AF695460

A Panorama of Singularities

CONTEMPORARY MATHEMATICS

742

A Panorama of Singularities

A Panorama on Singular Varieties
Conference to Celebrate Lê Dũng Tráng's 70th Birthday
February 7–10, 2017
University of Seville, IMUS, Spain

Francisco-Jesús Castro-Jiménez
David Bradley Massey
Bernard Teissier
Meral Tosun
Editors

American Mathematical Society
Real Sociedad Matemática Española

2010 *Mathematics Subject Classification.* Primary 14-xx, 32-xx, 57R45, 58K60.

Library of Congress Cataloging-in-Publication Data

Names: Panorama of Singularities: A Conference in Celebration of Lê Dũng Tráng's 70th birthday (2017 : Universidad de Sevilla), author. | Lê, Dũng Tráng, honoree. | Castro-Jiménez, Francisco-Jesús, 1957– editor. | Massey, David B., 1959–editor. | Teissier, Bernard, editor. | Tosun, Meral, 1967– editor.

Title: A panorama of singularities : a conference in celebration of Lê Dũng Tráng's 70th birthday, February 7–10, 2017, University of Seville, IMUS, Seville, Spain / Francisco-Jesús Castro-Jiménez, David Bradley Massey, Bernard Teissier, Meral Tosun, editors.

Description: [Providence, Rhode Island] : American Mathematical Society ; [Madrid, Spain] : Real Sociedad Matemática Espanõla, [2020] | Series: Contemporary mathematics, 0271-4132 ; volume 742 | Includes bibliographical references.

Identifiers: LCCN 2019039418 | ISBN 9781470447922 (paperback) | ISBN 9781470454524 (ebook)

Subjects: LCSH: Lê, Dũng Tráng–Congresses. | Singularities (Mathematics)–Congresses. | Geometry, Algebraic–Congresses. | AMS: Algebraic geometry. | Several complex variables and analytic spaces {For infinite-dimensional holomorphy, see 46G20, 58B12}. | Manifolds and cell complexes {For complex manifolds, see 32Qxx} – Differential topology {For foundational questions of differentiable manifolds, see 58Axx; for infinite-dimensional manifolds, see 58B} | Global analysis, analysis on manifolds [See also 32Cxx, 32Fxx, 32Wxx, 46-XX, 47Hxx, 53Cxx] {For geometric integration theory, see 49Q15} – Theory of singularities and catastrophe theory [See also 32S]

Classification: LCC QA614.58 .P36 2020 | DDC 516.3/5–dc23
LC record available at https://lccn.loc.gov/2019039418

Contemporary Mathematics ISSN: 0271-4132 (print); ISSN: 1098-3627 (online)

DOI: https://doi.org/10.1090/conm/742

Color graphic policy. Any graphics created in color will be rendered in grayscale for the printed version unless color printing is authorized by the Publisher. In general, color graphics will appear in color in the online version.

♾ The paper used in this book is acid-free and falls within the guidelines
established to ensure permanence and durability.
Visit the AMS home page at `https://www.ams.org/`

10 9 8 7 6 5 4 3 2 1 25 24 23 22 21 20

Dedicated to Lê Dũng Tráng on the occasion of his 70th birthday

Contents

Preface

This book contains texts dedicated to Lê Dũng Tráng on the occasion of his 70th Birthday, which was celebrated by a conference held in Sevilla from February 7–10, 2017.

The existence of this book is therefore due to the importance of Lê's work in an area of mathematics whose development is comparatively recent, the study of the geometry and topology of singular spaces in any dimension, mostly in the complex analytic framework.

This work spans a wide spectrum, from the celebrated constant Milnor number condition for the constancy of the embedded topological type for a family of hypersurfaces with isolated singularity of dimension $\neq 2$ (with C.P Ramanujam) to the JSJ decomposition of the complement of the link in $\mathbf{S}^3$ of a plane curve singularity and to the theory of $\mathcal{D}$-modules and the geometry of mixed Hodge structures. Within this spectrum lie many important theorems, for example on the monodromy and (with Hamm) Lefschetz type theorems on the homotpy of singular projective varieties, and constructions which have become widely used, such as carousels, the swing and, with the author of these lines, the theory of local polar varieties which provided, among other things, a better understanding of Whitney stratifications of complex analytic spaces and Thom-Mather's topological triviality theorem.

One should also mention the generosity evidenced by Lê's 20 students as well as his important contributions to the development of mathematics in Vietnam and other countries by his work as Mathematics Director of the ICTP in Trieste.

The texts gathered in this volume partly reflect the variety of interests of Lê and the width of his influence. To give an idea of their diversity we have tried to group them loosely by topics, as follows:

- Characteristic classes and divisor classes for singular varieties: Aluffi and Goresky, Hamm, Massey.
- Geometrical study of maps mixing holomorphic and anti-holomorphic functions: Cisneros-Molina and Romano-Velázquez, Oka.
- Algebraic and topological invariants of normal complex surface singularities: László and Némethi.
- Monodromy for complex projective hypersurfaces: Dimca and Sticlaru.
- Equisingularity of holomorphic maps: Sampaio.
- Equisingular deformations of plane curves and Hilbert schemes: Greuel.

- Hasse-Schmidt derivations: Narváez.
- Morphisms of Teichmüller spaces of marked Riemann spheres: E. Hironaka.
- Linear differential equations and the Stokes phenomenon: Teyssier.

Bernard Teissier
Directeur de Recherches émérite au CNRS

Contemporary Mathematics
Volume **742**, 2020
https://doi.org/10.1090/conm/742/14935

Pfaffian integrals and invariants of singular varieties

Paolo Aluffi and Mark Goresky

Dedicated to Lê Dũng Tráng on the occasion of his 70th birthday

1. Introduction

1.1. Chern classes of singular varieties have enjoyed an enormous development over the last forty years with hundreds of published papers and a number of books dedicated to the subject. The simple nature of the Gauss Bonnet theorem has received less attention, after a short announcement in [**8**]; see also [**24**], [**26**], [**9**]. In [**6, 7**], Alexandr Buryak studies the more general case of characteristic numbers for singular varieties, showing that any collection of characteristic numbers can be realized by a singular complex projective variety. In this paper we reformulate some of the statements in [**7, 8**] and give a number of extensions, applications and examples as we now describe.

Let $\overline{X}$ be an n dimensional complex projective algebraic variety and let X be the nonsingular part. Let $\Omega^r(X;\mathbb{C})$ denote the smooth differential r-forms on X. Endow X with the Kähler metric that is induced from the Fubini Study metric on the ambient projective space. Associated to this metric there is a unique torsion-free (Levi Civita) connection ∇_X and its curvature 2-form $\Theta \in \Omega^2(X;\mathrm{End}(TX))$ with values in the endomorphism bundle of the tangent bundle TX. Let

$$\mathrm{Pf}(\Theta_{\mathbb{R}}) \in \Omega^{2n}(X;\mathbb{R})$$

be the Pfaffian $2n$-form of the underlying real vector bundle with connection. If $X = \overline{X}$ is compact and nonsingular, then the Gauss Bonnet theorem states that the integral

$$(2\pi)^{-n}\int_X \mathrm{Pf}(\Theta_{\mathbb{R}}) = \chi(X)$$

is equal to the Euler characteristic of X. In the noncompact case, however, this is far from true. It does not even seem obvious, a priori, that the above integral is finite, or that it is independent of the embedding in projective space.

1.2. In [**18**], R. MacPherson defines, for each point $y \in \overline{X}$, the local Euler obstruction $\mathrm{Eu}_y(\overline{X}) \in \mathbb{Z}$. It is constant as the point y varies within any stratum of a Whitney stratification of X and has become a well studied local invariant of singular spaces.

2010 *Mathematics Subject Classification.* Primary: 14B05, 32S20; Secondary: 53C65, 57R20.

1.3. THEOREM. *Choose a Whitney stratification of $\overline{X}$ so that X is the largest stratum. Then the Gauss Bonnet integral is equal to the following sum,*

$$(2\pi)^{-n}\int_X \mathrm{Pf}(\Theta_{\mathbb{R}}) = \chi(X) + \sum_{Y<X} \mathrm{Eu}_y(\overline{X})\chi(Y) \tag{1.3.1}$$

*(where the sum is over singular strata Y and where $y \in Y$). It is an integer, is independent of the embedding, and is constant within any (Whitney) equisingular ([**16**]) family.*

This result is extended to higher Chern classes in Proposition 2.5 and proven in §3. Our argument relies on the Nash blow-up (the same strategy is also used in [**7**]). In §4 we give an intrinsic interpretation of the integral, relying on *controlled differential forms* rather than Nash blow-up. The more general case of characteristic numbers is treated in §5, and extended to degrees of products of Mather Chern classes. We prove that the information carried by the degrees of Mather Chern classes is equivalent to a suitable collection of Pfaffian integrals. In §6 we answer a query of Robert Langlands concerning the behavior of the Gauss Bonnet integral in a flat family. In §7 we briefly mention natural extensions.

1.4. We are grateful to Robert Langlands for asking these questions and for prompting us to publish these results. We wish to thank Jörg Schürmann for reviewing an earlier version of this note and for informing us of the work of Alexandr Buryak ([**6, 7**]).

We are pleased to dedicate this paper to Lê Dũng Tráng in recognition of his tremendous influence on singularity theory. In [**15**], Langevin and Lê consider curvature integrals near a singular point of a complex analytic variety X. If X is a hypersurface, they give an integral formula for the Milnor number of a generic hyperplane section. This result anticipates formulas of the type obtained in §6 in the present note.

2. Higher Chern classes

2.1. Suppose, as in the introduction, that X is the nonsingular part of a complex projective variety $\overline{X}$ equipped with the Fubini Study metric and associated Levi Civita connection ∇ and curvature form Θ. The *total Chern form*

$$c^*(\Theta) = \det\left(I + \frac{\sqrt{-1}}{2\pi}\Theta\right) \in \Omega^*(X;\mathbb{C}) \tag{2.1.1}$$

is a sum of homogeneous terms of even degree, and the degree $2r$ part, denoted $c^r(\Theta)$, is equal to $(\sqrt{-1}/2\pi)^r$ times the r-th elementary symmetric function of the eigenvalues of Θ.

2.2. Let $b : \widehat{X} \to \overline{X}$ be the Nash blowup of $\overline{X}$; it is defined to be the closure of the image of $X \to \mathrm{Gr}_n(T\mathbb{P}^n)$ in the Grassmann bundle of n dimensional subspaces of the tangent bundle of $\mathbb{P}^n$, under the Gauss map $x \mapsto T_xX$. The tautological n-plane bundle on $\mathrm{Gr}_n(T\mathbb{P}^n)$ when restricted to $\widehat{X}$ is a vector bundle, denoted ξ, that extends the tangent bundle $TX \to X$. Let $c^r(\xi) \in H^{2r}(\widehat{X};\mathbb{Z})$ denote its r-th Chern class. Recall [**18**] that the *Mather Chern class* of $\overline{X}$ is the homology class $c^M(\overline{X})$ whose component of dimension $n-r$ is

$$c^M_{n-r}(\overline{X}) = b_*\left(c^r(\xi) \cap [\widehat{X}]\right) \in H_{2n-2r}(\overline{X};\mathbb{Z})$$

where $[\widehat{X}] \in H_{2n}(\widehat{X};\mathbb{Z})$ denotes the fundamental (orientation) class of $\widehat{X}$.

2.3. Each point $y \in \widehat{X}$ corresponds to an n-dimensional subspace ξ_y of $T_{\pi(y)}\mathbb{P}^n$ and so it inherits a Hermitian metric from the Fubini Study metric on $\mathbb{P}^n$. Thus, not only does the tangent bundle of X extend to a vector bundle on $\widehat{X}$ but also the Kähler metric on X extends to a Hermitian metric on the budle ξ. We would like to say that the canonical connection ∇ and its curvature 2-form Θ also extends to $\widehat{X}$ but unfortunately this variety may still be singular (although it may be possible, with some work, to make sense of these notions in this setting). We avoid these technical difficulties by passing to a resolution of singularities

$$\widetilde{X} \xrightarrow{\ \pi\ } \widehat{X} \xrightarrow{\ b\ } \overline{X}$$

of the Nash blowup $\widehat{X}$.

2.4. The Hermitian metric on ξ pulls back to a Hermitian metric on the vector bundle $\pi^*(\xi)$. Let $\widetilde{\nabla}$ be the associated connection; it therefore extends the Levi Civita connection ∇ on TX. Let $\widetilde{\Theta}$ denote its curvature 2-form (with values in $\mathrm{End}(\pi^*(\xi))$) and let

$$c^*(\widetilde{\Theta}) = \det\left(I + \frac{\sqrt{-1}}{2\pi}\widetilde{\Theta}\right)$$

be the corresponding total Chern form on $\widetilde{X}$ with its associated cohomology class $[c^*(\widetilde{\Theta})] \in H^*(\widetilde{X};\mathbb{C})$. Let $\deg : H_0(\overline{X};\mathbb{Z}) \to \mathbb{Z}$ denote the augmentation.

2.5. PROPOSITION. *The homology class*

$$b_*\pi_*\left([c^*(\widetilde{\Theta})] \cap [\widetilde{X}]\right) = c^M(\overline{X}) \tag{2.5.1}$$

coincides with the (total) Mather Chern class of $\overline{X}$. Consequently the Gauss Bonnet integral

$$(2\pi)^{-n}\int_X \mathrm{Pf}(\Theta_{\mathbb{R}}) = \deg c_0^M(\overline{X}) \tag{2.5.2}$$

is equal to the degree zero part of the Mather-Chern class.

PROOF. The usual Chern Weil theorem describes the Chern class of $\pi^*(\xi)$ as follows:

$$c^*(\pi^*(\xi)) = [c^*(\widetilde{\Theta})].$$

Therefore

$$\begin{aligned} b_*\pi_*([c^*(\widetilde{\Theta})] \cap [\widetilde{X}]) &= b_*\pi_*(\pi^*c^*(\xi) \cap [\widetilde{X}]) \\ &= b_*(c^*(\xi) \cap [\widehat{X}]) \\ &= c^M(\overline{X}) \end{aligned}$$

which proves equation (2.5.1). For a complex rank n vector bundle (such as $\pi^*(\xi)$) the top Chern form agrees with the Pfaffian (see, for example, [**2**] p. 186), that is,

$$c^n(\widetilde{\Theta}) = \det\left(\frac{\sqrt{-1}}{2\pi}\widetilde{\Theta}\right) = \mathrm{Pf}\left(\frac{1}{2\pi}\widetilde{\Theta}_{\mathbb{R}}\right).$$

Moreover, the differential form $\mathrm{Pf}(\widetilde{\Theta}_{\mathbb{R}})$ on $\widetilde{X}$ is smooth. It agrees with $\mathrm{Pf}(\Theta_{\mathbb{R}})$ on X, and $\widetilde{X} - X$ has measure zero. So the degree zero part of equation (2.5.1) becomes

$$\int_{\widetilde{X}} c^n(\widetilde{\Theta}) = (2\pi)^{-n} \int_{\widetilde{X}} \mathrm{Pf}(\widetilde{\Theta}_{\mathbb{R}}) = (2\pi)^{-n} \int_X \mathrm{Pf}(\Theta_{\mathbb{R}}). \qquad \square$$

3. Proof of Theorem 1.3

3.1. As in [**18**], the Euler obstruction determines an isomorphism T from the group of algebraic cycles on $\overline{X}$ to the group of constructible functions on $\overline{X}$ by

$$T(\sum_i a_i V_i)(p) = \sum_i a_i \,\mathrm{Eu}_p(V_i).$$

The (MacPherson) total Chern class of a constructible function F is defined to be $c_*(F) = c^M T^{-1}(F)$. It follows that the Mather Chern class of $\overline{X}$ is the MacPherson Chern class of the constructible function that is $F(p) = \mathrm{Eu}_p(\overline{X})$. Proposition 2.5 says that the Gauss Bonnet integral is equal to the degree zero part of this class which is therefore the Euler characteristic of this constructible function F, giving equation (1.3.1).

3.2. The Euler obstruction is constant on strata of a Whitney stratification ([**4**]) and it is independent of the embedding so the right side of (1.3.1) is independent of the embedding and is constant within any Whitney-equisingular family. (In fact, the Nash blowup is independent of the embedding and can be defined intrinsically on $\overline{X}$ according to [**20**].) This completes the proof of Theorem 1.3. $\square$

4. The map to homology

4.1. It is possible to give an intrinsic description of the manner in which the differential form $c^*(\Theta) \in \Omega^*(X)$ gives rise to a homology class in $H_*(\overline{X};\mathbb{C})$ in equation (2.5.1), without referring to the Nash blowup. We give an outline only because the result uses various technicalities involving Whitney stratifications, and it is not essential for the development in this paper.

4.2. For any Whitney stratification of $\overline{X}$ and for an associated choice of control data ([**19, 27**]) there is a collection of *controlled differential forms* ([**10, 28**]). A controlled differential r-form (with real coefficients) is a differential r-form η_A on each stratum A with the property that, whenever $A < B$ are strata, then $\eta_B|T_A = \pi^*_{BA}(\eta_A)$ where $\pi_{BA} : (T_A \cap B) \to A$ is the projection function of a tubular neighborhood of A. The controlled differential forms constitute a complex $\Omega^*_{con}(\overline{X};\mathbb{R})$ whose cohomology is naturally isomorphic to the ordinary cohomology $H^*(\overline{X};\mathbb{R})$.

4.3. LEMMA. *Fix a Whitney stratification and control data on $\overline{X}$. Fix r with $0 \le r \le n$ and let $c^r(\Theta)$ be the r-th Chern form on X. For any controlled differential form $\eta \in \Omega^{2n-r}_{con}(\overline{X};\mathbb{C})$,*

$$\int_X c^r(\Theta) \wedge \eta < \infty$$

PROOF. Each controlled differential form η extends canonically to a smooth differential form in a neighborhood of $\overline{X}$. It follows that if $\widetilde{X} \to \overline{X}$ is a resolution of singularities then η pulls back to a smooth differential form $\tilde{\eta}$ on $\widetilde{X}$. Using the resolution of the Nash blowup as described in Proposition 2.5 above, we conclude that

$$\int_X c^r(\Theta)\wedge\eta = \int_{\widetilde{X}} c^r(\widetilde{\Theta})\wedge\tilde{\eta} < \infty. \qquad \square$$

4.4. Thus, each Chern form $c^r(\Theta)$ determines a homomorphism $\Omega^{2n-r}_{con}(\overline{X}) \to \mathbb{C}$ that vanishes on boundaries, and hence defines a class in the dual space $H^*(\overline{X};\mathbb{C})^* = H_*(\overline{X};\mathbb{C})$, which is easily seen to agree with the Mather Chern class.

5. Other curvature integrals

5.1. For any partition $I = i_1 + i_2 + \cdots + i_r = n$, the *Mather Chern number* is the integer $c^{M,I}(\overline{X}) = \deg\left(c^{i_1}(\xi)\cup\cdots\cup c^{i_r}(\xi)\cap[\widehat{X}]\right)$.

5.2. PROPOSITION. *Let $I = i_1+\cdots+i_r = n$ be a partition of n. Then*

$$\int_X c^{i_1}(\Theta)\wedge\cdots\wedge c^{i_r}(\Theta) = c^{M,I}(\overline{X}). \tag{5.2.1}$$

that is, the integral of this product of Chern forms equals the corresponding Mather Chern number.

The proof is the same as that of Proposition 2.5. □

Proposition 5.2 is also proven in [**7**].

One can also define a product of any collection $c^M_{i_1}(\overline{X}),\ldots,c^M_{i_r}(\overline{X})$ of Mather Chern classes, as the push-forward

$$b_*\left(c^{i_1}(\xi)\cup\cdots\cup c^{i_r}(\xi)\cap[\widehat{X}]\right)$$

in $H_*(\overline{X};\mathbb{C})$.

5.3. The (algebraic) degree $\deg(\alpha)$ of an integral homology class $\alpha\in H_{2r}(\mathbb{P}^N;\mathbb{Z})$ is the unique integer such that $\alpha = \deg(\alpha)[L^r]\in H_{2r}(\mathbb{P}^N;\mathbb{Z})$ where L^r denotes a codimension r linear subspace of $\mathbb{P}^N$.

5.4. PROPOSITION. *Let ω denote the Kähler form on X that is induced from the ambient projective space. Then the degree of the Mather Chern class $c^M_i(\overline{X})$ is given by the integral,*

$$\deg(c^M_i(\overline{X})) = \int_X c^{n-i}(\Theta)\wedge\omega^i. \tag{5.4.1}$$

Equation (5.4.1) follows from the fact that the class $[\omega]\in H^2(\mathbb{P}^N)$ represented by the Kähler form ω is the Poincaré dual of the homology class $[L]\in H_2(\mathbb{P}^N)$ represented by a hyperplane. □

5.5. The numbers $\deg(c^M_i(\overline{X}))$ may be used to determine the Gauss Bonnet integrals

$$\beta_r = \int_{X\cap L^r}\mathrm{Pf}(\Theta_r)$$

over a generic codimension r linear section $\overline{X}\cap L^r$, and vice versa. (Here, Θ_r is the curvature form on $X\cap L^r$.) Let

$$\mathrm{CM}(t) = \sum_{r\geq 0}\deg(c^M_r(\overline{X}))t^r$$

be the Mather Chern (Poincaré) polynomial and let

$$\mathrm{Pf}(t) = \sum_{r\geq 0} \beta_r(-t)^r$$

be the Pfaffian polynomial. Define an involution on the set of polynomials of fixed degree, $p \mapsto \mathcal{I}(p)$ by

$$\mathcal{I}(p)(t) = \frac{p(0) + t\,p(-1-t)}{1+t}.$$

5.6. PROPOSITION. *The involution $\mathcal{I}$ interchanges the Mather Chern polynomial with the Pfaffian polynomial:*

$$\mathrm{CM} = \mathcal{I}(\mathrm{Pf}) \quad \textit{and} \quad \mathrm{Pf} = \mathcal{I}(\mathrm{CM}).$$

The proof follows the same lines as that of Theorem 1.1 in [**1**]; the key technical step is the good behavior of the Mather Chern class with respect to generic linear sections, which follows for example from Lemma 1.2 in [**21**]. □

6. Behavior in families

6.1. We now consider the behavior of the Pfaffian integral along a smoothing family for $\overline{X}$. Localizing the integral in tubular neighborhoods of components of the singular locus of $\overline{X}$ in a smoothing family leads to expressions for invariants of the singularity in terms of Pfaffian integrals.

As above, $\overline{X}$ denotes a projective variety of dimension n, with nonsingular part X. We will let $S := \overline{X} \smallsetminus X$ be the singular locus of $\overline{X}$. Suppose that $\overline{X} = \overline{X}_0$ is the central fiber of a flat family $\overline{\mathcal{X}} \to D$ over a disk $D \subseteq \mathbb{C}$ centered at 0. We let $\overline{X}_\delta = \pi^{-1}(\delta)$ denote the fiber over $\delta \in D$ and we assume that $\overline{X}_\delta$ is nonsingular for all $\delta \neq 0$; thus $\overline{\mathcal{X}}$ is a smoothing of X. We will further assume that $\overline{\mathcal{X}} \subseteq \mathbb{P}^N$ is projective. It follows from the first isotopy lemma of R. Thom ([**19**]) that π is topologically locally trivial over $D \smallsetminus \{0\}$.

A metric is induced on the tangent spaces to X as above, and to every $\overline{X}_\delta$ for $\delta \neq 0$ in a neighborhood of 0. We have an induced connection ∇_δ on each $\overline{X}_\delta$, with curvature $\Theta_\delta \in \Omega^2(\overline{X}_\delta; \mathrm{End}(T\overline{X}_\delta))$ and Pfaffian $\mathrm{Pf}(\Theta_{\mathbb{R},\delta}) \in \Omega^{2n}(\overline{X}_\delta; \mathbb{R})$.

6.2. PROPOSITION. *Denote by $N_\epsilon(S)$ the ϵ-tubular neighborhood of S. Then*

$$\lim_{\epsilon\to 0}\lim_{\delta\to 0}(2\pi)^{-\dim\overline{X}}\int_{\overline{X}_\delta\cap N_\epsilon(S)} \mathrm{Pf}(\Theta_{\mathbb{R},\delta}) = \sum_{Y<X}\left(\sigma_y(\overline{\mathcal{X}}) - \mathrm{Eu}_y(\overline{X})\right)\chi(Y) \quad .$$

As above, the sum is over the strata Y contained in S, and $\mathrm{Eu}_y(\overline{X})$ denotes the (common) value of the local Euler obstruction at points y of the stratum Y. Similarly, $\sigma_y(\overline{\mathcal{X}})$ denotes the value at $y \in Y$ of Verdier's *specialization function* $\sigma(\overline{\mathcal{X}}) = \sigma_*(1\!\!1_{\overline{\mathcal{X}}})$, see [**29**, §3]. (We assume that $\sigma(\overline{\mathcal{X}})$ is locally constant along strata.) Explicitly,

$$\sigma_y(\overline{\mathcal{X}}) = \lim_{\epsilon\to 0}\lim_{\delta\to 0}\chi(\overline{X}_\delta \cap B_\epsilon(y))$$

where $B_\epsilon(y)$ is the ϵ-ball centered at y. Note that both $\mathrm{Eu}(\overline{X})$ and $\sigma(\overline{\mathcal{X}})$ equal 1 away from S. Therefore their difference is supported on S.

If $\overline{\mathcal{X}} \subseteq \overline{\mathcal{M}}$ may be realized as a hypersurface in a flat family $\overline{\mathcal{M}} \to D$ with nonsingular fibers, then every fiber $\overline{X}_\delta$, $\delta \in D$ is itself a hypersurface of a nonsingular variety. If $\overline{X} = \overline{X}_0$ has isolated singularities, then Proposition 6.2 admits the following particularly explicit formulation, which recovers a formula of R. Langevin ([**14**, Theorem 1]):

6.3. COROLLARY. *In the situation considered above, assume that* $\overline{X}$, $\overline{X}_\delta$ *are hypersurfaces of nonsingular varieties and that* $\dim S = 0$. *Then*

$$\lim_{\epsilon\to 0}\lim_{\delta\to 0}(2\pi)^{-\dim\overline{X}}\int_{\overline{X}_\delta\cap N_\epsilon(S)}\mathrm{Pf}(\Theta_{\mathbb{R},\delta}) = (-1)^{\dim\overline{X}}\sum_{p\in S}\left(\mu_{\overline{X}}(p)+\mu_{\overline{X}\cap H}(p)\right) \quad ,$$

where $\mu(p)$ *denotes the Milnor number at* p *and* H *is a general hyperplane through* p.

In particular, the left-hand-side is independent of the smoothing.

6.4. We prove Proposition 6.2 and Corollary 6.3.

PROOF OF PROPOSITION 6.2. Consider the complements $\overline{X}_\delta \smallsetminus N_\epsilon(S)$ of the tubular neighborhood. For all positive ϵ near 0 the integral

$$\int_{\overline{X}_\delta\smallsetminus N_\epsilon(S)}\mathrm{Pf}(\Theta_{\mathbb{R},\delta})$$

is a continuous function of δ near 0, therefore

$$\lim_{\epsilon\to 0}\lim_{\delta\to 0}(2\pi)^{-\dim\overline{X}}\int_{\overline{X}_\delta\smallsetminus N_\epsilon(S)}\mathrm{Pf}(\Theta_{\mathbb{R},\delta}) = \lim_{\epsilon\to 0}(2\pi)^{-\dim\overline{X}}\int_{\overline{X}\smallsetminus N_\epsilon(S)}\mathrm{Pf}(\Theta_{\mathbb{R},0}) = \deg c_0^M(\overline{X})$$

by Proposition 2.5. On the other hand, let $\chi = \chi(X_\delta)$ for any δ near 0. (By hypothesis, this number does not depend on δ.) Then

$$\lim_{\delta\to 0}(2\pi)^{-\dim\overline{X}}\int_{\overline{X}_\delta}\mathrm{Pf}(\Theta_{\mathbb{R},\delta}) = \chi$$

by the ordinary Gauss Bonnet formula. Since $\overline{X}_\delta\cap N_\epsilon(S) = \overline{X}_\delta \smallsetminus \left(\overline{X}_\delta \smallsetminus N_\epsilon(S)\right)$, this gives

$$\lim_{\epsilon\to 0}\lim_{\delta\to 0}(2\pi)^{-\dim\overline{X}}\int_{\overline{X}_\delta\cap N_\epsilon(S)}\mathrm{Pf}(\Theta_{\mathbb{R},\delta}) = \chi - \deg c_0^M(X)$$

Arguing as in §3.1,

$$\deg c_0^M(\overline{X}) = \sum_Y \mathrm{Eu}_y(\overline{X})\chi(Y) \quad . \tag{6.4.1}$$

By Verdier's specialization theorem [**29**, Théorème 5.1], we have, for δ near 0:
(6.4.2)

$$\chi = \deg\sigma_* c_*(1\!\!1_{\overline{X}_\delta}) = \deg c_*\sigma_*(1\!\!1_{\overline{\mathfrak{X}}}) = \deg c_*(\sum_Y \sigma_y(\overline{\mathfrak{X}})1\!\!1_Y) = \sum_Y \sigma_y(\overline{\mathfrak{X}})\chi(Y) \quad .$$

Identities (6.4.1) and (6.4.2) complete the proof of Proposition 6.2. □

PROOF OF COROLLARY 6.3. For an isolated hypersurface singularity,

$$\sigma_{\overline{X}}(p) = 1 + (-1)^{\dim\overline{X}}\mu_{\overline{X}}(p)$$

as a consequence of [**22**, Proposition 5.1]. On the other hand,

$$\mathrm{Eu}_{\overline{X}}(p) = 1 - (-1)^{\dim\overline{X}}\mu_{\overline{X}\cap H}(p)$$

as proven in [**23**] (also cf. [**8**, Remarque, p. 240]). The formula in Corollary 6.3 follows then immediately from Proposition 6.2. □

6.5. Assume now that $\overline{X}$ is a hypersurface in a nonsingular projective variety, with a single singular point p, and that $\overline{\mathfrak{X}} \to D$ is a smoothing family as above. If H is a general hyperplane through p, then $\overline{\mathfrak{X}} \cap H$ is a smoothing family for $\overline{X} \cap H$ over (a possibly smaller disk) D, satisfying the same hypotheses as $\overline{\mathfrak{X}}$. (In fact the set of parameter values $z \in D \smallsetminus \{0\}$ such that H fails to be transversal to the fiber $\overline{X}_z = \pi^{-1}(z)$ is algebraic, hence finite. By shrinking D if necessary we may assume that it contains no such points.) Iterating, we can find general hyperplanes $H_1, \dots, H_{\dim \overline{X}-1}$ through p such that

$$\overline{\mathfrak{X}}^{(r)} := \overline{\mathfrak{X}} \cap H_1 \cap \cdots \cap H_r$$

satisfies the same hypotheses for $r = 1, \dots, \dim \overline{X} - 1$ (perhaps at the price of further reducing the radius of the disk D).

Assume this is the case and let $\overline{\mathfrak{X}}^{(0)} = \overline{\mathfrak{X}}$. We have fibers $\overline{X}_\delta^{(r)}$ and corresponding Pfaffians $\mathrm{Pf}(\Theta_{\mathbb{R},\delta}^{(r)})$ for $r = 0, \dots, \dim \overline{X} - 1$. We let H be one further general hyperplane; the intersection

$$\overline{X}^{(\dim \overline{X})} = \overline{X} \cap H_1 \cap \cdots \cap H_{\dim \overline{X}-1} \cap H$$

is then zero-dimensional. It consists of an m-tuple point at p, where m is the multiplicity of $\overline{X}$ at p, and of a set of reduced points.

6.6. Corollary. *Let $\overline{X}$ be a hypersurface in a projective nonsingular variety; assume $\overline{X}$ has an isolated singularity p, of multiplicity m and Milnor number $\mu_{\overline{X}}(p)$. Then with notation as above:*
(6.6.1)

$$\lim_{\epsilon \to 0} \lim_{\delta \to 0} \sum_{i=1}^{\dim \overline{X}} (2\pi)^{-\dim \overline{X}+i-1} \int_{\overline{X}_\delta^{(i-1)} \cap N_\epsilon(S)} \mathrm{Pf}(\Theta_{\mathbb{R},\delta}^{(i-1)}) = (-1)^{\dim \overline{X}} \mu_{\overline{X}}(p) - m + 1 \quad .$$

Proof. We can evaluate each summand by using Corollary 6.3. This shows that the left-hand side equals

$$(-1)^{\dim \overline{X}} (\mu_{\overline{X}}(p) + \mu_{\overline{X}^{(1)}}(p)) + (-1)^{\dim \overline{X}-1} (\mu_{\overline{X}^{(1)}}(p) + \mu_{\overline{X}^{(2)}}(p)) \\ + \cdots + (-1)^1 (\mu_{\overline{X}^{(\dim \overline{X}-1)}}(p) + \mu_{\overline{X}^{(\dim \overline{X})}}(p))$$

that is ('telescoping') $(-1)^{\dim \overline{X}} \mu_{\overline{X}}(p) - \mu_{\overline{X}^{(\dim \overline{X})}}(p) = (-1)^{\dim \overline{X}} \mu_{\overline{X}}(p) - m + 1$, as stated. □

An integral formula for the Milnor number was obtained by Phillip Griffiths ([**11**], [**12**]; and see [**17**, p. 207]). It would be interesting to establish a direct relation between the two formulas.

6.7. Example. The simplest example illustrating the situation considered here is probably a family of nonsingular conics degenerating to the union $\overline{X}$ of two lines, $xy = \delta z^2$ in $\mathbb{P}^2$. Topologically, smooth conics and complex projective lines are 2-spheres; as $\delta \to 0$, the degeneration may be pictured as follows.

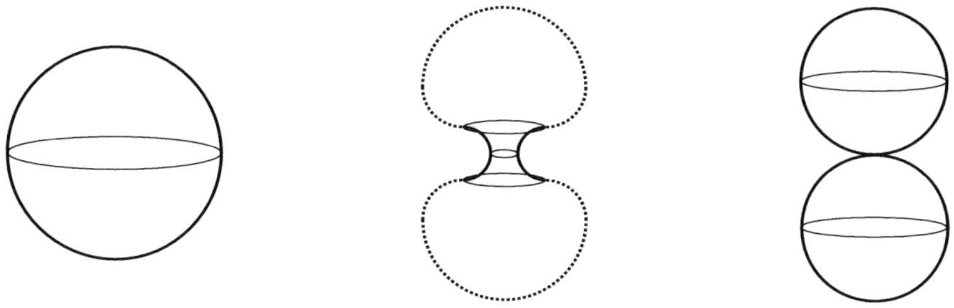

where the saddle in the middle depicts the intersection with an ϵ-polydisk centered at p. In this example, the integral in the left-hand side of (6.6.1) may be evaluated explicitly; as $\delta \to 0$, it converges to

$$\frac{2(\epsilon^2 - 1)}{(\epsilon^2 + 1)}$$

and hence to -2 as $\epsilon \to 0$. According to Corollary 6.6,

$$-2 = (-1)^1 \mu_{\overline{X}}(p) - 2 + 1 \quad ,$$

i.e., $\mu_{\overline{X}}(p) = 1$ (as it should).

7. Extensions and further questions

7.1. For simplicity of exposition we assumed that $\overline{X}$ is projective, but Theorem 1.3 applies to subvarieties of any algebraic Kähler manifold.

7.2. In fact, an analogue of Proposition 2.5 holds (with the same proof) for the Mather Chern class of any coherent sheaf ([**25**], [**13**]) realized as a quotient of a locally free sheaf $\mathcal{E}$, with respect to any hermitian metric defined on the vector bundle of sections of the dual $\mathcal{E}^\vee$.

7.3. Homology classes cannot, in general, be multiplied together. Therefore, multiplicativity properties of Mather Chern classes are subtle. It follows from [**3**] and [**5**] that these classes lift canonically to middle intersection homology, giving a well defined homology-valued product of any two such classes. Proposition 5.2 provides an invariant interpretation of any top-degree product of Mather Chern classes and, as observed there, a product of any collection of Mather-Chern classes may be defined by means of a corresponding product in the Nash blow-up. We do not know whether the vector space spanned by these products of Mather Chern classes forms a natural well-defined ring within the homology $H_*(\overline{X})$.

7.4. Pontrjagin classes of compact complex manifolds also admit Chern-Weil descriptions, and the signature of such a manifold may be expressed as a (rational) linear combination of Chern numbers, hence as a curvature integral. If $\overline{X}$ is a singular variety, the same linear combination of Mather-Chern numbers is defined and by Proposition 5.2 it is given by the same curvature integral over the nonsingular part, X. Is this number an integer? What is the relation between this number and the (intersection homological) signature of $\overline{X}$?

References

[1] P. Aluffi, *Euler characteristics of general linear sections and polynomial Chern classes*, Rend. Circ. Mat. Palermo (2) **62** (2013), no. 1, 3–26, DOI 10.1007/s12215-013-0106-x. MR3031565

[2] W. Ballmann, *Lectures on Kähler manifolds*, ESI Lectures in Mathematics and Physics, European Mathematical Society (EMS), Zürich, 2006. MR2243012

[3] G. Barthel, J.-P. Brasselet, K.-H. Fieseler, O. Gabber, and L. Kaup, *Relèvement de cycles algébriques et homomorphismes associés en homologie d'intersection* (French), Ann. of Math. (2) **141** (1995), no. 1, 147–179, DOI 10.2307/2118630. MR1314034

[4] J.-P. Brasselet and M.-H. Schwartz, *Sur les classes de Chern d'un ensemble analytique complexe* (French), The Euler-Poincaré characteristic (French), Astérisque, vol. 82, Soc. Math. France, Paris, 1981, pp. 93–147. MR629125

[5] J. P. Brasselet and A. Weber, A canonical lift of Chern-Mather classes, preprint, http://arXiv.org/alg-geom/9712023

[6] A. Y. Buryak, *Existence of a singular projective variety with an arbitrary set of characteristic numbers*, Math. Res. Lett. **17** (2010), no. 3, 395–400, DOI 10.4310/MRL.2010.v17.n3.a2. MR2653677

[7] A. Buryak, *Bott's residue formula for singular varieties*, TWMS J. Pure Appl. Math. **2** (2011), no. 1, 17–21. MR2828855

[8] A. S. Dubson, *Classes caractéristiques des variétés singulières* (French, with English summary), C. R. Acad. Sci. Paris Sér. A-B **287** (1978), no. 4, A237–A240. MR0499290

[9] J. H. G. Fu, *Curvature measures and Chern classes of singular varieties*, J. Differential Geom. **39** (1994), no. 2, 251–280. MR1267891

[10] M. Goresky and W. Pardon, *Chern classes of automorphic vector bundles*, Invent. Math. **147** (2002), no. 3, 561–612, DOI 10.1007/s002220100184. MR1893006

[11] P. A. Griffiths, *Complex differential and integral geometry and curvature integrals associated to singularities of complex analytic varieties*, Duke Math. J. **45** (1978), no. 3, 427–512. MR507455

[12] G. Kennedy, *Griffiths' integral formula for the Milnor number*, Duke Math. J. **48** (1981), no. 1, 159–165. MR610181

[13] M. Kwieciński. *Sur le transformé de Nash et la construction du graphe de MacPherson.* Thèse, Doctorat de l'Université de Provence, 1994.

[14] R. Langevin, *Courbure et singularités complexes* (French), Comment. Math. Helv. **54** (1979), no. 1, 6–16, DOI 10.1007/BF02566253. MR522029

[15] R. Langevin and Lê Dũng Tráng, *Courbure au voisinage d'une singularité* (French, with English summary), C. R. Acad. Sci. Paris Sér. A-B **290** (1980), no. 2, A95–A98. MR563947

[16] Lê Dũng Tráng and B. Teissier, *Cycles evanescents, sections planes et conditions de Whitney. II* (French, with English summary), Singularities, Part 2 (Arcata, Calif., 1981), Proc. Sympos. Pure Math., vol. 40, Amer. Math. Soc., Providence, RI, 1983, pp. 65–103. MR713238

[17] F. Loeser, *Formules intégrales pour certains invariants locaux des espaces analytiques complexes* (French), Comment. Math. Helv. **59** (1984), no. 2, 204–225, DOI 10.1007/BF02566346. MR749105

[18] R. D. MacPherson, *Chern classes for singular algebraic varieties*, Ann. of Math. (2) **100** (1974), 423–432, DOI 10.2307/1971080. MR0361141

[19] J. Mather, **Notes on Topological Stability**, mimeographed notes, Harvard University 1970. Reprinted in Bull. Amer. Math. Soc. **49** (2012), 475–506.

[20] A. Nobile, *Some properties of the Nash blowing-up*, Pacific J. Math. **60** (1975), no. 1, 297–305. MR0409462

[21] A. Parusiński and P. Pragacz, *Chern-Schwartz-MacPherson classes and the Euler characteristic of degeneracy loci and special divisors*, J. Amer. Math. Soc. **8** (1995), no. 4, 793–817, DOI 10.2307/2152829. MR1311826

[22] A. Parusiński and P. Pragacz, *Characteristic classes of hypersurfaces and characteristic cycles*, J. Algebraic Geom. **10** (2001), no. 1, 63–79. MR1795550

[23] R. Piene, *Cycles polaires et classes de Chern pour les variétés projectives singulières* (French), Introduction à la théorie des singularités, II, Travaux en Cours, vol. 37, Hermann, Paris, 1988, pp. 7–34. MR1074588

[24] I. Satake, *The Gauss-Bonnet theorem for V-manifolds*, J. Math. Soc. Japan **9** (1957), 464–492, DOI 10.2969/jmsj/00940464. MR0095520

[25] M.-H. Schwartz, *Classes et caractères de Chern-Mather des espaces linéaires* (French, with English summary), C. R. Acad. Sci. Paris Sér. I Math. **295** (1982), no. 5, 399–402. MR684735
[26] D. Siersma and M. Tibăr, *The Gauss-Bonnet defect of complex affine hypersurfaces*, Bull. Sci. Math. **130** (2006), no. 2, 110–122, DOI 10.1016/j.bulsci.2005.08.003. MR2200641
[27] R. Thom, *Ensembles et morphismes stratifiés* (French), Bull. Amer. Math. Soc. **75** (1969), 240–284, DOI 10.1090/S0002-9904-1969-12138-5. MR239613
[28] A. Verona, *Le théorème de de Rham pour les préstratifications abstraites* (French), C. R. Acad. Sci. Paris Sér. A-B **273** (1971), A886–A889. MR0290375
[29] J.-L Verdier, *Spécialisation des classes de Chern* (French), The Euler-Poincaré characteristic (French), Astérisque, vol. 82, Soc. Math. France, Paris, 1981, pp. 149–159. MR629126

Department of Mathematics, Florida State University, Tallahassee FL
Email address: aluffi@math.fsu.edu

School of Mathematics, Institute for Advanced Study, Princeton NJ
Email address: goresky@ias.edu

Contemporary Mathematics
Volume **742**, 2020
https://doi.org/10.1090/conm/742/14936

The real embedding method to study mixed functions

José Luis Cisneros-Molina and Agustín Romano-Velázquez

To Professor Lê Dũng Tráng on the occasion of his seventieth birthday.

Abstract. In this work, we describe the *embedding method* to study mixed functions. The idea of the method is to embed in a natural way a real singularity in a complex one, and to use properties and results of the complex singularity to study the real one. To ilustrate the method, we give a new proof of an *Isotopy Theorem* proved by Oka and later completed by Inaba, Kawashima and Oka.

Introduction

Let $\mathbf{a} = (a_1, \dots, a_n)$ and $\mathbf{b} = (b_1, \dots, b_n)$, with $a_i, b_i \in \mathbb{N}$. In [**6**] Oka considers the following families of mixed polynomials

$$f_A(\mathbf{z}) = z_1^{a_1}\bar{z}_1^{b_1} + \cdots + z_{n-1}^{a_{n-1}}\bar{z}_{n-1}^{b_{n-1}} + z_n^{a_n}\bar{z}_n^{b_n}, \tag{0.1}$$

$$f_B(\mathbf{z}) = z_1^{a_1}\bar{z}_1^{b_1}z_2 + \cdots + z_{n-1}^{a_{n-1}}\bar{z}_{n-1}^{b_{n-1}}z_n + z_n^{a_n}\bar{z}_n^{b_n}, \tag{0.2}$$

$$f_C(\mathbf{z}) = z_1^{a_1}\bar{z}_1^{b_1}z_2 + \cdots + z_{n-1}^{a_{n-1}}\bar{z}_{n-1}^{b_{n-1}}z_n + z_n^{a_n}\bar{z}_n^{b_n}z_1, \tag{0.3}$$

where $a_i > b_i \geq 0$, $i = 1, \dots, n$. In [**3**] to each f on the previous families, it is associated a graph. By the form of the graph, polynomials f_B are called of **B**amboo type and polynomials f_C are called of **C**yclic type. Since the graph associated to polynomials f_A consist of n isolated vertices we call them of **A**lone type.

These polynomials are polar weighted homogeneous polynomials and in the special case when $n = 3$ they appear in our previous work [**2**, Corollary 4.8] as polynomials of class **I**, **IV.a** and **V**. The mixed polynomials f_A given in (0.1) with $a_i \geq 2$ and $b_i = 0$ specialize to Brieskorn-Pham polynomials, which are weighted homogeneous with isolated critical point. When $a_i > b_i \geq 1$, these polynomials where studied by Seade in [**9**] and by Ruas, Seade and Verjovsky in [**8**], in particular, when $b_i = 1$ they are called twisted Briskorn-Pham polynomials. They proved that f_A is polar weighted homogeneous with isolated critical point at the origin, that the singular variety $f_A^{-1}(0)$ defined by f_A is homeomorphic to the Brieskorn variety defined by the Brieskorn-Pham polynomial $z_1^{a_1-b_1} + \cdots + z_n^{a_n-b_n}$, hence,

2010 *Mathematics Subject Classification.* Primary 14J17; Secondary 32C18.
Key words and phrases. Embedding method, isotopy theorem.
The first author is Regular Associate of the Abdus Salam International Centre for Theoretical Physics, Trieste, Italy. Both authors were supported by projects UNAM-DGAPA-PAPIIT IN106614 and CONACYT 253506.

the corresponding links are also homeomorphic, and that the corresponding Milnor fibrations are topologically equivalent [**8**, Theorem 4.1].

In [**6**] Oka improved the last assertion, he proved that the polynomials f_A and $z_1^{a_1-b_1}+\cdots+z_n^{a_n-b_n}$, and also the polynomials f_B and $z_1^{a_1-b_1}z_2+\cdots+z_{n-1}^{a_{n-1}-b_{n-1}}z_n+z_n^{a_n-b_n}$ have a Milnor fibration and the fibrations are C^∞-equivalent. In [**6**] Oka also conjectured the analogous result for the polynomials f_C and $z_1^{a_1-b_1}z_2+\cdots+z_{n-1}^{a_{n-1}-b_{n-1}}z_n+z_n^{a_n-b_n}z_1$; this was later proved by Inaba, Kawashima and Oka in [**3**].

In this article we introduce the *embedding method*. The idea of the method is to embed in a natural way a real singularity in a complex one, and to use properties and results of the complex singularity to study the real one. First we present the embedding method and we give some basics properties, later, we use this method to give a new proof[1] of the aforementioned results given in [**6**] and [**3**].

The organization of the article is as follows. In Section 1 we recall some basic facts about mixed functions, in particular, when a mixed polynomial is simplicial and the definition of polar weighted homogeneous polynomials. In Section 2 we introduce the embedding method and we define the holomorphic function associated to a mixed function (Definition 2.2). In Section 3 we use the embedding method to give another proof of the Isotopy Theorem (Theorem 3.4) given by Oka in [**6**] for polynomials f_A and f_B and by Inaba, Kawashima and Oka in [**3**] for polynomials f_C. In a separate article we plan to present other applications of the embedding method.

1. Mixed functions

Consider $\mathbb{C}^n$ with coordinates $z_1,\dots,z_n$. Let $\bar{z}_j$ be the complex conjugate of z_j. We will write $z_j=x_j+iy_j$ with $x_j,y_j\in\mathbb{R}$, $\mathrm{Re}(z_j)=x_j$ the real part and $\mathrm{Im}(z_j)=y_j$ the imaginary part of z_j. To simplify notation we shall write $\mathbf{z}=(z_1,\dots,z_n)$, $\bar{\mathbf{z}}=(\bar{z}_1,\dots,\bar{z}_n)$, $\mathbf{x}=(x_1,\dots,x_n)$ and $\mathbf{y}=(y_1,\dots,y_n)$. We also denote by $\mathbf{0}$ the origin in $\mathbb{C}^n$, by $\mathbb{C}^*$ the non-zero complex numbers and by $\mathbb{R}^+$ the positive real numbers. Considering $\mathbb{C}^n$ with coordinates $z_1,\dots,z_n,\bar{z}_1,\dots,\bar{z}_n$ is equivalent to consider it as a $2n$-dimensional *real* vector space with coordinates $x_1,y_1,\dots,x_n,y_n$.

DEFINITION 1.1. Let $\mu=(\mu_1,\dots,\mu_n)$ and $\nu=(\nu_1,\dots,\nu_n)$ with $\mu_j,\nu_j\in\mathbb{N}\cup\{0\}$, set $\mathbf{z}^\mu=z_1^{\mu_1}\dots z_n^{\mu_n}$ and $\bar{\mathbf{z}}^\nu=\bar{z}_1^{\nu_1}\dots\bar{z}_n^{\nu_n}$.

Following Oka [**5**], we call f a *mixed analytic function* if f is a complex valued function $f\colon\mathbb{C}^n\to\mathbb{C}$ such that f has a convergent power series of variables $\mathbf{z}$ and $\bar{\mathbf{z}}$,

$$f(\mathbf{z})=\sum_{\mu,\nu}c_{\mu,\nu}\mathbf{z}^\mu\bar{\mathbf{z}}^\nu.$$

We call f *mixed polynomial*, if f is a complex valued function $f\colon\mathbb{C}^n\to\mathbb{C}$ such that f is a polynomial in the variables $\mathbf{z}$ and $\bar{\mathbf{z}}$.

We consider f as a function $f\colon\mathbb{R}^{2n}\to\mathbb{R}^2$ in the $2n$ real variables $(\mathbf{x},\mathbf{y})$ writing $f(\mathbf{z})=g(\mathbf{x},\mathbf{y})+ih(\mathbf{x},\mathbf{y})$, taking $z_j=x_j+iy_j$ where $g,h\colon\mathbb{C}^n\cong\mathbb{R}^{2n}\to\mathbb{R}$ are real

[1]This proof of Oka's conjecture was presented at the "Brazil-Mexico 2nd meeting on Singularities" held in Salvador, Bahía, Brazil in 2015. Later this result was proved in [**3**] using different ideas.

analytic functions. Recall that for any real analytic function $k\colon \mathbb{R}^{2n} \to \mathbb{R}$ we have

$$\frac{\partial k}{\partial z_j} = \frac{1}{2}\left(\frac{\partial k}{\partial x_j} - i\frac{\partial k}{\partial y_j}\right), \qquad \frac{\partial k}{\partial \bar{z}_j} = \frac{1}{2}\left(\frac{\partial k}{\partial x_j} + i\frac{\partial k}{\partial y_j}\right).$$

So we have

$$\frac{\partial f}{\partial z_j} = \frac{\partial g}{\partial z_j} + i\frac{\partial h}{\partial z_j}, \qquad \frac{\partial f}{\partial \bar{z}_j} = \frac{\partial g}{\partial \bar{z}_j} + i\frac{\partial g}{\partial \bar{z}_j}.$$

As usual, we define the real gradients of g and h by

$$\mathrm{d}_{\mathbb{R}}g(\mathbf{x},\mathbf{y}) = \left(\frac{\partial g}{\partial x_1},\dots,\frac{\partial g}{\partial x_n},\frac{\partial g}{\partial y_1},\dots,\frac{\partial g}{\partial y_n}\right),$$

$$\mathrm{d}_{\mathbb{R}}h(\mathbf{x},\mathbf{y}) = \left(\frac{\partial h}{\partial x_1},\dots,\frac{\partial h}{\partial x_n},\frac{\partial h}{\partial y_1},\dots,\frac{\partial h}{\partial y_n}\right).$$

Following Oka [**4**] set

$$\mathrm{d}f(\mathbf{z}) = \left(\frac{\partial f(\mathbf{z})}{\partial z_1},\dots,\frac{\partial f(\mathbf{z})}{\partial z_n}\right), \qquad \bar{\mathrm{d}}f(\mathbf{z}) = \left(\frac{\partial f(\mathbf{z})}{\partial \bar{z}_1},\dots,\frac{\partial f(\mathbf{z})}{\partial \bar{z}_n}\right).$$

The following proposition is an useful criterium to determine whether a point $\mathbf{z} \in \mathbb{C}^n$ is a critical point of a mixed function f.

PROPOSITION 1.2 (Oka's Criterium [**4**, Proposition 1]). *Let $\mathbf{z} \in \mathbb{C}^n$. The following two conditions are equivalent,*

(1) *The vectors $\mathrm{d}_{\mathbb{R}}g(\mathbf{z})$ and $\mathrm{d}_{\mathbb{R}}h(\mathbf{z})$ are linearly dependent over $\mathbb{R}$.*
(2) *There exists a complex number $\alpha \in \mathbb{S}^1$ such that $\overline{\mathrm{d}f(\mathbf{z})} = \alpha\bar{\mathrm{d}}f(\mathbf{z})$.*

The following proposition gives the differential of a mixed function in the variables $z_1,\dots,z_n,\bar{z}_1,\dots,\bar{z}_n$.

PROPOSITION 1.3. *Let $f\colon \mathbb{C}^n \to \mathbb{C}$ be a mixed function. Let $\mathbf{v} = (v_1,\dots,v_n)$ be a tangent vector at $\mathbf{z} \in \mathbb{C}^n$. Then*

$$df_{\mathbf{z}}(\mathbf{v}) = \sum_{j=1}^{n}\left(\frac{\partial f}{\partial z_j}(\mathbf{z})v_j + \frac{\partial f}{\partial \bar{z}_j}(\mathbf{z})\bar{v}_j\right).$$

PROOF. Write f in terms of its real and imaginary part as

$$f(\mathbf{z}) = g(\mathbf{z},\bar{\mathbf{z}}) + ih(\mathbf{z},\bar{\mathbf{z}}).$$

Thus, the differential of f at $\mathbf{z}$ is given by the matrix

$$df_{\mathbf{z}}(\mathbf{v}) = \begin{pmatrix} \frac{\partial g}{\partial z_1}(\mathbf{z}) & \cdots & \frac{\partial g}{\partial z_n}(\mathbf{z}) & \frac{\partial g}{\partial \bar{z}_1}(\mathbf{z}) & \cdots & \frac{\partial g}{\partial \bar{z}_n}(\mathbf{z}) \\ \frac{\partial h}{\partial z_1}(\mathbf{z}) & \cdots & \frac{\partial h}{\partial z_n}(\mathbf{z}) & \frac{\partial h}{\partial \bar{z}_1}(\mathbf{z}) & \cdots & \frac{\partial h}{\partial \bar{z}_n}(\mathbf{z}) \end{pmatrix}.$$

The vector $\mathbf{v}$ written in the variables $z_1,\dots,z_n,\bar{z}_1,\dots,\bar{z}_n$ is given by the vector $\mathbf{v} = (v_1,\dots,v_n,\bar{v}_1,\dots,\bar{v}_n)$. Hence $df_{\mathbf{z}}(\mathbf{v})$ is the vector in $\mathbb{R}^2$ given by

$$df_{\mathbf{z}}(\mathbf{v}) = \left(\sum_{j=1}^{n}\left(\frac{\partial g}{\partial z_j}(\mathbf{z})v_j + \frac{\partial g}{\partial \bar{z}_j}(\mathbf{z})\bar{v}_j\right), \sum_{j=1}^{n}\left(\frac{\partial h}{\partial z_j}(\mathbf{z})v_j + \frac{\partial h}{\partial \bar{z}_j}(\mathbf{z})\bar{v}_j\right)\right).$$

Now, identifying $(u,v) \in \mathbb{R}^2$ with the complex number $u + iv \in \mathbb{C}$ we have that

$$df_{\mathbf{z}}(\mathbf{v}) = \sum_{j=1}^{n}\left(\frac{\partial g}{\partial z_j}(\mathbf{z})v_j + \frac{\partial g}{\partial \bar{z}_j}(\mathbf{z})\bar{v}_j\right) + i\left(\sum_{j=1}^{n}\left(\frac{\partial h}{\partial z_j}(\mathbf{z})v_j + \frac{\partial h}{\partial \bar{z}_j}(\mathbf{z})\bar{v}_j\right)\right), \tag{1.1}$$

but we have that

$$g = \frac{1}{2}\left(f + \bar{f}\right), \qquad h = \frac{1}{2i}\left(f - \bar{f}\right),$$

and substituting these expressions into (1.1) we get the desired formula. □

1.1. Simplicial mixed polynomials. Following Oka [**4**, §2.3] we have the following definition.

DEFINITION 1.4. Let $\mu_j = (\mu_{j,1}, \dots, \mu_{j,n})$ and $\nu_j = (\nu_{j,1}, \dots, \nu_{j,n})$ be multi-indices and let $f\colon \mathbb{C}^n \to \mathbb{C}$ be a mixed polynomial written as

$$f(\mathbf{z}) = \sum_{j=1}^{m} c_j \mathbf{z}^{\mu_j} \bar{\mathbf{z}}^{\nu_j},$$

where $c_1, \dots, c_m$ are non-zero. Consider the following matrices

$$P = \begin{pmatrix} \mu_{1,1}+\nu_{1,1} & \cdots & \mu_{1,n}+\nu_{1,n} \\ \vdots & \vdots & \vdots \\ \mu_{m,1}+\nu_{m,1} & \cdots & \mu_{m,n}+\nu_{m,n} \end{pmatrix}, \qquad Q = \begin{pmatrix} \mu_{1,1}-\nu_{1,1} & \cdots & \mu_{1,n}-\nu_{1,n} \\ \vdots & \vdots & \vdots \\ \mu_{m,1}-\nu_{m,1} & \cdots & \mu_{m,n}-\nu_{m,n} \end{pmatrix}.$$

We say that f is *simplicial* if $m \leq n$ and the rank of the matrices P and Q are equal to m. We say that f is *radial full* (respectivelly *angular full*) if $n = m$ and P (respectively, Q) has rank m. If f is radial and angular full, then we say that f is *full*. We call the matrix P the *radial matrix* and Q the *angular matrix* of f.

1.2. Polar weighted homogeneous polynomials. Polar weighted homogeneous polynomials are real analytic maps which generalize complex weighted homogeneous polynomials, they are mixed polynomials which are weighted homogeneous with respect to an $\mathbb{R}^+$-action and also with respect to a $\mathbb{S}^1$-action. Let $p_1, \dots, p_n$ and $q_1, \dots, q_n$ be non-zero integers such that $\gcd(p_1, \dots, p_n) = 1$ and $\gcd(q_1, \dots, q_n) = 1$. Let $w \in \mathbb{C}^*$ written in its polar form $w = t\tau$, with $t \in \mathbb{R}^+$ and $\tau \in \mathbb{S}^1$. A *polar* $\mathbb{C}^*$*-action* on $\mathbb{C}^n$ with *radial weights* $(p_1, \dots, p_n)$ and *angular weights* $(q_1, \dots, q_n)$ is given by:

$$t\tau \bullet \mathbf{z} = (t^{p_1}\tau^{q_1} z_1, \dots, t^{p_n}\tau^{q_n} z_n). \tag{1.2}$$

DEFINITION 1.5. A mixed function $f\colon \mathbb{C}^n \to \mathbb{C}$ is *polar weighted homogeneous* if there exists $p_1, \dots, p_n$ positive integers, $q_1, \dots, q_n$ non-zero integers, a, c positive integers, and a polar $\mathbb{C}^*$-action given by (1.2) such that f satisfies the following functional equation

$$f(t\tau \bullet \mathbf{z}) = t^a \tau^c f(\mathbf{z}). \tag{1.3}$$

We say that the polar weighted homogeneous function f has *radial weight type* $(p_1, \dots, p_n; a)$ and *angular weight type* $(q_1, \dots, q_n; c)$.

EXAMPLES 1.6. As examples of polar weighted homogeneous polynomials we have the following:

I. Complex weighted homogeneous polynomials are a particular case of polar weighted homogeneous polynomials with no $\bar{z}_j$ for $j = 1, \dots, n$ and with $p_j = q_j$ and $a = c$. In particular, Brieskorn-Pham polynomials

$$z_1^{a_1} + \cdots + z_n^{a_n}, \quad n > 0,\ a_j \in \mathbb{N},\ a_j \geq 2,\ j = 1, \dots, n.$$

II. A mixed polynomial in $\mathbb{C}^n$ of the form

$$f(\mathbf{z}) = c_1 z_1^{a_1} \bar{z}_{\sigma(1)} + \cdots + c_n z_n^{a_n} \bar{z}_{\sigma(n)},$$

is called a *twisted Brieskorn-Pham polynomial* of class $\{a_1, \dots, a_n; \sigma\}$, where each $a_j \geq 2$, $j = 1, \dots, n$, the c_j are non-zero complex numbers and σ is a permutation of the set $\{1, \dots, n\}$ called the *twisting*.

Twisted Brieskorn-Pham polynomials were the first examples of polar weighted homogeneous polynomials and they were defined and studied by Seade in [**9**] and by Ruas, Seade and Verjovsky in [**8**]. Inspired by these examples, polar weighted homogeneous polynomials were introduced by Cisneros-Molina in [**1**] and later they were studied by Oka in [**4, 5**].

2. The embedding method

In this section we define the real embedding, we define the holomorphic function associated to a mixed function and show some of its properties.

Consider the map given by

$$\begin{aligned} c_n \colon \mathbb{C}^n &\to \mathbb{C}^n, \\ \mathbf{z} &\mapsto \bar{\mathbf{z}}. \end{aligned} \tag{2.1}$$

Let $\mathrm{id}_n \colon \mathbb{C}^n \to \mathbb{C}^n$ be the identity map and consider the following embedding

$$\begin{aligned} e_n = (\mathrm{id}_n, c_n) \colon \mathbb{C}^n &\to \mathbb{C}^n \oplus \mathbb{C}^n, \\ \mathbf{z} &\mapsto (\mathbf{z}, \bar{\mathbf{z}}). \end{aligned} \tag{2.2}$$

REMARK 2.1. Notice that the image $H = e_n(\mathbb{C}^n)$ is a $2n$-dimensional real subspace of $\mathbb{C}^n \oplus \mathbb{C}^n$.

We are going to generalize the map e_n. Set $\underline{n} = \{1, 2, \dots, n\}$, so the cardinality $|\underline{n}|$ of $\underline{n}$ is n. Let K be a subset of $\underline{n}$, such that $|K| = k$. Define the projection

$$\begin{aligned} p_K \colon \mathbb{C}^n &\to \mathbb{C}^k, \\ (z_1, \dots, z_n) &\mapsto (z_{i_1}, z_{i_2}, \dots, z_{i_k}), \qquad \text{where } i_r \in K, \text{ for } r = 1, \dots, k. \end{aligned} \tag{2.3}$$

Let $K, L \subset \underline{n}$ such that $|K| = k$ and $|L| = l$. Define the map

$$\begin{aligned} e_{K,L} = (p_K, c_l \circ p_L) \colon \mathbb{C}^n &\to \mathbb{C}^k \oplus \mathbb{C}^l \\ (z_1, \dots, z_n) &\mapsto \big((z_{i_1}, z_{i_2}, \dots, z_{i_k}), (\bar{z}_{j_1}, \bar{z}_{j_2}, \dots, \bar{z}_{j_l})\big). \end{aligned}$$

So we have that $e_n = e_{\underline{n},\underline{n}}$.

Let $f \colon \mathbb{C}^n \to \mathbb{C}$ be a mixed function. We define the following two subsets of $\underline{n}$

$$K = \{ i \in \underline{n} \mid \text{the variable } z_i \text{ appears in } f \},$$
$$L = \{ j \in \underline{n} \mid \text{the variable } \bar{z}_j \text{ appears in } f \}.$$

We also ask that $K \cup L = \underline{n}$, otherwise, let J be the complement of $K \cup L$ in $\underline{n}$ and suppose $|J| = j$, hence we can consider the function f defined on $\mathbb{C}^{n-j}$ instead of $\mathbb{C}^n$ using the coordinates with indices in $K \cup L$.

DEFINITION 2.2. Let $f \colon \mathbb{C}^n \to \mathbb{C}$ be a mixed function. Let K and L be as before. Suppose that $|K| = k$, $|L| = l$ and consider the complex space $\mathbb{C}^k \oplus \mathbb{C}^l$ with variables $z_1, \dots, z_{i_k}, w_1, \dots, w_{j_l}$. Define a function $F \colon \mathbb{C}^k \oplus \mathbb{C}^l \to \mathbb{C}$ by replacing in f each variable $\bar{z}_{j_s}$ with $j_s \in L$ by the corresponding complex variable w_{j_s}. The function F is called the *holomorphic function associated to the mixed function f*.

From the definition it is easy to see that the following diagram is commutative

$$\begin{array}{ccc} \mathbb{C}^n & \xrightarrow{e_{K,L}} & \mathbb{C}^k \oplus \mathbb{C}^l \\ & \searrow^{f} & \downarrow F \\ & & \mathbb{C} \end{array}$$

so we have that

$$f = F \circ e_{K,L}. \tag{2.4}$$

Set

$$\begin{aligned} V_f &= f^{-1}(0), & V_f^* &= V_f - \operatorname{Sing} V_f, \\ V_F &= F^{-1}(0), & V_F^* &= V_F - \operatorname{Sing} V_F. \end{aligned}$$

Let $\mathbf{z} \in V_f^*$. As usual denote by $T_{\mathbf{z}} V_f^*$ the tangent space of V_f^* at the point $\mathbf{z}$. We use the analogous notation for the points in V_F^*.

As before, denote by H the image $e_{K,L}(\mathbb{C}^n)$. Notice that

$$V_f \cong V_F \cap H. \tag{2.5}$$

Lemma 2.3. *Let $\mathbf{z} \in V_f^*$. Then $e_{K,L}(\mathbf{z}) \in V_F^*$.*

Proof. Since $\mathbf{z}$ belongs to the set V_f^* we have that $\operatorname{rank}_{\mathbb{R}} Df_{\mathbf{z}} = 2$. Suppose $e_{K,L}(\mathbf{z}) \in \operatorname{Sing} V_F$, then $\operatorname{rank}_{\mathbb{R}} DF_{e_{K,L}(\mathbf{z})} < 2$. By (2.4) and the chain rule we have that $Df_{\mathbf{z}} = DF_{e_{K,L}(\mathbf{z})} \circ e_{K,L}$. Hence

$$2 = \operatorname{rank}_{\mathbb{R}} Df_{\mathbf{z}} = \operatorname{rank}_{\mathbb{R}}(DF_{e_{K,L}(\mathbf{z})} \circ e_{K,L}) \leq \operatorname{rank}_{\mathbb{R}} DF_{e_{K,L}(\mathbf{z})} < 2,$$

which clearly is a contradiction. □

Lemma 2.4. *Let $\mathbf{z} \in V_f$ such that $e_{K,L}(\mathbf{z}) \in V_F^*$. Then, $\mathbf{z} \in V_f^*$ if and only if $T_{e_{K,L}(\mathbf{z})} V_F^* \pitchfork H$.*

Proof. Suppose $\mathbf{z} \in V_f^*$, then $\dim_{\mathbb{R}} T_{\mathbf{z}} V_f^* = 2n - 2$. Since $e_{K,L}(\mathbf{z}) \in V_F^*$ we have that $\dim_{\mathbb{R}} T_{e_{K,L}(\mathbf{z})} V_F^* = 2(k+l) - 2$. We also have that $\dim_{\mathbb{R}} H = 2n$. On the other hand we have that

$$T_{\mathbf{z}} V_f^* \cong T_{e_{K,L}(\mathbf{z})} V_F^* \cap H. \tag{2.6}$$

Hence, $\operatorname{codim}_{\mathbb{R}} T_{e_{K,L}(\mathbf{z})} V_F^* \cap H = \operatorname{codim}_{\mathbb{R}} T_{e_{K,L}(\mathbf{z})} V_F^* + \operatorname{codim}_{\mathbb{R}} H$, and this is the case only if $T_{e_{K,L}(\mathbf{z})} V_F^* \pitchfork H$.

Conversely, suppose that $T_{e_{K,L}(\mathbf{z})} V_F^* \pitchfork H$. We have that $f = F \circ e_{K,L}$ and by the chain rule $Df_{\mathbf{z}} = DF_{e_{K,L}(\mathbf{z})} \circ e_{K,L}$. The rank of the map $e_{K,L}$ is $2n$, since $e_{K,L}(\mathbf{z}) \in V_F^*$, we have that $\dim_{\mathbb{R}} T_{e_{K,L}(\mathbf{z})} V_F^* = 2(k+l) - 2$. Since $T_{e_{K,L}(\mathbf{z})} V_F^* \pitchfork H$ we have that $\dim_{\mathbb{R}} T_{e_{K,L}(\mathbf{z})} V_F^* \cap H = 2n - 2$. Hence, by (2.6)

$$\dim_{\mathbb{R}} T_{\mathbf{z}} V_f^* = \ker Df_{\mathbf{z}} = 2n - 2.$$

Therefore the rank of $Df_{\mathbf{z}}$ is 2 and $\mathbf{z} \in V_f^*$. □

Consider the standard Hermitian product $\langle\ ,\ \rangle$ on $\mathbb{C}^n$; the Hermitian inner product on the direct sum $\mathbb{C}^k \oplus \mathbb{C}^l$ gives also the standard Hermitian product

$$\langle\ ,\ \rangle : \mathbb{C}^k \oplus \mathbb{C}^l \to \mathbb{C},$$

$$\langle (\mathbf{z}, \mathbf{w}), (\mathbf{z}', \mathbf{w}') \rangle = \sum_{j=1}^{k} z_j \bar{z}'_j + \sum_{j=1}^{l} w_j \bar{w}'_j.$$

Recall that the Euclidean inner product on $\mathbb{C}^n$ is given by the real part of the Hermitian inner product.

Denote by $\mathbb{S}_\varepsilon^{2n-1}$ the $(2n-1)$-sphere in $\mathbb{C}^n$ of radius $\varepsilon > 0$ with centre at the origin. Let $\mathbf{z} \in \mathbb{C}^n$ and consider the following tangent spaces:

$$T_{\mathbf{z}}\mathbb{S}_{\|\mathbf{z}\|}^{2n-1} = \{v \in \mathbb{C}^n \mid \mathrm{Re}\langle \mathbf{z}, v\rangle = 0\},$$
$$T_{e_n(\mathbf{z})}\mathbb{S}_{\|e_n(\mathbf{z})\|}^{4n-1} = \{w \in \mathbb{C}^n \oplus \mathbb{C}^n \mid \mathrm{Re}\langle e_n(\mathbf{z}), w\rangle = 0\}.$$

We have the following proposition:

PROPOSITION 2.5. *Let* $\mathbf{z} \in \mathbb{C}^n$. *Then the map*

$$e_n = (\mathrm{id}_n, c_n)\colon \mathbb{C}^n \to \mathbb{C}^n \oplus \mathbb{C}^n,$$
$$\mathbf{z} \mapsto (\mathbf{z}, \bar{\mathbf{z}}),$$

induces the map

$$e_n\colon T_{\mathbf{z}}\mathbb{S}_{\|\mathbf{z}\|}^{2n-1} \to T_{e_n(\mathbf{z})}\mathbb{S}_{\|e_n(\mathbf{z})\|}^{4n-1}.$$

PROOF. For $v \in T_{\mathbf{z}}\mathbb{S}_{\|\mathbf{z}\|}^{2n-1}$ we have $\mathrm{Re}\langle \mathbf{z}, v\rangle = \mathrm{Re}\left(\sum_{j=1}^n z_j\bar{v}_j\right) = 0$, then

$$\mathrm{Re}\langle e(\mathbf{z}), e(v)\rangle = \mathrm{Re}\left(\sum_{j=1}^n z_j\bar{v}_j + \sum_{j=1}^n \bar{z}_j v_j\right) = 2\mathrm{Re}\left(\sum_{j=1}^n z_j\bar{v}_j\right) = 0.$$

□

3. The isotopy theorem

In this section we use the embedding method in order to give another proof of the results given in [**3**, **6**]. Our work relays in a more geometrical way of thinking the mixed singularities, this allow us to give another interpretation of some properties.

Following [**3**, **6**] consider the following families of mixed polynomials.

Let $\mathbf{a} = (a_1, a_2, a_3)$ and $\mathbf{b} = (b_1, b_2, b_3)$, with $a_i, b_i \in \mathbb{N}$ and $a_i > b_i \geq 0$, $i = 1, \ldots, 3$.

(3.1)
$$f_A(\mathbf{z}) = z_1^{a_1}\bar{z}_1^{b_1} + z_2^{a_2}\bar{z}_2^{b_2} + z_3^{a_3}\bar{z}_3^{b_3},$$

(3.2)
$$f_B(\mathbf{z}) = z_1^{a_1}\bar{z}_1^{b_1}z_2 + z_2^{a_2}\bar{z}_2^{b_2}z_3 + z_3^{a_3}\bar{z}_3^{b_3}, \qquad (a_2, a_3) \neq (b_2 + 2, b_3 - 1),$$

(3.3)
$$f_C(\mathbf{z}) = z_1^{a_1}\bar{z}_1^{b_1}z_2 + z_2^{a_2}\bar{z}_2^{b_2}z_3 + z_3^{a_3}\bar{z}_3^{b_3}z_1, \quad \begin{cases} (a_1 - b_1)(a_2 - b_2)(a_3 - b_3) \neq -1, \\ (a_{i-1}, a_{i+1}) \neq (b_{i-1} + 1, b_{i+1}), \\ (a_{i-1}, a_{i+1}) \neq (b_{i-1} - 1, b_{i+1} + 2), \\ i = 1, 2, 3. \end{cases}$$

With the given conditions on the exponents, these correspond respectively (up to a change of coordinates) to classes **I**, **IV.a** and **V** of polar weighted homogeneous polynomials given in [**2**, Theorem 4.5], which generalize Orlik and Wagreich classes of complex weighted homogeneous polynomials with isolated critical point, given in [**7**, §3.1]. These mixed polynomials are simplicial full [**2**, Corollary 4.7] and have isolated critical point [**2**, Theorem 4.10].

When $b_i = 1$ for $i = 1, \dots, n$, the maps of the form given in (3.1) are *twisted Brieskorn-Pham polynomials* with *twisting* the identity permutation.

Now define

$$h_j(z_j) = \begin{cases} z_j & \text{if} \quad a_j - b_j > 0, \\ \bar{z}_j & \text{if} \quad a_j - b_j < 0, \end{cases}$$

for $j = 1, 2, 3$ and consider the maps

$$\begin{aligned} g_A(\mathbf{z}) &= h_1(z_1)^{|a_1-b_1|} + h_2(z_2)^{|a_2-b_2|} + h_3(z)_3^{|a_3-b_3|}, \\ g_B(\mathbf{z}) &= h_1(z_1)^{|a_1-b_1|} z_2 + h_2(z_2)^{|a_2-b_2|} z_3 + h_3(z_3)^{|a_3-b_3|}, \\ g_C(\mathbf{z}) &= h_1(z_1)^{|a_1-b_1|} z_2 + h_2(z_2)^{|a_2-b_2|} z_3 + h_3(z_3)^{|a_3-b_3|} z_1. \end{aligned}$$

Consider the map $f_{l,t} \colon \mathbb{C}^3 \to \mathbb{C}$ given by $f_{l,t}(\mathbf{z}) := (1-t) f_l(\mathbf{z}) + t g_l(\mathbf{z})$ for $0 \le t \le 1$ and consider the family $V_{l,t} = f_{l,t}^{-1}(0)$ where $l = A, B, C$.

LEMMA 3.1. *The function $f_{l,t}(\mathbf{z}) \colon \mathbb{C}^3 \to \mathbb{C}$ has an isolated critical point for any $0 \le t \le 1$ and $l = A, B, C$.*

PROOF. **Case $l = A$:** Under a change of coordinates we can assume that

$$g_A(\mathbf{z}) = z_1^{a_1-b_1} + z_2^{a_2-b_2} + z_3^{a_3-b_3},$$

and this case was proved by Oka [**6**, Lemma 1].

Case $l = B$: Under a change of coordinates we can assume that

$$g_B(\mathbf{z}) = z_1^{a_1-b_1} z_2 + h_2(z_2)^{|a_2-b_2|} z_3 + h_3(z_3)^{|a_3-b_3|}.$$

The case $a_2 > b_2$ and $a_3 > b_3$ was proved by Oka [**6**, Lemma 9].

Suppose that $a_2 < b_2$ and $a_3 > b_3$, then

$$g_B(\mathbf{z}) = z_1^{a_1-b_1} z_2 + \bar{z}_2^{b_2-a_2} z_3 + z_3^{a_3-b_3}.$$

For $t = 0, 1$ the assertion is true by [**2**, Theorem 4.10], therefore suppose $t \in (0, 1)$. We have that

$$f_{B,t}(\mathbf{z}) = (1-t)\left(z_1^{a_1} \bar{z}_1^{b_1} z_2 + z_2^{a_2} \bar{z}_2^{b_2} z_3 + z_3^{a_3} \bar{z}_3^{b_3}\right) + t\left(z_1^{a_1-b_1} z_2 + \bar{z}_2^{b_2-a_2} z_3 + z_3^{a_3-b_3}\right).$$

Suppose that (z_1, z_2, z_3) is a critical point of $f_{B,t}$, then by Proposition 1.2 exist $\alpha \in \mathbb{S}^1$ such that

$$\overline{\mathrm{d} f_{B,t}(\mathbf{z})} = \alpha \bar{\mathrm{d}} f_{B,t}(\mathbf{z}).$$

This equality gives us the following system

$$\bar{z}_1^{a_1-b_1-1} \bar{z}_2 \left((1-t) a_1 |z_1|^{2b_1} + t(a_1 - b_1)\right) = \alpha (1-t) b_1 z_1^{a_1} \bar{z}_1^{b_1-1} z_2, \tag{3.4}$$

$$\begin{aligned} (1-t)(z_1^{b_1} \bar{z}^{a_1} + a_2 z_2^{b_2} \bar{z}^{a_2-1} \bar{z}_3) + t \bar{z}_1^{a_1-b_1} \\ = \alpha \bar{z}_2^{b_2-a_2-1} z_3((1-t) b_2 |z_2|^{2a_2} + t(b_2 - a_2)), \end{aligned} \tag{3.5}$$

$$\begin{aligned} \bar{z}_3^{a_3-b_3-1} \left((1-t) a_3 |z_3|^{2b_3} + t(a_3 - b_3)\right) + z_2^{b_2-a_2} \left((1-t)|z_2|^{2a_2} + t\right) \\ = \alpha (1-t) b_3 z_3^{a_3} \bar{z}_3^{b_3-1}. \end{aligned} \tag{3.6}$$

Suppose $z_3 \neq 0$ and let $s = \min\{j \mid z_k \neq 0,\ k \ge j\}$.

If $s=1$, multiplying (3.4) by $\bar{z}_1$ we get

$$\bar{z}_1^{a_1-b_1}\bar{z}_2\left((1-t)a_1|z_1|^{2b_1}+t(a_1-b_1)\right)=z_1^{a_1-b_1}z_2(\alpha(1-t)b_1|z_1|^{2b_1}),$$

computing the norm and since $|z_1|,|z_2|\neq 0$ we have

$$(1-t)a_1|z_1|^{2b_1}+t(a_1-b_1)=(1-t)b_1|z_1|^{2b_1},$$

therefore

$$(a_1-b_1)\left((1-t)|z_1|^{2b_1}+t\right)=0,$$

and this can not happen since $(a_1-b_1)\neq 0$, $t\in(0,1)$ and $|z_1|>0$.

If $s=2$, using (3.5) we get a contradiction in the same way as in the previous case.

If $s=3$, using (3.6) again we get a contradiction using the same idea.

Finally, suppose that $z_3=0$ and define $k=\min\{j\mid z_k=0,\ k\geq j\}$.

If $k=3$, by (3.6) and the fact that $a_3-b_3\neq 1$ (see conditions of family **IV.a**)

$$(1-t)|z_2|^{2a_2}+t=0,$$

and this can not happen.

If $k=2$, by (3.5)

$$(1-t)|z_1|^{2b_1}+t=0,$$

but again, this can not happen.

Therefore $k=1$, so the origin is the only critical point.

The cases $a_2>b_2$, $a_3<b_3$ and $a_2<b_2$, $a_3<b_3$ are analogous.

Case $l=C$: For $t=0,1$ the assertion is true by [**2**, Theorem 4.10], therefore suppose t belongs to the set $(0,1)$. If $a_j>b_j$ for $j=1,2,3$ then,

$$f_{C,t}(\mathbf{z})=(1-t)\left(z_1^{a_1}\bar{z}_1^{b_1}z_2+z_2^{a_2}\bar{z}_2^{b_2}z_3+z_3^{a_3}\bar{z}_3^{b_3}z_1\right)+t\left(z_1^{a_1-b_1}z_2+z_2^{a_2-b_2}z_3+z_3^{a_3-b_3}z_1\right). \tag{3.7}$$

Suppose that (z_1,z_2,z_3) is a critical point of $f_{C,t}$, then by Proposition 1.2 exist $\alpha\in\mathbb{S}^1$ such that

$$\overline{\mathrm{d}f_{C,t}(\mathbf{z})}=\alpha\bar{\mathrm{d}}f_{C,t}(\mathbf{z}).$$

Recall that by definition of f_A, f_B and f_C we have that $a_i>b_i$. This fact and the previous equality give us the following system

$$\bar{z}_1^{a_1-b_1-1}\bar{z}_2\left((1-t)a_1|z_1|^{2b_1}+t(a_1-b_1)\right)+\bar{z}_3^{a_3-b_3}\left((1-t)|z_3|^{2b_3}+t\right)=\alpha(1-t)b_1z_1^{a_1}\bar{z}_1^{b_1-1}z_2, \tag{3.8}$$

$$\bar{z}_2^{a_2-b_2-1}\bar{z}_3\left((1-t)a_2|z_2|^{2b_2}+t(a_2-b_2)\right)+\bar{z}_1^{a_1-b_1}\left((1-t)|z_1|^{2b_1}+t\right)=\alpha(1-t)b_2z_2^{a_2}\bar{z}_2^{b_2-1}z_3, \tag{3.9}$$

$$\bar{z}_3^{a_3-b_3-1}\bar{z}_1\left((1-t)a_3|z_3|^{2b_3}+t(a_3-b_3)\right)+\bar{z}_2^{a_2-b_2}\left((1-t)|z_2|^{2b_2}+t\right)=\alpha(1-t)b_3z_3^{a_3}\bar{z}_3^{b_3-1}z_1. \tag{3.10}$$

If $z_1=0$, then by (3.10)

$$\bar{z}_2^{a_2-b_2}\left((1-t)|z_2|^{2b_2}+t\right)=0,$$

therefore $z_2 = 0$ and then by (3.8)

$$\bar{z}_3^{a_3-b_3}\left((1-t)|z_3|^{2b_3}+t\right)=0,$$

hence $z_3 = 0$.

Suppose that $z_3 = 0$, then by (3.9)

$$z_1^{b_1-a_1}\left((1-t)|z_1|^{2a_1}+t\right)=0,$$

therefore $z_1 = 0$.

Finally suppose that $z_j \neq 0$ for $j = 1, 2, 3$. Here is were we use the embedding method.

Consider the holomorphic function $F_{C,t}\colon \mathbb{C}^3 \oplus \mathbb{C}^3 \to \mathbb{C}$ associated to $f_{C,t}$ given by

$$\begin{aligned} F_{C,t}(\mathbf{z},\mathbf{w}) = z_1^{a_1-b_1} z_2 \left((1-t)z_1^{b_1}w_2^{b_1}+t\right) + z_2^{a_2-b_2} z_3 \left((1-t)z_2^{b_2}w_2^{b_2}+t\right) \\ + z_3^{a_3-b_3} z_1 \left((1-t)z_3^{b_3}w_3^{b_3}+t\right) \end{aligned}$$

and to simplify notation we denote by e the map e_3 defined in (2.2)

$$\begin{aligned} e\colon \mathbb{C}^3 &\to \mathbb{C}^3 \oplus \mathbb{C}^3 \\ \mathbf{z} &\mapsto (\mathbf{z}, \bar{\mathbf{z}}), \end{aligned}$$

therefore

$$f_{C,t}(\mathbf{z}) = F_{C,t}(e(\mathbf{z})).$$

Since e is a $\mathbb{R}$-linear embedding, its differential is e itself. Let $\mathbf{z} \in \mathbb{C}^3$ and consider $\mathbf{v} \in \mathbb{C}^3$ as a *real* tangent vector of $\mathbb{C}^3$ at $\mathbf{z}$, hence we have that

$$d(f_{C,t})_{\mathbf{z}}(\mathbf{v}) = \boldsymbol{\nabla} F_{C,t}(e(\mathbf{z})) \cdot e(\mathbf{v}),$$

where $d(f_{C,t})_{\mathbf{z}}(\mathbf{v})$ is given by Proposition 1.3. Notice that even though $\boldsymbol{\nabla} F_{C,t}(e(\mathbf{z}))$ is $\mathbb{C}$-linear, the map $d(f_{C,t})_{\mathbf{z}}$ is only $\mathbb{R}$-linear since e is only $\mathbb{R}$-linear. Now we have that

$$\begin{aligned} &\boldsymbol{\nabla} F_{C,t}(\mathbf{z},\mathbf{w}) = \\ &\quad\Big((1-t)\left(a_1 z_1^{a_1-1} w_1^{b_1} z_2 + z_3^{a_3} w_3^{b_3}\right) + t\left((a_1-b_1)z_1^{a_1-b_1-1} z_2 + z_3^{a_3-b_3}\right), \\ &\quad(1-t)\left(z_1^{a_1} w_1^{b_1} + a_2 z_2^{a_2-1} w_2^{b_2} z_3\right) + t\left(z_1^{a_1-b_1} + (a_2-b_2) z_2^{a_2-b_2-1} z_3\right), \\ &\quad(1-t)\left(z_2^{a_2} w_2^{b_2} + a_3 z_3^{a_3-1} w_3^{b_3} z_1\right) + t\left(z_2^{a_2-b_2} + (a_3-b_3) z_3^{a_3-b_3-1} z_1\right), \\ &\quad(1-t) b_1 z_1^{a_1} w_1^{b_1-1} z_2, \\ &\quad(1-t) b_2 z_2^{a_2} w_2^{b_2-1} z_3, \\ &\quad(1-t) b_3 z_3^{a_3} w_3^{b_3-1} z_1\Big). \end{aligned}$$

Therefore

$$\begin{aligned}
\nabla F_{C,t}(e(\mathbf{z})) = \\
&\Big(z_1^{a_1-b_1-1} z_2 \left((1-t)a_1|z_1|^{2b_1} + t(a_1-b_1)\right) + z_3^{a_3-b_3}\left((1-t)|z_3|^{2b_3}+t\right), \\
&z_2^{a_2-b_2-1} z_3 \left((1-t)a_2|z_2|^{2b_2} + t(a_2-b_2)\right) + z_1^{a_1-b_1}\left((1-t)|z_1|^{2b_1}+t\right), \\
&z_3^{a_3-b_3-1} z_1 \left((1-t)a_3|z_3|^{2b_3} + t(a_3-b_3)\right) + z_2^{a_2-b_2}\left((1-t)|z_2|^{2b_2}+t\right), \\
&(1-t)b_1 z_1^{a_1}\bar{z}_1^{b_1-1} z_2, \\
&(1-t)b_2 z_2^{a_2}\bar{z}_2^{b_2-1} z_3, \\
&(1-t)b_3 z_3^{a_3}\bar{z}_3^{b_3-1} z_1\Big).
\end{aligned}$$

Since $z_j \neq 0$ for $j = 1,2,3$, then $e(\mathbf{z})$ is a regular point of the holomorphic map $F_{C,t}$ and the kernel of $\nabla F_{C,t}(e(\mathbf{z}))$ is a complex subspace of complex codimension 1, that is, of real codimension 2. In order to prove that $\mathbf{z}$ is a regular point of $f_{C,t}$ we need to find a 2-dimensional *real* subspace in the image $e(\mathbb{C}^3)$ which is complementary to the kernel of $\nabla F_{C,t}(e(\mathbf{z}))$.

Consider the following *real* basis of $\mathbb{C}^3$:

$$\begin{aligned}
v_1 &= (z_1, 0, 0), & iv_1 &= (iz_1, 0, 0), \\
v_2 &= (0, z_2, 0), & iv_2 &= (0, iz_2, 0), \\
v_3 &= (0, 0, z_3), & iv_3 &= (0, 0, iz_3),
\end{aligned}$$

and define

$$\begin{aligned}
P_j &= (1-t)(a_j+b_j)|z_j|^{2b_j} + t(a_j-b_j), \\
Q_j &= (1-t)(a_j-b_j)|z_j|^{2b_j} + t(a_j-b_j), \\
R_j &= (1-t)|z_{j-1}|^{2b_{j-1}} + t.
\end{aligned} \tag{3.11}$$

Notice that

$$0 < R_{j+1} < Q_j < P_j. \tag{3.12}$$

Also by the definition of the function f_C we have that $a_j - b_j \neq 1$ for any $j = 1,2,3$. Therefore we have that

$$0 < (a_j - b_j - 1)\left((1-t)|z_j|^{2b_j} + t\right). \tag{3.13}$$

We have that

$$\begin{aligned}
\nabla F_{C,t}(e(\mathbf{z})) \cdot e(v_1)^\top &= z_1^{a_1-b_1} z_2 P_1 + z_3^{a_3-b_3} z_1 R_1, \\
\nabla F_{C,t}(e(\mathbf{z})) \cdot e(iv_1)^\top &= i\left(z_1^{a_1-b_1} z_2 Q_1 + z_3^{a_3-b_3} z_1 R_1\right), \\
\nabla F_{C,t}(e(\mathbf{z})) \cdot e(v_2)^\top &= z_2^{a_2-b_2} z_3 P_2 + z_1^{a_1-b_1} z_2 R_2, \\
\nabla F_{C,t}(e(\mathbf{z})) \cdot e(iv_2)^\top &= i\left(z_2^{a_2-b_2} z_3 Q_2 + z_1^{a_1-b_1} z_2 R_2\right), \\
\nabla F_{C,t}(e(\mathbf{z})) \cdot e(v_3)^\top &= z_3^{a_3-b_3} z_1 P_3 + z_2^{a_2-b_2} z_3 R_3, \\
\nabla F_{C,t}(e(\mathbf{z})) \cdot e(iv_3)^\top &= i\left(z_3^{a_3-b_3} z_1 Q_3 + z_2^{a_2-b_2} z_3 R_3\right).
\end{aligned} \tag{3.14}$$

We want to check that $e(v_j)$ and $e(iv_j)$, $j = 1,2,3$, are not in the kernel of $\nabla F_{C,t}(e(\mathbf{z}))$.

Suppose that

$$\nabla F_{C,t}(e(\mathbf{z}))\cdot e(v_1)^\top = z_1^{a_1-b_1}z_2P_1 + z_3^{a_3-b_3}z_1R_1 = 0. \tag{3.15}$$

Multiplying equations (3.8), (3.9) and (3.10) by $\bar{z}_1$, $\bar{z}_2$ and $\bar{z}_3$ respectively we get

$$\bar{z}_1^{a_1-b_1}\bar{z}_2\left((1-t)a_1|z_1|^{2b_1}+t(a_1-b_1)\right)+\bar{z}_3^{a_3-b_3}\bar{z}_1R_1 = \alpha(1-t)b_1z_1^{a_1-b_1}z_2|z_1|^{2b_1}, \tag{3.16}$$

$$\bar{z}_2^{a_2-b_2}\bar{z}_3\left((1-t)a_2|z_2|^{2b_2}+t(a_2-b_2)\right)+\bar{z}_1^{a_1-b_1}\bar{z}_2R_2 = \alpha(1-t)b_2z_2^{a_2-b_2}z_3|z_2|^{2b_2}, \tag{3.17}$$

$$\bar{z}_3^{a_3-b_3}\bar{z}_1\left((1-t)a_3|z_3|^{2b_3}+t(a_3-b_3)\right)+\bar{z}_2^{a_2-b_2}\bar{z}_3R_3 = \alpha(1-t)b_3z_3^{a_3-b_3}z_1|z_3|^{2b_3}, \tag{3.18}$$

using equation (3.16) and the conjugate of (3.15) we have

$$\begin{aligned}\alpha(1-t)b_1z_1^{a_1-b_1}|z_1|^{2b_1}z_2 &= \bar{z}_1^{a_1-b_1}\bar{z}_2\left((1-t)a_1|z_1|^{2b_1}+t(a_1-b_1)-P_1\right)\\ &= -\bar{z}_1^{a_1-b_1}\bar{z}_2(1-t)b_1|z_1|^{2b_1},\end{aligned}$$

therefore

$$\alpha z_1^{a_1-b_1}z_2 = -\bar{z}_1^{a_1-b_1}\bar{z}_2.$$

Now multiplying (3.17) by $z_1^{a_1-b_1}z_2$ we get

$$\begin{aligned}&\alpha z_1^{a_1-b_1}z_2(1-t)b_2z_2^{a_2-b_2}|z_2|^{2b_2}z_3 = -z_2^{a_2-b_2}z_3\bar{z}_1^{a_1-b_1}\bar{z}_2(1-t)b_2|z_2|^{2b_2}\\ &= \bar{z}_2^{a_2-b_2}\bar{z}_3z_1^{a_1-b_1}z_2\left((1-t)a_2|z_2|^{2b_2}+t(a_2-b_2)\right)+|z_1|^{2(a_1-b_1)}|z_2|^2R_2.\end{aligned}$$

Hence we get

$$\begin{aligned}&-\operatorname{Re}\left(z_2^{a_2-b_2}z_3\bar{z}_1^{a_1-b_1}\bar{z}_2\right)(1-t)b_2|z_2|^{2b_2}\\ &= \operatorname{Re}\left(\bar{z}_2^{a_2-b_2}\bar{z}_3z_1^{b_1-a_1}z_2\right)\left((1-t)a_2|z_2|^{2b_2}+t(a_2-b_2)\right)+|z_1|^{2(a_1-b_1)}|z_2|R_2\\ &= \operatorname{Re}\left(z_2^{a_2-b_2}z_3\bar{z}_1^{a_1-b_1}\bar{z}_2\right)\left((1-t)a_2|z_2|^{2b_2}+t(a_2-b_2)\right)+|z_1|^{2(a_1-b_1)}|z_2|R_2,\end{aligned}$$

so

$$\operatorname{Re}\left(z_2^{a_2-b_2}z_3\bar{z}_1^{a_1-b_1}\bar{z}_2\right)P_2+|z_1|^{2(a_1-b_1)}|z_2|R_2 = 0. \tag{3.19}$$

Also

$$\begin{aligned}&-\operatorname{Im}\left(z_2^{a_2-b_2}z_3\bar{z}_1^{a_1-b_1}\bar{z}_2\right)(1-t)b_2|z_2|^{2b_2}\\ &= \operatorname{Im}\left(\bar{z}_2^{a_2-b_2}\bar{z}_3z_1^{b_1-a_1}z_2\left((1-t)a_2|z_2|^{2b_2}+t(a_2-b_2)\right)+|z_1|^{2(a_1-b_1)}|z_2|R_2\right)\\ &= -\operatorname{Im}\left(z_2^{a_2-b_2}z_3\bar{z}_1^{a_1-b_1}\bar{z}_2\right)\left((1-t)a_2|z_2|^{2b_2}+t(a_2-b_2)\right),\end{aligned}$$

so

$$\operatorname{Im}\left(z_2^{a_2-b_2}z_3\bar{z}_1^{a_1-b_1}\bar{z}_2\right)Q_2 = 0,$$

and since $Q_2\neq 0$, then $z_2^{a_2-b_2}z_3\bar{z}_1^{a_1-b_1}\bar{z}_2\in\mathbb{R}$ and using (3.19) we get

$$z_2^{a_2-b_2}z_3P_2+z_1^{a_1-b_1}z_2R_2 = 0. \tag{3.20}$$

Using equation (3.17) we get

$$\bar{z}_2^{a_2-b_2}\bar{z}_3\left((1-t)a_2|z_2|^{2b_2}+t(a_2-b_2)\right)-\bar{z}_2^{a_2-b_2}\bar{z}_3P_2 = \alpha(1-t)b_2z_2^{a_2-b_2}z_3|z_2|^{2b_2},$$

hence

$$-\bar{z}_2^{a_2-b_2}\bar{z}_3 = \alpha z_2^{a_2-b_2} z_3.$$

Therefore equation (3.17) is just

$$\begin{aligned}
0 &= \bar{z}_2^{a_2-b_2}\bar{z}_3\left((1-t)a_2|z_2|^{2b_2} + t(a_2-b_2) + (1-t)b_2|z_2|^{2b_2}\right) + \bar{z}_1^{a_1-b_1}\bar{z}_2 R_2\\
&= \bar{z}_2^{a_2-b_2}\bar{z}_3 P_2 + \bar{z}_1^{a_1-b_1}\bar{z}_2 R_2\\
&= z_2^{a_2-b_2} z_3 P_2 + z_1^{a_1-b_1} z_2 R_2\\
&= \nabla F_{C,t}(e(\mathbf{z}))\cdot e(v_2)^\top.
\end{aligned}$$

In the same way we get

$$\nabla F_{C,t}(e(\mathbf{z}))\cdot e(v_3)^\top = 0.$$

This give us the system

$$\begin{aligned}
z_1^{a_1-b_1} z_2 P_1 + z_3^{a_3-b_3} z_1 R_1 &= 0,\\
z_2^{a_2-b_2} z_3 P_2 + z_1^{a_1-b_1} z_2 R_2 &= 0,\\
z_3^{a_3-b_3} z_1 P_3 + z_2^{a_2-b_2} z_3 R_3 &= 0,
\end{aligned}$$

therefore

$$z_3^{a_3-b_3} z_1\left(R_1R_2R_3 + P_1P_2P_3\right) = 0,$$

but this can not happen since $z_3^{a_3-b_3} z_1 \neq 0$ and $R_j, P_j > 0$ for all $j = 1,2,3$.

Using a similiar argument we can deduce that

$$\nabla F_{C,t}(e(\mathbf{z}))\cdot e(v_j)^\top \neq 0, \qquad \nabla F_{C,t}(e(\mathbf{z}))\cdot e(iv_j)^\top \neq 0, \tag{3.21}$$

for all $j = 1,2,3$.

Finally suppose that for all $j = 1,2,3$, there exists $s_j \in \mathbb{R}$ such that

$$\nabla F_{C,t}(e(\mathbf{z}))\cdot e(v_j + s_j i v_j) = 0.$$

This gives us the system

$$\begin{aligned}
z_1^{a_1-b_1} z_2\left(P_1 + is_1Q_1\right) + z_3^{a_3-b_3} z_1 R_1\left(1+is_1\right) &= 0,\\
z_2^{a_2-b_2} z_3\left(P_2 + is_2Q_2\right) + z_1^{a_1-b_1} z_2 R_2\left(1+is_2\right) &= 0,\\
z_3^{a_3-b_3} z_1\left(P_3 + is_3Q_3\right) + z_2^{a_2-b_2} z_3 R_3\left(1+is_3\right) &= 0,
\end{aligned}$$

therefore

$$\begin{aligned}
z_3^{a_3-b_3} z_1(&R_1(1+is_1)R_2(1+is_2)R_3(1+is_3)\\
&+ (P_1+is_1Q_1)(P_2+is_2Q_2)(P_3+is_3Q_3) = 0,
\end{aligned}$$

and since $z_j \neq 0$ for all $j = 1,2,3$, then

$$R_1(1+is_1)R_2(1+is_2)R_3(1+is_3) + (P_1+is_1Q_1)(P_2+is_2Q_2)(P_3+is_3Q_3) = 0,$$

and using the norm we get

$$\left|R_1(1+is_1)R_2(1+is_2)R_3(1+is_3)\right| = \left|(P_1+is_1Q_1)(P_2+is_2Q_2)(P_3+is_3Q_3)\right|. \tag{3.22}$$

We have that $0 < R_{j+1} < Q_j < P_j$ (3.12) therefore

$$|(1+is_j)R_{j+1}|^2 = R_{j+1}^2 + s_j^2 R_{j+1}^2 < P_j^2 + s_j^2 Q_j^2 = |P_j + is_j Q_j|^2,$$

hence

$$|R_1(1+is_1)R_2(1+is_2)R_3(1+is_3)|^2 < |(P_1+is_1Q_1)(P_2+is_2Q_2)(P_3+is_3Q_3)|^2,$$

and this can not happen by (3.22).

Therefore for some $j \in \{1,2,3\}$ we have that $\nabla F_{C,t}(e(\mathbf{z})) \cdot e(v_j + siv_j) \neq 0$ for all $s \in \mathbb{R}$. Thus we have found a 2-dimensional *real* subspace in the image $e(\mathbb{C}^3)$ which is complementary to the kernel of $\nabla F_{C,t}(e(\mathbf{z}))$ and therefore $\mathbf{z}$ is a regular point of $f_{C,t}$.

So the origin is the only critical point.

Using the same method we can prove the assertion for the rest of the cases.

□

Again consider the map $f_{l,t}\colon \mathbb{C}^3 \to \mathbb{C}$ given by $f_{l,t}(\mathbf{z}) := (1-t)f_l(\mathbf{z}) + tg_l(\mathbf{z})$ for $0 \leq t \leq 1$ and the family $V_{l,t} = f_{l,t}^{-1}(0)$ where $l = A, B, C$.

PROPOSITION 3.2. *For any $0 \leq t \leq 1$ and $r > 0$, the sphere $\mathbb{S}_r^5$ intersects transversely $V_{l,t}$ for $l = A, B, C$.*

PROOF. (1) The case $l = A$ was proved by Oka [**6**, Lemma 2].

(2) Let $l = B$. Under a change of coordinates we can suppose that $a_1 > b_1$. The case $a_2 > b_2$, $a_3 > b_3$ was proved by Oka [**6**, Lemma 10]. Suppose that $a_2 < b_2$, $a_3 > b_3$, then

$$f_{B,t}(\mathbf{z}) = (1-t)\left(z_1^{a_1}\bar{z}_1^{b_1}z_2 + z_2^{a_2}\bar{z}_2^{b_2}z_3 + z_3^{a_3}\bar{z}_3^{b_3}\right) + t\left(z_1^{a_1-b_1}z_2 + \bar{z}_2^{b_2-a_2}z_3 + z_3^{a_3-b_3}\right).$$

For $t = 0, 1$ the assertion is true since f_B and g_B are polar weighted homogeneous with isolated singularity. Therefore suppose $t \in (0,1)$.

Consider the holomorphic associated function $F_{B,t}\colon \mathbb{C}^3 \oplus \mathbb{C}^3 \to \mathbb{C}$ of $f_{B,t}$ given by

$$\begin{aligned} F_{B,t}(\mathbf{z},\mathbf{w}) = z_1^{a_1-b_1}z_2\left((1-t)z_1^{b_1}w_1^{b_1} + t\right) + w_2^{b_2-a_2}z_3\left((1-t)z_2^{b_2}w_2^{b_2} + t\right) \\ + z_3^{a_3-b_3}\left((1-t)z_3^{b_3}w_3^{b_3} + t\right) \end{aligned}$$

and the map

$$\begin{aligned} e\colon \mathbb{C}^3 &\to \mathbb{C}^3 \oplus \mathbb{C}^3 \\ \mathbf{z} &\mapsto (\mathbf{z}, \bar{\mathbf{z}}), \end{aligned}$$

therefore

$$f_{B,t}(\mathbf{z}) = F_{B,t}(e(\mathbf{z})).$$

Now we have that

$$\begin{aligned}\nabla F_{B,t}(\mathbf{z},\mathbf{w}) = \Big(&(1-t)a_1 z_1^{a_1-1}w_1^{b_1}z_2 + t(a_1-b_1)z_1^{a_1-b_1-1}z_2,\\ &(1-t)\left(z_1^{a_1}w_1^{b_1} + a_2 z_2^{a_2-1}w_2^{b_2}z_3\right) + tz_1^{a_1-b_1},\\ &(1-t)\left(z_2^{a_2}w_2^{b_2} + a_3 z_3^{a_3-1}w_3^{b_3}\right) + t\left(w_2^{b_2-a_2} + (a_3-b_3)z_3^{a_3-b_3-1}\right),\\ &(1-t)b_1 z_1^{a_1}w_1^{b_1-1}z_2,\\ &(1-t)\left(b_2 z_2^{a_2}w_2^{b_2-1}z_3\right) + t(b_2-a_2)w_2^{b_2-a_2-1}z_3,\\ &(1-t)b_3 z_3^{a_3}w_3^{b_3-1}\Big).\end{aligned}$$

Therefore

$$\begin{aligned}&\nabla F_{B,t}(e(\mathbf{z})) =\\ &\quad\Big(z_1^{a_1-b_1-1}z_2\left((1-t)a_1|z_1|^{2b_1}+t\right),\\ &\quad(1-t)\left(a_2 z_2^{a_2-1}\bar{z}_2^{b_2}z_3\right) + z_1^{a_1-b_1}\left((1-t)|z_1|^{2b_1}+t\right),\\ &\quad z_3^{a_3-b_3-1}\left((1-t)a_3|z_3|^{2b_3}+t(a_3-b_3)\right) + \bar{z}_2^{b_2-a_2}\left((1-t)|z_2|^{2a_2}+t\right),\\ &\quad(1-t)b_1 z_1^{a_1}\bar{z}_1^{b_1-1}z_2,\\ &\quad\bar{z}_2^{b_2-a_2-1}z_3\left((1-t)b_2|z_2|^{2a_2}+t(b_2-a_2)\right),\\ &\quad(1-t)b_3 z_3^{a_3}\bar{z}_3^{b_3-1}\Big).\end{aligned}$$

By Proposition 2.5 we have that

$$e\colon T_{\mathbf{z}}\mathbb{S}^5_{\|\mathbf{z}\|} \to T_{e(\mathbf{z})}\mathbb{S}^{11}_{\|e(\mathbf{z})\|}. \tag{3.23}$$

Let $\mathbf{z} = (z_1, z_2, z_3) \in V_{B,t}$ with $t \neq 0, 1$. We want to find a tangent vector to $e(V_{B,t})$ at $e(\mathbf{z})$ which is not in $T_{e(\mathbf{z})}\mathbb{S}^{11}_{\|e(\mathbf{z})\|}$.

Consider the following vectors in $\mathbb{C}^3$

$$v_1 = (z_1, 0, 0), \quad v_2 = (0, z_2, 0), \quad v_3 = (0, 0, z_3),$$

and define

$$\begin{aligned}P_j &= (1-t)(a_j+b_j)|z_j|^{2b_j} + t(a_j-b_j),\\ \tilde{P}_2 &= (1-t)(a_2+b_2)|z_2|^{2a_2} + t(b_2-a_2),\\ R_{(j+1)\bmod 3} &= (1-t)|z_j|^{2b_j} + t,\\ \tilde{R}_3 &= (1-t)|z_2|^{2a_2} + t,\end{aligned}$$

for $j = 1, 3$.

We have that

$$\begin{aligned}\nabla F_{B,t}(e(\mathbf{z}))\cdot e(v_1)^\top &= z_1^{a_1-b_1}z_2P_1,\\ \nabla F_{B,t}(e(\mathbf{z}))\cdot e(v_2)^\top &= \bar{z}_2^{b_2-a_2}z_3\tilde{P}_2 + z_1^{a_1-b_1}z_2R_2,\\ \nabla F_{B,t}(e(\mathbf{z}))\cdot e(v_3)^\top &= z_3^{a_3-b_3}P_3 + \bar{z}_2^{b_2-a_2}z_3\tilde{R}_3.\end{aligned}$$

Let $r_1, r_2, r_3 \in \mathbb{R}$, then

$$\begin{aligned}
&\boldsymbol{\nabla} F_{B,t}(e(\mathbf{z})) \cdot e(r_1 v_1 + r_2 v_2 + r_3 v_3)^\top \\
&= z_1^{a_1-b_1} z_2 P_1 r_1 + r_2\left(\bar{z}_2^{b_2-a_2} z_3 \tilde{P}_2 + z_1^{a_1-b_1} z_2 R_2\right) + r_3\left(z_3^{a_3-b_3} P_3 + \bar{z}_2^{b_2-a_2} z_3 \tilde{R}_3\right) \\
&= z_1^{a_1-b_1} z_2 (P_1 r_1 + r_2 R_2) + \bar{z}_2^{b_2-a_2} z_3 (r_2 \tilde{P}_2 + r_3 \tilde{R}_3) + z_3^{a_3-b_3} r_3 P_3,
\end{aligned}$$

also we have

$$f_{B,t}(\mathbf{z}) = z_1^{a_1-b_1} z_2 R_2 + \bar{z}_2^{b_2-a_2} z_3 \tilde{R}_3 + z_3^{a_3-b_3} R_1 = 0.$$

Consider the system

$$\begin{aligned}
P_1 r_1 + r_2 R_2 &= R_2, \\
r_2 \tilde{P}_2 + r_3 \tilde{R}_3 &= \tilde{R}_3, \\
r_3 P_3 &= R_1,
\end{aligned}$$

whose solution is given by

$$r_1 = \frac{R_2(1-r_2)}{P_1}, \quad r_2 = \frac{\tilde{R}_3(1-r_3)}{\tilde{P}_2}, \quad r_3 = \frac{R_1}{P_3}. \tag{3.24}$$

By (3.12) we have that $0 < R_{(j+1) \bmod 3} < P_j$ for $j = 1, 3$ and $\tilde{R}_3 < \tilde{P}_2$, therefore r_1, r_2 and r_3 are all positive real numbers.

Therefore using (3.24) we get that

$$\boldsymbol{\nabla} F_{B,t}(e(\mathbf{z})) \cdot e(r_1 v_1 + r_2 v_2 + r_3 v_3)^\top = 0,$$

and also

$$\operatorname{Re}\langle \mathbf{z}, (r_1 z_1, r_2 z_2, r_3 z_3)\rangle = r_1|z_1|^2 + r_2|z_2|^2 + r_3|z_3|^2 \neq 0,$$

therefore we have that $r_1 v_1 + r_2 v_2 + r_3 v_3 \notin T_{\mathbf{z}}\mathbb{S}^5_{\|\mathbf{z}\|}$, so the intersection with $V_{B,t}$ is transversal.

The other cases are just analogous.

(3) Suppose $l = C$. For $t = 0, 1$ the assertion is true since f_C and g_C are polar weighted homogeneous with isolated singularity. Suppose $t \in (0,1)$.

If $a_j > b_j$ for $j = 1, 2, 3$ then,

$$\begin{aligned}
f_{C,t}(\mathbf{z}) = (1-t)&\left(z_1^{a_1}\bar{z}_1^{b_1} z_2 + z_2^{a_2}\bar{z}_2^{b_2} z_3 + z_3^{a_3}\bar{z}_3^{b_3} z_1\right) \\
&+ t\left(z_1^{a_1-b_1} z_2 + z_2^{a_2-b_2} z_3 + z_3^{a_3-b_3} z_1\right).
\end{aligned}$$

Using again the definitions given in (3.11) we obtain the first three equations in (3.14)

$$\begin{aligned}
\boldsymbol{\nabla} F_{C,t}(e(\mathbf{z})) \cdot e(v_1)^\top &= z_1^{a_1-b_1} z_2 P_1 + z_3^{a_3-b_3} z_1 R_1, \\
\boldsymbol{\nabla} F_{C,t}(e(\mathbf{z})) \cdot e(v_2)^\top &= z_2^{a_2-b_2} z_3 P_2 + z_1^{a_1-b_1} z_2 R_2, \\
\boldsymbol{\nabla} F_{C,t}(e(\mathbf{z})) \cdot e(v_3)^\top &= z_3^{a_3-b_3} z_1 P_3 + z_2^{a_2-b_2} z_3 R_3.
\end{aligned}$$

Since

$$f_{C,t}(\mathbf{z}) = z_1^{a_1-b_1} z_2 R_2 + z_2^{a_2-b_2} z_3 R_3 + z_3^{a_3-b_3} z_1 R_1 = 0,$$

then using the same idea we can prove that there exist $r_1, r_2, r_3 \in \mathbb{R}^+$ such that

$$\nabla F_{C,t}(e(\mathbf{z})) \cdot e(r_1 v_1 + r_2 v_2 + r_3 v_3)^\top = 0,$$
$$\operatorname{Re}\langle \mathbf{z}, r_1 v_1 + r_2 v_2 + r_3 v_3 \rangle \neq 0,$$

therefore the intersection is transversal. The other cases are just analogous. □

REMARK 3.3. This proof generalizes in a straightforward way to the case of n variables.

Fix $\varepsilon > 0$ and let $l \in \{A, B, C\}$. Since $f_{l,t}$ has isolated singularity at the origin, there exist $\delta > 0$ such that for any $\eta \in \mathbb{C}$ such that $|\eta| \leq \delta$ the fibers $f_{l,t}^{-1}(\eta)$ and the sphere $\mathbb{S}^5_\varepsilon$ intersect transversely. Let

$$\mathcal{N}(\varepsilon, \delta) := \{(\mathbf{z}, t) \in \mathbb{C}^3 \times I \,|\, |f_{l,t}(\mathbf{z})| = \delta, \|\mathbf{z}\| \leq r\},$$
$$\mathcal{N}_t(\varepsilon, \delta) := \{\mathbf{z} \in \mathbb{C}^3 \,|\, |f_{l,t}(\mathbf{z})| = \delta, \|\mathbf{z}\| \leq r\}.$$

Note that $f_{l,t} \colon \mathcal{N}_t(\varepsilon, \delta) \to \mathbb{S}_\delta$ is the Milnor fibration of $f_{l,t}$ on the Milnor tube. Applying Ehresmann's fibration theorem to the projections

$$\pi \colon \mathbb{S}^5_r \times I \to I, \qquad \pi' \colon \mathcal{N}(\varepsilon, \delta) \to I,$$

we obtain:

THEOREM 3.4 (Isotopy Theorem). *Let $\varepsilon > 0$ small enough. Choose $\delta > 0$ sufficiently small such that for any $\eta \in \mathbb{C}$ with $|\eta| \leq \delta$, the fibre $f_{l,t}^{-1}(\eta)$ and $\mathbb{S}^5_r$ intersect transversely. Then,*

(1) *There exists an isotopy $h_t \colon (\mathbb{S}^5_\varepsilon, K_{0,\varepsilon}) \to (\mathbb{S}^5_\varepsilon, K_{t,\varepsilon})$ such that $f_{l,t}(h_t(\mathbf{z})) = f_{l,0}(\mathbf{z})$ for any $\mathbf{z}$ with $|f_{l,0}(\mathbf{z})| \leq \delta$.*
(2) *The Milnor fibrations*

$$f_{l,0} \colon \mathcal{N}_0(\varepsilon, \delta) \to \mathbb{S}^1_\delta,$$
$$f_{l,t} \colon \mathcal{N}_t(\varepsilon, \delta) \to \mathbb{S}^1_\delta,$$

are C^∞ equivalent for any $t \in [0, 1]$ and with $l = A, B, C$.

COROLLARY 3.5. *The Milnor fibrations of $f_l(\mathbf{z})$ and $g_l(\mathbf{z})$ are C^∞ equivalent.*

References

[1] José Luis Cisneros-Molina, *Join theorem for polar weighted homogeneous singularities*, Singularities II, Contemp. Math., vol. 475, Amer. Math. Soc., Providence, RI, 2008, pp. 43–59, DOI 10.1090/conm/475/09274. MR2454370

[2] José Luis Cisneros-Molina and Augustín Romano-Velázquez, *Classification of isolated polar weighted homogeneous singularities*, Singularities in geometry, topology, foliations and dynamics, Trends Math., Birkhäuser/Springer, Cham, 2017, pp. 77–104. MR3706214

[3] Kazumasa Inaba, Masayuki Kawashima, and Mutsuo Oka, *Topology of mixed hypersurfaces of cyclic type*, J. Math. Soc. Japan **70** (2018), no. 1, 387–402, DOI 10.2969/jmsj/07017538. MR3750281

[4] Mutsuo Oka, *Topology of polar weighted homogeneous hypersurfaces*, Kodai Math. J. **31** (2008), no. 2, 163–182, DOI 10.2996/kmj/1214442793. MR2435890

[5] Mutsuo Oka, *Non-degenerate mixed functions*, Kodai Math. J. **33** (2010), no. 1, 1–62, DOI 10.2996/kmj/1270559157. MR2732230

[6] Mutsuo Oka, *On mixed Brieskorn variety*, Topology of algebraic varieties and singularities, Contemp. Math., vol. 538, Amer. Math. Soc., Providence, RI, 2011, pp. 389–399, DOI 10.1090/conm/538/10615. MR2777832

[7] Peter Orlik and Philip Wagreich, *Isolated singularities of algebraic surfaces with C* action*, Ann. of Math. (2) **93** (1971), 205–228, DOI 10.2307/1970772. MR0284435

[8] Maria Aparecida Soares Ruas, José Seade, and Alberto Verjovsky, *On real singularities with a Milnor fibration*, Trends in singularities, Trends Math., Birkhäuser, Basel, 2002, pp. 191–213. MR1900787

[9] José Seade, *Open book decompositions associated to holomorphic vector fields*, Bol. Soc. Mat. Mexicana (3) **3** (1997), no. 2, 323–335. MR1679312

Instituto de Matemáticas, Unidad Cuernavaca, Universidad Nacional Autónoma de México, Avenida Universidad s/n, Colonia Lomas de Chamilpa, CP62210, Cuernavaca, Morelos Mexico

Email address: jlcisneros@im.unam.mx

Email address: agustin.rom5@gmail.com

Contemporary Mathematics
Volume **742**, 2020
https://doi.org/10.1090/conm/742/14937

Computing Milnor fiber monodromy for some projective hypersurfaces

Alexandru Dimca and Gabriel Sticlaru

This paper is dedicated to Lê Dung Tráng 70th birthday.

ABSTRACT. We describe an algorithm computing the monodromy and the pole order filtration on the top Milnor fiber cohomology of hypersurfaces in $\mathbb{P}^n$ whose pole order spectral sequence partially degenerates at the second page. In the case of hyperplane arrangements and free, locally quasi-homogeneous hypersurfaces, and assuming a key conjecture, this algorithm is much faster than for a hypersurface as above. Our conjecture is based on some results due to L. Narváez-Macarro and M. Saito on the roots of Bernstein-Sato polynomials of such hypersurfaces, and is supported by certain computed examples and by one partial result. For hyperplane arrangements coming from reflection groups, a surprising symmetry of their pole order spectra on top cohomology is displayed in our examples. We also improve our previous results in the case of plane curves.

1. Introduction

The monodromy of singularities or of families of algebraic varieties is a fundamental object of study in Mathematics. Lê Dung Tráng has contributed many deep and fruitful ideas to this area of research, see for instance [**31**–**33**].

Let $V : f = 0$ be a reduced hypersurface in the complex projective space $\mathbb{P}^n$, defined by a homogeneous polynomial $f \in S = \mathbb{C}[x_0, ..., x_n]$, of degree d. Consider the corresponding complement $M = \mathbb{P}^n \setminus V$, and the global Milnor fiber F defined by $f(x_0, ..., x_n) = 1$ in $\mathbb{C}^{n+1}$, with monodromy action $h : F \to F$, $h(x) = \exp(2\pi i/d) \cdot x$. A special case of great interest is when f is a product of linear forms, and then V is a hyperplane arrangement $\mathcal{A}$, and the corresponding complement is traditionally denoted by $M(\mathcal{A})$. A lot of efforts were made, in the case of hyperplane arrangements most of the time, to determine the eigenvalues of the monodromy operators

$$h^m : H^m(F, \mathbb{C}) \to H^m(F, \mathbb{C}) \tag{1.1}$$

2000 *Mathematics Subject Classification.* Primary 32S40; Secondary 32S22, 32S55.

Key words and phrases. hyperplane arrangements, hypersurfaces, Milnor fiber, monodromy, pole order filtration, b-function.

The first author was supported in part by the French government, through the UCA^{JEDI} Investments in the Future project managed by the National Research Agency (ANR) with the reference number ANR-15-IDEX-01.

with $1 \leq m \leq n$, see for instance [**1–4, 8–10, 12, 21, 22, 36, 37, 40, 51, 54**]. However, in most of these papers, either only the monodromy action on $H^1(F, \mathbb{C})$ is considered, or the results are just sufficient conditions for the vanishing of some eigenspaces $H^m(F, \mathbb{C})_\lambda$. These conditions are usually not necessary, see Example 5.7 below. For complexified real arrangements, an approach to compute the monodromy operators using the associated Salvetti complex is explained in [**9, 10, 49, 51**]. However, note that the Milnor fibers considered in [**9, 10**] are not the same as the Milnor fibers in our note, but they correspond to the discriminants of some reflection groups.

In this paper we explain an approach working for *some* hypersurfaces, namely in technical terms for hypersurfaces $V : f = 0$ whose pole order spectral sequence $E_*(f)$ described below degenerates at the E_2-term. For hyperplane arrangements and free locally quasi-homogeneous hypersurfaces, modulo a basic conjecture that is *one of the main contribution of this paper*, see Conjecture 3.9 below, this algorithm is quite efficient.

As a warning, note that our notion of a *free projective hypersurface* might not be the usual one, but it rather corresponds to the notion of a *projectively free hypersurface* in $\mathbb{P}^n$, as explained in Remark 3.3.

Conjecture 3.9 is suggested by the fact that, for these latter hypersurfaces, the roots of their Bernstein-Sato polynomials enjoy special properties, as proved by L. Narváez-Macarro [**38**] and M. Saito [**45**]. In fact our results are either *conjectural*, depending on whether the Conjecture 3.9 holds (as for instance in Examples 5.3, 5.4, 5.5 and 5.7, and for k a resonant value), or *certain*, but based on additional information coming from other viewpoints (as for instance in the previous examples, but for k non-resonant, see Remark 3.8 on this point, or in Examples 5.1, 5.2, 5.8, 5.9 and 5.10).

Our computation gives not only the dimensions of the eigenspaces $H^n(F, \mathbb{C})_\lambda$ of the monodromy, but also the dimensions of the graded pieces $Gr_P^p H^n(F, \mathbb{C})_\lambda$, where P denotes the pole order filtration on the cohomology group $H^m(F, \mathbb{C})$, see section 2 below for the definition. The dimensions of the eigenspaces $H^m(F, \mathbb{C})_\lambda$ for $m < n$ can then be computed by decreasing induction on n, using a generic linear section and the formula (1.4) below.

In the case $n = 2$, this approach was already described in [**27, 30**] in the case of a reduced plane curve $C : f = 0$. However, even in this case, we bring here valuable new information, see Proposition 2.2 and Corollary 2.4. These two results have short and rather direct proofs, but their consequences for the practical computations are substantial, and hence we regard them as *main results* of our paper.

Assume now that $n > 2$ and let $H \subset \mathbb{C}^{n+1}$ be a generic hyperplane with respect to the hypersurface V, passing through the origin. Let $V_H = V \cap H$ be the corresponding hyperplane section of V in $\mathbb{P}(H) = \mathbb{P}^{n-1}$, and denote by F_H the corresponding Milnor fiber in $H = \mathbb{C}^n$ and by

$$h_H^m : H^m(F_H, \mathbb{C}) \to H^m(F_H, \mathbb{C})$$

the associated monodromy operators. Then it is known that the obvious inclusion $\iota_H : F_H \to F$ induces isomorphisms $H^m(F, \mathbb{C}) = H^m(F_H, \mathbb{C})$ for $m = 1, 2, ..., n-2$, as well as a monomorphism

$$\iota_H^* : H^{n-1}(F, \mathbb{C}) \to H^{n-1}(F_H, \mathbb{C}), \tag{1.2}$$

see for instance [**17**], which are compatible with the monodromy operators. Consider the Alexander polynomials of V, which are just the characteristic polynomials of the monodromy, namely

$$\Delta^j(V)(t) = \det(t \cdot Id - h^j | H^j(F,\mathbb{C})), \tag{1.3}$$

for $j = 0, 1, ..., n$, denoted by $\Delta^j(\mathcal{A})(t)$ in the case $V = \mathcal{A}$. It is clear that one has $\Delta^0(V)(t) = t - 1$, and moreover

$$\Delta^0(V)(t)\Delta^1(V)(t)^{-1} \cdots \Delta^n(V)(t)^{(-1)^n} = (t^d - 1)^{\chi(M)}, \tag{1.4}$$

where $\chi(M)$ denotes the Euler characteristic of the complement M, see for instance [**17**, Proposition 4.1.21]. When V is a hyperplane arrangement $\mathcal{A}$, the Euler characteristic $\chi(M(\mathcal{A}))$ is easily computable from the intersection lattice $L(\mathcal{A})$, see [**21, 39**]. By induction, assume that we know how to compute the characteristic polynomials $\Delta^j(V_H)(t)$ for $j = 0, 1, ..., n - 1$. It follows that

$$\Delta^j(V)(t) = \Delta^j(V_H)(t), \tag{1.5}$$

for $j = 0, 1, ..., n - 2$, and hence, in view of the formula (1.4), it is enough to determine the top degree Alexander polynomial $\Delta^n(V)(t)$ and the Euler characteristic $\chi(M) = n + 1 - \chi(V)$. *The computation of this Alexander polynomial $\Delta^n(V)(t)$ is the main aim of this paper.* For the computation of the Euler characteristic $\chi(V)$, see [**35**, Corollary 2].

Here is in short how we proceed. Let Ω^j denote the graded S-module of (polynomial) differential j-forms on $\mathbb{C}^{n+1}$, for $0 \leq j \leq n+1$. The complex $K^*_f = (\Omega^*, \mathrm{d} f\wedge)$ is just the Koszul complex in S of the partial derivatives $f_0, f_1 ... f_n$ of the polynomial f with respect to $x_0, x_1, ..., x_n$. The general theory says that there is a spectral sequence $E_*(f)$, whose first term $E_1(f)$ is computable from the cohomology of the Koszul complex K^*_f and whose limit $E_\infty(f)$ gives us the action of the monodromy operator on the graded pieces of the reduced cohomology $\tilde{H}^*(F,\mathbb{C})$ of the Milnor fiber with respect to the pole order filtration P, see [**16, 24, 47, 48**] as well as [**17**, Chapter 6]. In this note we present an algorithms to compute the second page of the spectral sequence $E_*(f)$. Several examples computed so far suggest the following.

CONJECTURE 1.1. The spectral sequences $E_*(f)$ degenerates at the E_2-term for any hyperplane arrangement $\mathcal{A} : f = 0$ and any free locally quasi-homogeneous hypersurface $V : f = 0$ in $\mathbb{P}^n$.

For the moment it is not clear how to prove this conjecture, not even how to check that it holds in a specific example. For a related property, *extremely useful for performing our computations*, see Definition 3.7, Remark 3.8 and Conjecture 3.9. This property holds in all the cases where we dispose of enough additional information to compute the monodromy operators, see Examples 5.1, 5.2, 5.8. It also holds for some irreducible non-free surfaces, see Examples 5.9 and 5.10. Theorem 3.10 gives some theoretical support for Conjecture 3.9, and is a final *main result* in our paper. Conjecture 1.1 can also be regarded as an extension of the following recent deep result due to M. Saito [**47**].

THEOREM 1.2. *If a hypersurface $V : f = 0$ in $\mathbb{P}^n$ has only isolated singularities, then the spectral sequence $E_*(f)$ degenerates at the E_2-term if and only if these singularities are weighted homogeneous. In particular, Conjecture 1.1 holds for $n = 2$.*

From a different point of view, Conjecture 1.1 can be regarded as a special case of a general conjecture for singular projective hypersurfaces going back to H. Terao [**57**], and saying that $E_2(f) = E_\infty(f)$ always holds. This conjecture is known to fail in general, e.g. by Theorem 1.2 above or by looking at the surface $V' : f' = 0$ introduced at the end of Example 5.9, see also [**16**]. The remarkable fact pointed out in our paper is that Terao's Conjecture seems to hold *in a stronger form* for any hyperplane arrangement.

In the final section several examples of plane arrangements in $\mathbb{P}^3$, as well as examples of (free, locally quasi-homogeneous or general) surfaces in $\mathbb{P}^3$, are considered to illustrate the method. For reflection arrangements, the pole order spectrum $Sp^0_P(f)$ of the top cohomology group $H^3(F, \mathbb{C})$ has a surprising symmetry property, see Remark 5.6, which is not present in the case of other arrangements considered in Examples 5.1 and 5.7. There is no explanation for this symmetry for the moment, just a possible analogy to the formula (3.1) verified by the Bernstein-Sato polynomial of a free, locally quasi-homogeneous hypersurface.

The computations in this note were made using the computer algebra system Singular [**14**]. The corresponding codes are available on request.

We thank Morihiko Saito for very useful discussions related to the subject and the presentation of this note, see in particular Remark 2.3 and Remark 4.1. We also thank the referee for his careful reading of the manuscript and his useful remarks.

2. Gauss-Manin complexes, Koszul complexes, and Milnor fiber cohomology

Let S be the polynomial ring $\mathbb{C}[x_0, ..., x_n]$ with the usual grading and consider a reduced homogeneous polynomial $f \in S$ of degree d. The graded Gauss-Manin complex C^*_f associated to f is defined by taking $C^j_f = \Omega^j[\partial_t]$, i.e. formal polynomials in ∂_t with coefficients in the space of differential forms Ω^j, where $\deg \partial_t = -d$ and the differential $\mathrm{d} : C^j_f \to C^{j+1}_f$ is $\mathbb{C}$-linear and given by

$$\mathrm{d}(\omega \partial_t^q) = (\mathrm{d}\,\omega)\partial_t^q - (\mathrm{d}\,f \wedge \omega)\partial_t^{q+1}, \tag{2.1}$$

see for more details [**24**, Section 4]. The complex C^*_f has a natural increasing filtration P'_* defined by

$$P'_q C^j_f = \oplus_{i \leq q+j} \Omega^j \partial_t^i. \tag{2.2}$$

If we set $P'^q = P'_{-q}$ in order to get a decreasing filtration, then one has

$$Gr^q_{P'} C^*_f = \sigma_{\geq q}(K^*_f((n+1-q)d)), \tag{2.3}$$

the truncation of a shifted version of the Koszul complex K^*_f. Moreover, this filtration P'^q yields a decreasing filtration P' on the cohomology groups $H^j(C^*_f)$ and a spectral sequence

$$E_1^{q,j-q}(f) \Rightarrow H^j(C^*_f). \tag{2.4}$$

On the other hand, the reduced cohomology $\tilde{H}^j(F, \mathbb{C})$ of the Milnor fiber $F : f(x_0, ..., x_n) = 1$ associated to f has a pole order decreasing filtration P, see [**24**, Section 3], such that there is a natural identification for any integers q, j and $k \in [1, d]$

$$P'^{q+1} H^{j+1}(C^*_f)_k = P^q \tilde{H}^j(F, \mathbb{C})_\lambda, \tag{2.5}$$

where $\lambda = \exp(-2\pi ik/d)$. Moreover, the E_1-term of the spectral sequence (2.4) is completely determined by the morphisms of graded $\mathbb{C}$-vector spaces

$$\mathrm{d}' : H^j(K_f^*) \to H^{j+1}(K_f^*), \tag{2.6}$$

induced by the exterior differentiation of forms, i.e. $\mathrm{d}' : [\omega] \mapsto [\mathrm{d}(\omega)]$. More precisely, this spectral sequence $E_*(f)$ is the direct sum of d spectral sequences $E_*(f)_k$, for $k \in [1, d]$, where

(2.7)

$$E_1^{s,t}(f)_k = H^{s+t+1}(K_f^*)_{td+k} \text{ and } \mathrm{d}_1 : E_1^{s,t}(f)_k \to E_1^{s+1,t}(f)_k, \quad \mathrm{d}_1 : [\omega] \mapsto [\mathrm{d}(\omega)].$$

With this notation, one has

$$E_\infty^{s,t}(f)_k = Gr_P^s \tilde{H}^{s+t}(F, \mathbb{C})_\lambda. \tag{2.8}$$

Since the Milnor fiber F is a smooth affine variety, its cohomology groups $H^m(F, \mathbb{C})$ have a decreasing Hodge filtration F coming from the mixed Hodge structure constructed by Deligne, see [**41**]. The two filtrations P and F are related by the inclusion

$$F^s H^m(F, \mathbb{C}) \subset P^s H^m(F, \mathbb{C}), \tag{2.9}$$

for any integers s, m, see formula (4.4.8) in [**24**]. This inclusion and the equality $F^0 H^m(F, \mathbb{C}) = H^m(F, \mathbb{C})$, imply the vanishing

$$E_\infty^{s,t}(f)_k = 0 \tag{2.10}$$

for any $s < 0$ or $t < 0$, in other words the limit page of the spectral sequence is contained in the first quadrant. Moreover, for $k = d$, one has in addition $E_\infty^{0,n}(f)_d = 0$, see [**6**, Proposition 5.2].

EXAMPLE 2.1. If F is the Milnor fiber associated to a smooth hypersurface V in $\mathbb{P}^n$, then the inclusion (2.9) becomes an equality, see for instance [**53**] and [**17**, Example 6.2.13]. Moreover one has

$$E_\infty^{s,t}(f)_k = 0 \text{ for } s + t \neq n \text{ and } E_\infty^{n-t,t}(f)_k = \mu(td + k - n - 1), \tag{2.11}$$

where $\mu(a)$ is the coefficient of t^a in the polynomial

$$\left(\frac{1 - t^{d-1}}{1 - t}\right)^{n+1}.$$

In particular, $\mu(a) \neq 0$ if and only if $1 \leq a \leq (n + 1)(d - 2)$.

One has the following result, the second part of which answers positively a conjecture made in Remark 2.9 (i) in [**30**].

PROPOSITION 2.2. *Let V be a hypersurface in $\mathbb{P}^n$ and H a generic hyperplane. Consider the linear inclusion $\iota_H : F_H \to F$ defined in the Introduction. Then the induced morphisms $\iota_H^* : H^m(F, \mathbb{C}) \to H^m(F_H, \mathbb{C})$ are strictly compatible with the Hodge filtration F^p and compatible with the pole order filtration P^p. Moreover, one has the following.*

(1) *If V has only isolated singularities, then the Hodge filtration F^p and the pole order filtration P^p coincide on $H^{n-1}(F, \mathbb{C})$.*
(2) *For any hypersurface V, in particular for any hyperplane arrangement $\mathcal{A}$, the Hodge filtration F^p and the pole order filtration P^p coincide on $H^1(F, \mathbb{C})$.*

PROOF. Since $\iota_H : F_H \to F$ is a regular mapping, the strict compatibility of ι_H^* with the Hodge filtration F^p is well known, see [**15**]. In particular, for $m = n-1$ and V with isolated singularities, since ι_H^* is injective, this means that a cohomology class $\alpha \in H^{n-1}(F, \mathbb{C})$ satisfies

$$\alpha \in F^p H^{n-1}(F, \mathbb{C}) \text{ if and only if } \iota_H^*(\alpha) \in F^p H^{n-1}(F_H, \mathbb{C}). \tag{2.12}$$

The compatibility of ι_H^* with the pole order filtration P^p means that

$$\alpha \in P^p H^m(F, \mathbb{C}) \text{ implies } \iota_H^*(\alpha) \in P^p H^m(F_H, \mathbb{C}). \tag{2.13}$$

This property comes from the fact that ι_H^* induces a morphism $\iota_H^* : C_f^* \to C_{f_H}^*$ between the corresponding Gauss-Manin complexes, where f_H denotes the restriction of the polynomial f to the hyperplane H, thought of here as a hyperplane in $\mathbb{C}^{n+1}$. This morphism preserves the P'_* filtrations introduced in (2.2), i.e. one clearly has $\iota_H^*(P'_q C_f^*) \subset P'_q C_{f_H}^*$, for any integer q. To prove the claim (1), it is enough in view of the inclusion in (2.9) to prove the converse inclusion $P^p H^{n-1}(F, \mathbb{C}) \subset F^p H^{n-1}(F, \mathbb{C})$ for any p. So take $\alpha \in P^p H^{n-1}(F, \mathbb{C})$. Using (2.13), it follows that

$$\iota_H^*(\alpha) \in P^p H^{n-1}(F_H, \mathbb{C}) = F^p H^{n-1}(F_H, \mathbb{C}).$$

The last equality is due to the fact that the hyperplane section V_H being smooth, the Hodge filtration F^p and the pole order filtration P^p coincide on $H^{n-1}(F_H, \mathbb{C})$, as seen in Example 2.1. We conclude using (2.12). The claim (2) follows in a similar way, by taking a generic $(n-2)$-codimensional linear section instead of the hyperplane H.

□

REMARK 2.3. (i) We thank Morihiko Saito for teaching us to be strict about the difference between strict compatibility and compatibility of morphisms of filtered objects in the above proof. This corrects a serious gap in our original presentation of the proof above. See also [**47**, Remark 4.4] for a similar property in a general context.
(ii) We do not know whether the morphism $\iota_H^* : H^m(F, \mathbb{C}) \to H^m(F_H, \mathbb{C})$ is strictly compatible with the pole order filtration P^p. Indeed, a morphism of filtered complexes, strictly compatible with the filtrations, does not induce in general a strictly compatible morphism when we pass to cohomology groups, see [**25**, Section 1.10]

If this strict compatibility holds for $\iota_H^* : H^m(F, \mathbb{C}) \to H^m(F_H, \mathbb{C})$, and if one knows the P-filtration on $H^*(F_H, \mathbb{C})$, in order to determine it on $H^*(F, \mathbb{C})$, it is enough to determine the P-filtration on the top cohomology $H^n(F, \mathbb{C})$ and to identify the image of $H^{n-1}(F, \mathbb{C})$ inside $H^{n-1}(F_H, \mathbb{C})$ under ι_H^*.
(iii) It is known that the equality $F^s = P^s$ does not hold on $H^2(F, \mathbb{C})$, even when F is the Milnor fiber of a line arrangement in $\mathbb{P}^2$, see for instance [**30**, Remark 2.9.(ii)].

The following result is a major improvement of Theorem 1.2 in [**27**].

COROLLARY 2.4. *For any curve $V : f = 0$ in $\mathbb{P}^2$, in order to compute the corresponding Alexander polynomial $\Delta^1(V)$, it is enough to compute the dimensions $\dim E_2^{1,0}(f)_k$ for any $k = 1, 2, ..., d$. More precisely, let $\lambda = \exp(-2\pi i k/d)$ and $m(\lambda)$ be the multiplicity of λ as a root of the Alexander polynomial $\Delta^1(V)$. Then one has*

$$m(\lambda) = \dim E_2^{1,0}(f)_k + \dim E_2^{1,0}(f)_{d-k},$$

for $1 \leq k < d$ *and* $m(1) = \dim E_2^{1,0}(f)_d$.

PROOF. It is well known, e.g. one can use the proof of Proposition 2.2 above, that $H^1(F, \mathbb{C})_{\neq 1}$ is a pure Hodge structure of weight 1. For any $\lambda = \exp(-2\pi i k/d) \neq 1$, and with obvious notation, it follows that

$$m(\lambda) = h^{1,0}(H^1(F,\mathbb{C})_\lambda) + h^{0,1}(H^1(F,\mathbb{C})_\lambda) = h^{1,0}(H^1(F,\mathbb{C})_\lambda) + h^{1,0}(H^1(F,\mathbb{C})_{\overline{\lambda}}),$$

where $\overline{\lambda}$ denotes the complex conjugate of λ. On the other hand, we have

$$h^{1,0}(H^1(F,\mathbb{C})_\lambda) = \dim Gr_P^1 \tilde{H}^1(F,\mathbb{C})_\lambda = \dim E_\infty^{1,0}(f)_k = \dim E_2^{1,0}(f)_k,$$

where the last equality is obvious. These equalities yield our claim for $\lambda \neq 1$. The claim for $\lambda = 1$ follows from the fact that $H^1(F,\mathbb{C})_1 = H^1(M,\mathbb{C})$ is a pure Hodge structure of weight $(1,1)$.

$\square$

REMARK 2.5. Consider the j-th *Hodge spectrum* of the plane curve $V : f = 0$, defined by

$$Sp_F^j(f) = \sum_{\alpha > 0} n_{F,f,\alpha}^j t^\alpha \tag{2.14}$$

for $j = 0, 1$, where

$$n_{F,f,\alpha}^j = \dim Gr_F^p H^{2-j}(F,\mathbb{C})_\lambda$$

with $p = [3 - \alpha]$ and $\lambda = \exp(-2\pi i \alpha)$. When V is a line arrangement, then simple formulas for the difference

$$Sp_F(f) = Sp_F^0(f) - Sp_F^1(f)$$

are given in [**7**]. It follows from Proposition 2.2 and the proof of Corollary 2.4, that once we know the dimensions $\dim E_2^{1,0}(f)_k$ for any $k = 1, 2, ..., d-1$, we can compute spectrum $Sp_F^1(f)$, and hence via [**7**], the spectrum $Sp_F^0(f)$ as well. This gives us precise information on the Hodge structure on $H^2(F,\mathbb{C})$ in this case.

3. Hyperplane arrangements, free locally quasi-homogeneous divisors, and Bernstein-Sato polynomials

In this section we explain why the limit page of the spectral sequences discussed above enjoy a very useful property in the case of hyperplane arrangements. Let $(D,0) : g = 0$ be a complex analytic hypersurface germ at the origin of $\mathbb{C}^{n+1}$ and denote by $b_{g,0}(s)$ the corresponding (local) Bernstein-Sato polynomial. If the analytic germ g is given by a homogeneous polynomial, then one can define also the global Bernstein-Sato polynomial $b_g(s)$ of g, and one has an equality $b_g(s) = b_{g,0}(s)$, see for more details [**43, 44, 46**]. Let $R_{g,0}$ be the set of roots of the polynomial $b_{g,0}(-s)$. When g is a homogeneous polynomial, we use the simpler notation $R_g = R_{g,0}$.

In this section we consider the case when $g = f$ is the defining equation of a hypersurface V in $\mathbb{P}^n$ and denote by $D = CV$ the affine cone over V, defined in $\mathbb{C}^{n+1}$ by the equation $f = 0$. Recall M. Saito's fundamental results in [**44**, Theorem 2] and [**45**, Theorem 1].

THEOREM 3.1. *Let* $V : f = 0$ *be a hypersurface in* $\mathbb{P}^n$, *let* $\alpha > 0$ *be a rational number and set* $\lambda = \exp(-2\pi i \alpha)$.

(1) *If* $Gr_P^p H^n(F,\mathbb{C})_\lambda \neq 0$, *where* $p = [n + 1 - \alpha]$, *then* $\alpha \in R_f$.

(2) *If the sets* $\alpha+\mathbb{N}$ *and* $\cup_{a\in D, a\neq 0} R_{f,a}$ *are disjoints, then the converse of the assertion* (1) *holds.*

THEOREM 3.2. *Let* $\mathcal{A}: f=0$ *be an arrangement of d hyperplanes in* $\mathbb{P}^n$. *Then*

$$\max R_f < 2 - \frac{1}{d}.$$

REMARK 3.3. If V is a hypersurface in a projective space $\mathbb{P}^n$, then there are two possible definitions of freeness for V, which are both in use, without being equivalent. We say that V is *projectively free* if the cone $D=CV$ over V is free at its vertex as in K. Saito's paper [**42**]. It is known that V is projectively free if and only if the sheaf $T(-\log V)$ of logarithmic vector fields along V is a direct sum of line bundles on $\mathbb{P}^n$, see for instance [**21**, Section 8.1]. Note that M. Saito calls V a strongly free divisor in [**48**] if it satisfies these equivalent conditions.

On the other hand, sometimes V is said to be *free* if for any point $p\in V$, the germ (V,p) is free as defined in K. Saito's paper [**42**], which is equivalent to asking the sheaf $T(-\log V)$ to be a locally free coherent sheaf, i.e. to come from a vector bundle on $\mathbb{P}^n$.

It is clear that V projectively free implies V free, but not the converse. Indeed, any curve C in $\mathbb{P}^2$ is free, but not necessarily projectively free. This terminology is similar to, and was suggested by, the normal versus the projectively normal subvarieties X in some $\mathbb{P}^N$. For the *sake of simplicity*, we call below V free when we actually mean projectively free. This simplified terminology is quite standard, see [**28, 29, 50, 52**].

REMARK 3.4. As shown by Narváez Macarro, the Bernstein-Sato polynomial b_f of a free arrangement $\mathcal{A}: f=0$ satisfies the equality

$$b_f(s-2) = \pm b_f(-s), \tag{3.1}$$

see [**38**]. This equality implies that the zero set $R_f \subset (0,2)$ is stable under the involution $\alpha \mapsto 2-\alpha$, including the multiplicities of the roots. In fact, the equation (3.1) holds for a larger class of free hypersurfaces, namely those of *linear Jacobian type*, see [**38**]. As noted in [**38**, Corollary 4.3], for any such free hypersurface one has $R_f \subset (0,2)$. A locally quasi-homogeneous divisor $V: f=0$ in $\mathbb{P}^n$ is of linear Jacobian type, see [**38**, Theorem (1.6)], and an example of such a surface in given below, see Example 5.8. Indeed, it is easy to see that the hypersurface V is locally a quasi-homogeneous divisor (with strictly positive weights) if and only if its affine cone $D=CV$ (regarded as a germ at the origin) has the same property.

Assume from now on in this section, except in Definition 3.7 *and in the final subsection* 3.1, *that* $V: f=0$ *is either a hyperplane arrangement in* $\mathbb{P}^n$, *or a free locally quasi-homogeneous divisor in* $\mathbb{P}^n$. *Let* $\delta_{k,d}$ *be 1 if* $k=d$ *and 0 otherwise, and* $\lambda=\exp(-2\pi i k/d)$.

COROLLARY 3.5. *With the above assumption, one has*

$$Gr_P^p H^n(F,\mathbb{C}) = 0$$

for any $p \le n-2$ *and* $Gr_P^{n-1} H^n(F,\mathbb{C})_1 = 0$. *In other words, for any* $k=1,\dots,d$ *one has*

$$P^{n-1+\delta_{k,d}} H^n(F,\mathbb{C})_\lambda = H^n(F,\mathbb{C})_\lambda.$$

In particular $E_\infty^{n-t,t}(f)_k = 0$ *for* $t > 1-\delta_{k,d}$.

PROOF. Assume $Gr_P^p H^n(F,\mathbb{C})_\eta \neq 0$ for some p and some $\eta = \exp(-2\pi i\alpha)$ with $p \leq n+1-\alpha < p+1$. Then Theorem 3.1 (1) implies that $\alpha \in R_f$ and hence using Theorem 3.2 or Remark 3.4, we get $2 > \alpha > n-p$. If $p \leq n-2$, or if $p = n-1$ and α is an integer, then we get a contradiction.

□

REMARK 3.6. If the morphism $\iota_H^* : H^m(F,\mathbb{C}) \to H^m(F_H,\mathbb{C})$ is strictly compatible with the pole order filtration P^p, then one has in addition the following property:

($\star$) for a hyperplane arrangement, $E_\infty^{s,t}(f)_k = 0$ for any $s \geq 0$ and $t \geq 2 - \delta_{k,d}$.

Indeed, the property ($\star$) is equivalent to $P^{m-1+\delta_{k,d}}H^m(F,\mathbb{C})_\lambda = H^m(F,\mathbb{C})_\lambda$ for any m. It follows that in this situation the limit page $E_\infty^{s,t}(f)_k$ is not only contained in the first quadrant, but in fact the non-zero terms are situated on just two horizontal lines when $k \neq d$, namely the lines $t = 0$ and $t = 1$ (and, respectively, only on the line $t = 0$ when $k = d$).

If we assume Conjecture 1.1 and proceed by induction, it remains to compute the dimension of the terms $E_2^{n-q,q}(f)_k = 0$ for any $q \in \{0,1\}$ and $k = 1,2,...,d$. It is this property that makes possible the computations in a reasonable amount of time. Since Conjecture 1.1 is difficult to check in practice, we introduce the following notion, motivated by Corollary 3.5.

DEFINITION 3.7. Let k,m be positive integers satisfying $1 \leq k \leq d$ and $1 \leq m \leq n$. We say that the hypersurface $V : f = 0$ is (k,m)-top-computable if $E_\infty^{n-t,t}(f)_k = 0$ for $m - \delta_{k,d} < t \leq n - \delta_{k,d}$, and

$$\sum_{t=0}^{m-\delta_{k,d}} \dim E_2^{n-t,t}(f)_k = \dim H^n(F,\mathbb{C})_\lambda.$$

The vanishing of $E_\infty^{0,n}(f)_d = 0$, which is essential for this definition, follows from [**6**, Proposition 5.2]. As an example, a smooth hypersurface $V : f = 0$ in $\mathbb{P}^n$ is (k,n)-top-computable for any k, but not $(k,n-1)$-top-computable. For any hypersurface $V : f = 0$ and any t, one has $\dim E_2^{n-t,t}(f)_k \geq \dim E_\infty^{n-t,t}(f)_k$. Hence, if V is (k,m)-top-computable then $E_2^{n-t,t}(f)_k = E_\infty^{n-t,t}(f)_k$ for $0 \leq t \leq m - \delta_{k,d}$. In particular, the information on the P-filtration on $H^n(F,\mathbb{C})$ given by the second term E_2 is also complete in this case. For a hypersurface $V : f = 0$ not covered by Corollary 3.5, the simplest way to check the vanishings $E_\infty^{n-t,t}(f)_k = 0$ for $t > m - \delta_{k,d}$ is to use the vanishings in (2.10) and to check by a direct computation whether $E_2^{n-t,t}(f)_k = 0$ for $m - \delta_{k,d} < t \leq n - \delta_{k,d}$. Using this approach, irreducible, non-free surfaces in $\mathbb{P}^3$ that are still $(k,1)$-top-computable are displayed in Examples 5.9 and 5.10.

REMARK 3.8. The conditions in Definition 3.7 are easy to check as soon as we know $\dim H^n(F,\mathbb{C})_\lambda$ in the case of an arrangement $\mathcal{A}$ of d hyperplanes. Note that the vanishings necessary for the $(k,1)$-top-computability hold by Corollary 3.5. Let $d' = d/e$ where $e = G.C.D.(d,k)$. Then λ is a d-root of unity of order d'. If there is a hyperplane $H \in \mathcal{A}$, such that for any dense edge $X \subset H$, the number n_X of hyperplanes in $\mathcal{A}$ containing X is not a multiple of d', then it is known that $H^m(F,\mathbb{C})_\lambda = 0$ for $m < n$ and $\dim H^n(F,\mathbb{C})_\lambda = |\chi(M(\mathcal{A})|$, see [**18**, Theorem

6.4.18], [**34**]. In this case we say that the eigenvalue λ and the integer k are *non-resonant with respect to the arrangement $\mathcal{A}$*. Hence, the answer given by the second page of the spectral sequence is correct as soon as we know that k is non-resonant and we have the equality

$$\sum_{q=0}^{1-\delta_{k,d}} \dim E_2^{n-q,q}(f)_k = |\chi(M)|.$$

The new information given in such a case by the second page of the spectral sequence concerns the pole order filtration on $H^n(F,\mathbb{C})_\lambda$. On the other hand, the fact that this equality holds in all the computed cases gives strong support for Conjecture 3.9 below.

The following is the main conjecture put forth in our paper.

CONJECTURE 3.9. For any arrangement $\mathcal{A} : f = 0$ of d hyperplanes in $\mathbb{P}^n$, and for any free locally quasi-homogeneous divisor $V : f = 0$ of degree d in $\mathbb{P}^n$, the defining polynomial f is $(k,1)$-top-computable for any positive integer k satisfying $1 \leq k \leq d$.

One has the following partial result, saying that Conjecture 3.9 holds for $k = d$.

THEOREM 3.10. *With the above assumption on $V : f = 0$, one has*

$$\dim E_2^{n,0}(f)_d = \dim E_\infty^{n,0}(f)_d = \dim H^n(F,\mathbb{C})_1 = \dim H^n(M,\mathbb{C}).$$

PROOF. Let us denote as above by D the affine cone in $X = \mathbb{C}^{n+1}$ over V. Then let us denote by $\Omega^*(*D)$ the de Rham complex of rational differential forms on X with poles of arbitrary orders along the hypersurface D. Consider the subcomplex $\Omega^*(\log D)$ of differential forms in $\Omega^*(*D)$ with logarithmic poles along the hypersurface D. A form $\omega \in \Omega^p(\log D)$ satisfies, by definition

$$f\omega \in \Omega^p \text{ and } f\,\mathrm{d}\,\omega \in \Omega^{p+1},$$

or, equivalently $\omega = \eta/f$ with $\eta \in \Omega^p$ and $\mathrm{d}\,f \wedge \eta$ divisible by f. Note also that $\Omega^{n+1}(\log D) = \Omega^{n+1}(D)$ by definition. Since the differential $\mathrm{d}\,f : \Omega^n_{k-d} \to \omega^n_k$ vanishes for $k < n+d$, we have an identification

$$\Omega^{n+1}(\log D)_0 = \Omega^{n+1}_d = H^{n+1}(K^*_f)_d = E_1^{n,0}(f)_d,$$

given by $\eta/f \mapsto \eta$, where the grading on the S-module $\Omega^p(\log D)$ is the usual one, i.e. $|\eta'/f| = |\eta'| - d$, for any $\eta' \in \Omega^p$. Next the homogeneous component $\Omega^n(\log D)_0$ can be identified in the same way as above to the internal direct sum

$$S_{d-n-1} \cdot \omega_n \oplus Syz^n_d \subset \Omega^n_d,$$

where

$$\omega_n = \sum_{i=0}^{n} (-1)^i x_i \,\mathrm{d}\,x_0 \wedge \mathrm{d}\,x_1 \wedge \ldots \wedge \widehat{\mathrm{d}\,x_i} \wedge \ldots \wedge \mathrm{d}\,x_n,$$

and

$$Syz^n_d = \ker\left(\mathrm{d}\,f\wedge : \Omega^n_d \to \Omega^{n+1}_{2d}\right) = H^n(K^*_f)_d = E_1^{n-1,0}(f)_d.$$

To see this, use the second description of logarithmic forms given above and the fact that $\mathrm{d}\,f \wedge \omega_n = d \cdot f\omega'_n$, where $\omega'_n = \mathrm{d}\,x_0 \wedge \mathrm{d}\,x_1 \wedge \ldots \wedge \mathrm{d}\,x_n$. If $h \in S_m$, a direct computation shows that

$$d\left(\frac{h\cdot\omega_n}{f}\right) = (m+n+1-d)\frac{h\cdot\omega'_n}{f}.$$

In particular the image of the differential $d : \Omega^n(\log D)_0 \to \Omega^{n+1}(\log D)_0$ coincides with the image of the differential $d_1 : E_1^{n-1,0}(f)_d \to E_1^{n,0}(f)_d$, and all the other differentials $d : \Omega^n(\log D)_q \to \Omega^{n+1}(\log D)_q$ for $q \neq d$ are surjective. It follows that

$$\dim E_2^{n,0}(f)_d = \dim H^{n+1}(\Omega^*(\log D)_0) = \dim H^{n+1}(\Omega^*(\log D)). \tag{3.2}$$

On the other hand, we have $\dim H^n(M, \mathbb{C}) = \dim H^{n+1}(X \setminus D, \mathbb{C})$, see for instance [**18**, Prop. 6.4.1].

It remains to show that $\dim H^{n+1}(\Omega^*(\log D)) = \dim H^{n+1}(X \setminus D, \mathbb{C})$. When $(D, 0)$ is a free locally quasi-homogeneous divisor, this follows from [**11**]. When D is a hyperplane arrangement, one has to use [**56**, Proposition 6.1].

□

3.1. The arbitrary hypersurface case. The algorithm presented below can be applied for any hypersurface $V : f = 0$ in $\mathbb{P}^n$ to compute the terms $E_2^{n-q,q}(f)_k$ of the second page of the above spectral sequences. However, in the general case $Q = qd + k$ takes values up to $(n+1)d$ and not only $2d - 1$, which increases dramatically the computer time. Once the second page is computed, then it is a difficult question to decide whether

$$E_2^{n-q,q}(f)_k = E_\infty^{n-q,q}(f)_k, \tag{3.3}$$

and hence whether we have obtained the correct results. In some cases, one can proceed as follows. Using (2.10), we see that

$$\dim H^n(F, \mathbb{C})_\lambda \leq \sum_{q=0,\dots,n} \dim E_2^{n-q,q}(f)_k \tag{3.4}$$

where $\lambda = \exp(-2\pi ik/d)$, and equality holds if and only if one has the equality (3.3) for any q.

In conclusion, without assuming Conjecture 3.9 one has only the inequality (3.4) after computing the second page of the spectral sequence. When a good lower estimate for the top Betti number of F is available, then one can get equalities in (3.4) for any k. Two examples of such ad hoc computations are given below in Examples 5.9 and 5.10.

Note that M. Saito in [**48**] has obtained a computer program for a *certified calculation* based on a computation of the pages E_r of the pole order spectral sequence for $r \leq 3$ and surfaces in $\mathbb{P}^3$, and for degrees up-to $4d - 1$ ($2d - 1$ for hyperplane arrangements). These results confirm the computations in our Examples 5.7-10 in our paper.

4. The algorithm

Consider the graded S–submodule $AR(f) \subset S^{n+1}$ of *all relations* involving the derivatives of f, namely

$$r = (r_0, r_1, ..., r_n) \in AR(f)_q$$

if and only if

$$r_0 f_0 + r_1 f_1 + ... + r_n f_n = 0 \tag{4.1}$$

and the polynomials $r_0, r_1, ..., r_n$ are in S_q. Since S is a noetherian ring, the graded S-module $AR(f)$ admits a (minimal) system of generators $r^{(j)}$ of Jacobian syzygies, where $j = 1, ..., g$. Assume that

$$r^{(j)} = (r_0^{(j)}, r_1^{(j)}, ..., r_n^{(j)})$$

for $j = 1, ..., g$ and let $d_j = \deg r_m^{(j)}$, for any $m \in [0, n]$ with $r_m^{(j)} \neq 0$. Assume moreover that

$$d_1 \leq d_2 \leq ... \leq d_g.$$

Such a system of generators can be determined using the software SINGULAR [**14**] or CoCoA [**13**], see Remark 4.2. To each syzygy $r = (r_0, r_1, ..., r_n) \in AR(f)_q$ we can associate an n-differential form in Ω^n_{q+n} by the formula

$$\omega(r) = \sum_{i=0}^{n} (-1)^i r_i \, \mathrm{d}\, x_0 \wedge \mathrm{d}\, x_1 \wedge ... \wedge \widehat{\mathrm{d}\, x_i} \wedge ... \wedge \mathrm{d}\, x_n. \tag{4.2}$$

Then equation (4.1) is equivalent to $\mathrm{d}\, f \wedge \omega(r) = 0$ and

$$\mathrm{d}\, \omega(r) = (\sum_{i=0}^{n} (r_i)_i) \, \mathrm{d}\, x_0 \wedge \mathrm{d}\, x_1 \wedge ... \wedge \mathrm{d}\, x_n, \tag{4.3}$$

where $(r_i)_i$ denotes the partial derivative of r_i with respect to x_i for $i = 0, 1, ..., n$. It follows that the dimension of $E_2^{n-k,q}(f)_k$, which is the cokernel of the differential

$$\mathrm{d}_1 : E_1^{n-1-q,q}(f)_k \to E_1^{n-q,q}(f)_k$$

can be computed as follows. Consider the linear mapping

$$\phi_Q : S_{Q-d_1-n} \times ... \times S_{Q-d_g-n} \times S^{n+1}_{Q-d-n} \to S_{Q-n-1}, \tag{4.4}$$

given by

$$((A_1, ..., A_g), (B_0, ..., B_n)) \mapsto \sum_{0 \leq i \leq n} ((\sum_{1 \leq j \leq g} A_j r_i^{(j)})_i + B_i f_i),$$

for $Q = qd + k$. This map can be described in a more compact way by using differential forms as follows. The formula (4.2) implies that the application ϕ_Q is nothing else but the map

$$S_{Q-d_1-n} \times ... \times S_{Q-d_n-n} \times \Omega^n_{Q-d} \to \Omega^{n+1}_Q$$

given by

$$((A_1, ..., A_n), \eta) \mapsto \mathrm{d}(\sum_{j=1,n} A_j \omega(r^{(j)})) + \mathrm{d}\, f \wedge \eta.$$

Let R_Q be the rank of this linear mapping, which is computed using the software SINGULAR for instance. Then one clearly has

$$\dim E_2^{n-q,q}(f)_k = \dim S_{Q-n-1} - R_Q = \binom{Q-1}{n} - R_Q, \tag{4.5}$$

for any $q = 0, ..., n$.

In the case of a hyperplane arrangement $\mathcal{A} : f = 0$, or of a free divisor $V : f = 0$ of linear Jacobian type, we can consider only the values $q \leq 1$ if we assume Conjecture 3.9, while $k \leq d$ by definition. It follows that it is enough to take $Q \leq 2d$ in this case. In fact, for a hyperplane arrangement, it is known that $F^n H^n(F, \mathbb{C})_1 = H^n(F, \mathbb{C})_1 = H^n(M(\mathcal{A}), \mathbb{C})$, which implies that in fact we need to consider only the values $Q \leq 2d - 1$ in such a case.

REMARK 4.1. In case one does not like to use the generating system of syzygies produced by a computer software, one can proceed in the following more direct way, already considered by us in [**30**] and by Morihiko Saito in [**48**]. Let $V : f = 0$ be a degree d hypersurface in $\mathbb{P}^n$, and for each $Q = qd + k$ consider the linear map

$$\phi_Q^1 : \Omega_{Q-d}^n \to \Omega_Q^{n+1}, \quad \eta \mapsto \mathrm{d} f \wedge \eta. \tag{4.6}$$

Let $\kappa^1(Q)$ be the dimension of the kernel $K^1(Q)$ of this map. Clearly $\kappa^1(Q) = 0$ for $Q < d + n$. Consider next the map

$$\phi_Q^2 : \Omega_{Q-d}^n \times \Omega_Q^n \to \Omega_Q^{n+1} \times \Omega_{Q+d}^{n+1}, \tag{4.7}$$

given by

$$(\eta_1, \eta_2) \mapsto (\mathrm{d} f \wedge \eta_1 + \mathrm{d}(\eta_2), \mathrm{d} f \wedge \eta_2).$$

Let $\kappa^2(Q)$ be the dimension of the kernel $K^2(Q)$ of this map. Note that one has

$$K^2(Q) \subset \Omega_{Q-d}^n \times K^1(Q + d),$$

and hence the dimension of the vector space $\phi_Q^2(\Omega_{Q-d}^n \times K^1(Q + d))$ is given by

$$R_Q = \dim(\Omega_{Q-d}^n \times K^1(Q+d)) - \kappa^2(Q) = \kappa^1(Q+d) - \kappa^2(Q) + (n+1)\binom{Q-d}{n}. \tag{4.8}$$

The dimensions $\kappa^1(Q + d)$ and $\kappa^2(Q)$ can be computed using the software SINGULAR for instance. Then one clearly has the same formula as above, namely

$$\dim E_2^{n-k,q}(f)_k = \dim S_{Q-n-1} - R_Q = \binom{Q-1}{n} - R_Q. \tag{4.9}$$

This approach seems to increase the necessary computing time as well as the necessary computer memory substantially. We thank Morihiko Saito for telling us that the algorithm to compute the second page of the spectral sequence using the system of generators is better not only in the case of a free hypersurface, when we had already used it, but also in the general case.

REMARK 4.2. (i) The command *syz(...)* in the software SINGULAR does not always give a minimal set of generators for the graded S-module $AR(f)$. For the quartic surface discussed in Example 5.9, it lists 6 generators for the order *rp*, 7 generators for the order *dp* and 8 generators for the orders *lp* and *Dp*. Here *rp* = reverse lexicographical ordering, *dp* = degree reverse lexicographical ordering, *lp* = lexicographical ordering, and *Dp* = degree lexicographical ordering. Note also that in some of these listings, the generators are not given with the degrees in increasing order. To get the minimal set of generators one should use the command *minbase(syz(...))*.

(ii) To get information on the complexity of computations in the algorithm, it would be useful to have an upper bound in terms of the geometry of the hypersurface $V : f = 0$ on g, the minimal number of generators for the S-module $AR(f)$, and a lower bound on the minimal degree d_1.

Note that for a smooth hypersurface we have the $g = n(n + 1)/2$ linear independent Koszul generators of degree $d_1 = d - 1$. The equality $g = n(n+1)/2$ holds also for the plane arrangement in Example 5.7 and the singular surface in Example 5.9 below. On the other hand, note that [**26**, formula (1.3) and Example 4.3 (i)] imply that for a hypersurface with just one A_1 singularity, one has

$$g = \frac{n(n + 1)}{2} + 1.$$

And [**26**, Theorem 1.4] implies that for a nodal curve C in $\mathbb{P}^2$, one has $g \geq r-1$, where r is the number of irreducible components of C. Lower bounds for d_1 are known for hypersurfaces having only isolated singularities, see [**25**, Theorem 9] for the case of weighted homogeneous singularities, and [**20**, Theorem 2.4] for arbitrary isolated singularities.

(iii) Note that [**26**, Theorem 1.5 and Example 4.3 (i)] imply that for a hypersurface with just one A_1 singularity, one has $d_1 = ...d_{g-1} = d-1$ and $d_g = n(d-2)$.

REMARK 4.3. If $V: f=0$ is a free hypersurface, then $g=n$ as the S-module $AR(f)$ is free. One can verify that the basis given by SINGULAR is correct using Saito's criterion, i.e. the $(n+1)$-square matrix having as the first row $x_0, ..., x_n$, and the $j+1$-st row given by $(r_0^{(j)}, r_1^{(j)}, ..., r_n^{(j)})$ for $j = 1, ..., n$, should have as determinant a constant, non-zero multiple of f, see [**39, 58, 60**]. In particular, in this case $d_1 + d_2 + ... + d_n = d-1$.

It is known that, in the case of a hyperplane arrangement $\mathcal{A}: f=0$, the exponents d_j determine the Betti numbers of the complement $M(\mathcal{A})$, see for instance [**39**],

$$\pi(\mathcal{A}, t) := \sum_{i=0}^{n} b_i(M(\mathcal{A}))t^i = \prod_{j=1}^{n}(1+d_jt). \tag{4.10}$$

5. Examples

In this section we consider plane arrangements in $\mathbb{P}^3$, except in Examples 5.8, 5.9 and 5.10 where irreducible quartic surfaces in $\mathbb{P}^3$ are considered, and we replace the coordinates x_0, x_1, x_2, x_3 by x, y, z, w. To state the results, we consider the pole order spectrum defined by

$$Sp_P^0(f) = \sum_{\alpha>0} n_{P,f,\alpha} t^\alpha \tag{5.1}$$

where

$$n_{P,f,\alpha} = \dim Gr_P^p H^3(F, \mathbb{C})_\lambda$$

with $p = [4-\alpha]$ and $\lambda = \exp(-2\pi i \alpha)$. In view of Corollary 3.5 and assuming Conjecture 3.9, in the case of a plane arrangement the exponents α with possibly non-zero coefficients $n_{P,f,\alpha}$ are of the form

$$\alpha = \frac{Q}{d} \text{ and } n_{P,f,\alpha} = \dim E_2^{n-q,q}(f)_k, \tag{5.2}$$

where $Q = qd + k$ as above, with $q = 0, 1$ and $k = 1, ..., d$. Note that one has the equality

$$b_n(F) = \sum_{\alpha>0} n_{P,f,\alpha}.$$

Recall from Remark 3.8 the notion of non-resonant eigenvalue λ and non-resonant integer k, for which one has without any computation

$$H^m(F, \mathbb{C})_\lambda = 0 \text{ for } m < n \text{ and } \dim H^n(F, \mathbb{C})_\lambda = |\chi(M(\mathcal{A})|,$$

as well as the notation $e = G.C.D.(k, d)$.

EXAMPLE 5.1 (A family of free arrangements). Consider the arrangement

$$\mathcal{A}(p,q) : (x^p + y^p)(z^q + w^q) = 0.$$

This arrangement is free with exponents $(1, p-1, q-1)$ and the monodromy operators $h^m : H^m(F, \mathbb{C}) \to H^m(F, \mathbb{C})$ can be easily computed using [**19**, Theorem 1.4] or [**55**]. For all the pairs (p,q) we have tested, i.e. $2 \leq p, q \leq d = p+q \leq 12$, the algorithm described above gives the correct result. In other words, the corresponding arrangements $\mathcal{A}(p,q)$ are $(k,1)$-top-computable for all the integers k with $1 \leq k \leq d$. For instance, for the arrangement $\mathcal{A}(4,8)$ we get the following spectrum

$$Sp^0_P(f) = 3t^{\frac{6}{12}} + 10t^{\frac{9}{12}} + 21t^{\frac{12}{12}} + 12t^{\frac{15}{12}} + 9t^{\frac{18}{12}} + 2t^{\frac{21}{12}}. \tag{5.3}$$

Note that this spectrum is not symmetric with respect to the monomial $21t = 21t^{\frac{12}{12}}$, and an equality similar to that in Corollary 2.4 does not hold for the multiplicities of the roots of

$$\Delta^3(\mathcal{A}(4,8)) = \Phi_1^{21} \cdot \Phi_2^{12} \cdot \Phi_4^{12}.$$

Here and in the sequel, Φ_j denotes the j-th cyclotomic polynomial.

EXAMPLE 5.2 (The braid arrangement A_4). The braid arrangement A_4 is defined in $\mathbb{C}^5$ by the equation

$$\prod_{0 \leq i < j \leq 4} (x_i - x_j) = 0.$$

However, this arrangement is not essential. Using the coordinate change $x = x_1 - x_0$, $y = x_2 - x_0$, $z = x_3 - x_0$, $w = x_4 - x_0$, $u = x_0 + x_1 + x_2 + x_3 + x_4$, we see that the essential version of this arrangement is given in $\mathbb{C}^4$, corresponding to $u = 0$, by the equation

$$\mathcal{A} : f = xyzw(x-y)(x-z)(x-w)(y-z)(y-w)(z-w) = 0.$$

Regarded as an arrangement in $\mathbb{P}^3$, this arrangement is known to be free with exponents $(d_1, d_2, d_3) = (2, 3, 4)$. Running the algorithm described above and using the fact that (3.4) is an equality in this case as implied by Settepanella's results in [**51**, Table 2], we get that $\mathcal{A}$ is $(k,1)$-top-computable for any $k \in [1, 10]$. In particular, we have

$$Sp^0_P(f) = t^{\frac{4}{10}} + 4t^{\frac{5}{10}} + 5t^{\frac{6}{10}} + 6t^{\frac{7}{10}} + 6t^{\frac{8}{10}} + 6t^{\frac{9}{10}} + 24t^{\frac{10}{10}} + \tag{5.4}$$
$$+ 6t^{\frac{11}{10}} + 6t^{\frac{12}{10}} + 6t^{\frac{13}{10}} + 5t^{\frac{14}{10}} + 4t^{\frac{15}{10}} + t^{\frac{16}{10}}.$$

Then the formula for the spectrum clearly implies the following formula for the Alexander polynomial $\Delta^3(\mathcal{A})$:

$$\Delta^3(\mathcal{A}) = \Phi_1^{24} \cdot \Phi_2^{8} \cdot \Phi_5^{6} \cdot \Phi_{10}^{6}, \tag{5.5}$$

which coincides of course with the formula given in [**51**, Table 2]. It is known that in this case $\Delta^1(\mathcal{A}) = \Phi_1^9$, see [**36**] or [**51**, Table 2]. Using the formula (1.4) and (4.10), we get $\chi(M(\mathcal{A})) = -6$ and it follows that

$$\Delta^2(\mathcal{A}) = \Phi_1^{26} \cdot \Phi_2^{2}.$$

This coincides again with the formula given in [**51**, Table 2]. Any integer k such that $e = 1, 2$ is non-resonant with respect to the arrangement A_4.

EXAMPLE 5.3 (The Coxeter arrangement D_4). The arrangement D_4 is defined in $\mathbb{C}^4$ by the equation

$$\mathcal{A} : f = (x^2 - y^2)(x^2 - z^2)(x^2 - w^2)(y^2 - z^2)(y^2 - w^2)(z^2 - w^2) = 0.$$

Regarded as an arrangement in $\mathbb{P}^3$, this arrangement is known to be free with exponents $(d_1, d_2, d_3) = (3, 3, 5)$, see [**39**]. Running the algorithm described above *and assuming Conjecture* 3.9 *true*, we get

$$(5.6)\quad Sp^0_P(f) = t^{\frac{4}{12}} + 4t^{\frac{5}{12}} + 10t^{\frac{6}{12}} + 12t^{\frac{7}{12}} + 23t^{\frac{8}{12}} + 16t^{\frac{9}{12}} + 20t^{\frac{10}{12}} + 16t^{\frac{11}{12}} + 45t^{\frac{12}{12}} + \\ + 16t^{\frac{13}{12}} + 20t^{\frac{14}{12}} + 16t^{\frac{15}{12}} + 23t^{\frac{16}{12}} + 12t^{\frac{17}{12}} + 10t^{\frac{18}{12}} + 4t^{\frac{19}{12}} + t^{\frac{20}{12}}.$$

This formula for the spectrum clearly implies the following formula for the Alexander polynomial $\Delta^3(\mathcal{A})$:

$$\Delta^3(\mathcal{A}) = \Phi_1^{45} \cdot \Phi_2^{20} \cdot \Phi_3^{24} \cdot \Phi_4^{16} \cdot \Phi_6^{20} \cdot \Phi_{12}^{16}. \tag{5.7}$$

It is known that in this case $\Delta^1(\mathcal{A}) = \Phi_1^{11}\Phi_3$, see [**36**]. Using the formula (1.4) and (4.10), we get $\chi(M(\mathcal{A})) = -16$ and it follows that

$$\Delta^2(\mathcal{A}) = \Phi_1^{39} \cdot \Phi_2^{4} \cdot \Phi_3^{9} \cdot \Phi_6^{4}.$$

Any integer k such that $e = 1, 3$ is non-resonant with respect to the arrangement D_4.

EXAMPLE 5.4 (The complex reflection arrangement $\mathcal{A}(3, 3, 4)$). The hyperplane arrangement $\mathcal{A}(3, 3, 4)$ is defined in $\mathbb{C}^4$ by the equation

$$\mathcal{A} : f = (x^3 - y^3)(x^3 - z^3)(x^3 - w^3)(y^3 - z^3)(y^3 - w^3)(z^3 - w^3) = 0.$$

Regarded as an arrangement in $\mathbb{P}^3$, this arrangement is known to be free with exponents $(d_1, d_2, d_3) = (4, 6, 7)$, see [**39**]. Running the algorithm described above *and assuming Conjecture* 3.9 *true*, we get

$$(5.8)\quad Sp^0_P(f) = t^{\frac{4}{18}} + 4t^{\frac{5}{18}} + 10t^{\frac{6}{18}} + 19t^{\frac{7}{18}} + 31t^{\frac{8}{18}} + 46t^{\frac{9}{18}} + 59t^{\frac{10}{18}} + 71t^{\frac{11}{18}} + 98t^{\frac{12}{18}} + \\ + 86t^{\frac{13}{18}} + 89t^{\frac{14}{18}} + 92t^{\frac{15}{18}} + 90t^{\frac{16}{18}} + 90t^{\frac{17}{18}} + 168t^{\frac{18}{18}} + 90t^{\frac{19}{18}} + 90t^{\frac{20}{18}} + 92t^{\frac{21}{18}} + 89t^{\frac{22}{18}} + \\ + 86t^{\frac{23}{18}} + 98t^{\frac{24}{18}} + 71t^{\frac{25}{18}} + 59t^{\frac{26}{18}} + 46t^{\frac{27}{18}} + 31t^{\frac{28}{18}} + 19t^{\frac{29}{18}} + 10t^{\frac{30}{18}} + 4t^{\frac{31}{18}} + t^{\frac{32}{18}}.$$

This formula for the spectrum clearly implies the following formula for the Alexander polynomial $\Delta^3(\mathcal{A})$:

$$\Delta^3(\mathcal{A}) = \Phi_1^{168} \cdot \Phi_2^{92} \cdot \Phi_3^{108} \cdot \Phi_6^{92} \cdot \Phi_9^{90} \cdot \Phi_{18}^{90}. \tag{5.9}$$

It is known that in this case $\Delta^1(\mathcal{A}) = \Phi_1^{17}\Phi_3$, see [**36**]. In fact, a generic plane section $\mathcal{A}_H$ has 42 triple points and 27 nodes, and the result follows also from [**8**]. Using the formula (1.4) and (4.10), we get $\chi(M(\mathcal{A})) = -90$ and it follows that

$$\Delta^2(\mathcal{A}) = \Phi_1^{94} \cdot \Phi_2^{2} \cdot \Phi_3^{19} \cdot \Phi_6^{2}.$$

Any integer k such that $e = 1, 2$ is non-resonant with respect to the arrangement $\mathcal{A}(3, 3, 4)$.

EXAMPLE 5.5 (The complex reflection arrangement $\mathcal{A}(2, 1, 4)$). The hyperplane arrangement $\mathcal{A}(2, 1, 4)$ is defined in $\mathbb{C}^4$ by the equation

$$\mathcal{A} : f = xyzw(x^2 - y^2)(x^2 - z^2)(x^2 - w^2)(y^2 - z^2)(y^2 - w^2)(z^2 - w^2) = 0.$$

Regarded as an arrangement in $\mathbb{P}^3$, this arrangement is known to be free with exponents $(d_1, d_2, d_3) = (3, 5, 7)$, see [**39**]. Running the algorithm described above *and assuming Conjecture* 3.9 *true*, we get

$$Sp^0_P(f) = t^{\frac{4}{16}} + 4t^{\frac{5}{16}} + 9t^{\frac{6}{16}} + 16t^{\frac{7}{16}} + 25t^{\frac{8}{16}} + 32t^{\frac{9}{16}} + 39t^{\frac{10}{16}} + 44t^{\frac{11}{16}} + \\ +47t^{\frac{12}{16}} + 48t^{\frac{13}{16}} + 48t^{\frac{14}{16}} + 48t^{\frac{15}{16}} + 105t^{\frac{16}{16}} + 48t^{\frac{17}{16}} + 48t^{\frac{18}{16}} + 48t^{\frac{19}{16}} + 47t^{\frac{20}{16}} + \\ +44t^{\frac{21}{16}} + 39t^{\frac{22}{16}} + 32t^{\frac{23}{16}} + 25t^{\frac{24}{16}} + 16t^{\frac{25}{16}} + 9t^{\frac{26}{16}} + 4t^{\frac{27}{16}} + t^{\frac{28}{16}}. \tag{5.10}$$

This formula for the spectrum clearly implies the following formula for the Alexander polynomial $\Delta^3(\mathcal{A})$:

$$\Delta^3(\mathcal{A}) = \Phi_1^{105} \cdot \Phi_2^{50} \cdot \Phi_4^{48} \cdot \Phi_8^{48} \cdot \Phi_{16}^{48}. \tag{5.11}$$

It is known that in this case $\Delta^1(\mathcal{A}) = \Phi_1^{15}$, see [**36**]. Using the formula (1.4) and (4.10), we get $\chi(M(\mathcal{A})) = -48$ and it follows that

$$\Delta^2(\mathcal{A}) = \Phi_1^{73} \cdot \Phi_2^2.$$

Any integer k such that $e = 1, 2, 4$ is non-resonant with respect to the arrangement $\mathcal{A}(2, 1, 4)$.

REMARK 5.6. Note that all the spectra coming from reflection groups in Examples 5.2, 5.3, 5.4, 5.5 above enjoy a perfect symmetry with respect the monomial containing t, i.e. the coefficients of t^α and $t^{2-\alpha}$ coincide for all $0 < \alpha < 1$. This symmetry might be related to the symmetry of the Bernstein-Sato polynomial b_f of a free arrangement $\mathcal{A} : f = 0$ recalled in (3.1). However, note that for some free arrangements as in Example 5.1, the pole order spectra are not symmetric, but such arrangements seem to be quite exceptional. Indeed, most of the free arrangements we have tested so far enjoy the above spectrum symmetry property.

EXAMPLE 5.7 (A non free arrangement). Consider the arrangement $\mathcal{A}$ defined in $\mathbb{C}^4$ by the equation

$$\mathcal{A} : f = xyzw(x + y + z)(y - z + w) = 0.$$

This arrangement is far from being free, the S-module $AR(f)$ has 6 generators of degrees 2, 2, 3, 3, 3, 3 respectively. Running the algorithm described above *and assuming Conjecture* 3.9 *true*, we get

$$Sp^0_P(f) = t^{\frac{4}{6}} + 2t^{\frac{5}{6}} + 8t^{\frac{6}{6}} + 2t^{\frac{7}{6}} + 2t^{\frac{8}{6}} + 2t^{\frac{9}{6}} + t^{\frac{10}{6}}. \tag{5.12}$$

This formula for the spectrum clearly implies the following formula for the Alexander polynomial $\Delta^3(\mathcal{A})$:

$$\Delta^3(\mathcal{A}) = \Phi_1^8 \cdot \Phi_2^2 \cdot \Phi_3^2 \cdot \Phi_6^2. \tag{5.13}$$

A generic plane section of $\mathcal{A}$ is a nodal line arrangement, and hence in this case $\Delta^1(\mathcal{A}) = \Phi_1^5$. It is easy to compute $\chi(M(\mathcal{A})) = -2$, and hence using the formula (1.4), we get

$$\Delta^2(\mathcal{A}) = \Phi_1^{10}.$$

Note that the two points $A = (0 : 0 : 0 : 1)$ and $B = (1 : 0 : 0 : 0)$ both correspond to dense edges X with $n_X = 4$, and any hyperplane in $\mathcal{A}$ contains at least one of these two points. It follows that the value $k = 3$ is resonant with respect to the arrangement $\mathcal{A}$, i.e. the defining property of non-resonancy in Remark 3.8 is not satisfied, but still there is no contribution to $H^m(F, \mathbb{C})_{-1}$ for $m < 3$. This fact suggests that Yoshinaga's results in [**59**] for real line arrangements might have a higher dimensional analogue.

EXAMPLE 5.8 (A free discriminant surface). Consider the surface in $\mathbb{P}^3$ given by

$$V : f = y^2z^2 - 4xz^3 - 4y^3w + 18xyzw - 27x^2w^2 = 0.$$

Then V is just the discriminant of cubic binary forms in $\mathbb{P}(\mathbb{C}[u,v]_3) = \mathbb{P}^3$, i.e. the set of cubic forms in u,v with a multiple linear factor. It is known that V is a free surface with exponents $d_1 = d_2 = d_3 = 1$ and V is homeomorphic to $\mathbb{P}^1 \times \mathbb{P}^1$, see [**29**]. Using the homogeneity under the obvious $G\ell_2(\mathbb{C})$-action on $\mathbb{P}(\mathbb{C}[u,v]_3)$, it is easy to see that V is locally quasi-homogeneous. Running the algorithm described above, we get the following for the terms occurring in the inequality (3.4)

$$E_2^{3-q,q}(f)_k = 0 \text{ for } (q,k) \neq (0,4) \text{ and } \dim E_2^{3,0}(f)_4 = 1. \tag{5.14}$$

It follows that $b_3(F) \leq 1$. On the other hand we have

$$\chi(F) = 4\chi(M) = 4(\chi(\mathbb{P}^3) - \chi(V)) = 4(4-4) = 0.$$

Note that a generic plane section of V is a quartic curve with 4 cusps, and hence $b_0(F) = 1$ and $b_1(F) = 0$, see for instance [**17**, Proposition 4.4.8]. It follows that

$$b_3(F) = b_2(F) + b_0(F) \geq 1.$$

It follows that $b_3(F) = 1$ and hence f is k-top-computable for any $k \in [1,4]$. The corresponding Alexander polynomials are $\Delta^3(V) = \Delta^0(V) = \Phi_1$ and $\Delta^2(V) = \Delta^1(V) = 1$.

EXAMPLE 5.9 (A non-free irreducible surface). Consider the surface in $\mathbb{P}^3$ given by

$$V : f = x^3z + x^2y^2 + y^2w(y+w) = 0,$$

and the corresponding Milnor fiber $F : f = 1$ in $\mathbb{C}^4$. Let H be the hyperplane in $\mathbb{C}^4$ given by $x = 0$ and note that $F_0 = F \cap H$ is given by $y^2w(y+w) = 1$ in $\mathbb{C}^3$, with coordinates y,z,w. It follows that F_0 is a smooth surface, homotopically equivalent to the affine curve $F_0' : y^2w(y+w) = 1$ in $\mathbb{C}^2$, with coordinates y,w. It is easy to see that the projective closure C of F_0' is a quartic irreducible curve, with a unique singular point of type A_3. It follows that $\chi(C) = 2 - (4-1)(4-2) + \mu(C) = -1$, see for instance [**17**, Corollary 5.4.4]. Then $\chi(F_0') = \chi(C) - 3 = -4$, since C has 3 points at infinity. It follows that $b_0(F_0) = 1$, $b_1(F_0) = 5$ and $b_j(F_0) = 0$ for $j \geq 2$. On the other hand, the projection on the x-coordinate induces a locally trivial fibration

$$\mathbb{C}^2 \to F \setminus F_0 \to \mathbb{C}^*,$$

and hence $F \setminus F_0$ is homotopy equivalent to $\mathbb{C}^*$. The Gysin sequence in homology,

$$\cdots \to H_k(F \setminus F_0) \to H_k(F) \to H_{k-2}(F_0) \to H_{k-1}(F \setminus F_0) \to \cdots,$$

see for instance [**17**, Equation (2.2.13)], yields the following Betti numbers for F: $b_0(F) = 1$, $b_1(F) = b_2(F) = 0$ and $b_3(F) = 5$. Indeed, note that a generic plane section $V_P = V \cap P$ of V is an irreducible curve in $P = \mathbb{P}^2$ having a point of multiplicity 3. It follows by [**17**, Corollary 4.3.8] that

$$\pi_1(P \setminus V_P) = \pi_1(\mathbb{P}^3 \setminus V) = \mathbb{Z}/d\mathbb{Z}.$$

Then [**17**, Corollary 4.1.10] implies that $b_1(F) = 0$.

Running the algorithm described above, we get the following non-zero terms among those occurring in the inequality (3.4)

$$\dim E_2^{2,1}(f)_k = 1 \text{ for } k = 1,2,3,4 \text{ and } \dim E_2^{3,0}(f)_4 = 1. \tag{5.15}$$

It follows that we get an equality in (3.4), with implies that $V : f = 0$ is $(k, 1)$-top-computable for all the integers k with $1 \leq k \leq 4$. In particular we have

$$Sp^0_P(f) = t^{\frac{4}{4}} + t^{\frac{5}{4}} + t^{\frac{6}{4}} + t^{\frac{7}{4}} + t^{\frac{8}{4}}. \tag{5.16}$$

The corresponding Alexander polynomials are $\Delta^3(V) = \Phi_1^2 \cdot \Phi_2 \cdot \Phi_4$, $\Delta^0(V) = \Phi_1$ and $\Delta^2(V) = \Delta^1(V) = 1$.

The value of the Alexander polynomial $\Delta^3(V)$ can also be obtained from the isomorphism $H_3(F) \to H_1(F_0)$ in the above Gysin sequence, using its functoriality, but not the spectrum $Sp^0_P(f)$. Note also that the surface in $\mathbb{P}^3$ given by

$$V' : f' = x^3z + x^2y^2 + y^2w(y + w) + x^2w^2 = f + x^2w^2 = 0,$$

has the same topological properties as $V : f = 0$, but this time the inequality in (3.4) is strict.

EXAMPLE 5.10 (Another non-free irreducible surface). Consider the surface in $\mathbb{P}^3$ given by

$$V : f = x^4 + x^3z + yz^2w = 0,$$

and the corresponding Milnor fiber $F : f = 1$ in $\mathbb{C}^4$. Consider the hyperplanes $H_y : y = 0$ and $H_z : z = 0$ in $\mathbb{C}^4$ and set $F_y = F \cap H_y$, $F_z = F \cap H_z$, $F_0 = F_y \cap F_z$ and $F' = F \setminus F_0$. It follows that F_0 is a smooth curve in F, homotopically equivalent to 4 points. The corresponding Gysin sequence in homology contains the sequence

$$0 = H_4(F) \to H_0(F_0) \to H_3(F') \to H_3(F) \to H_{-1}(F_0) = 0,$$

which implies $b_3(F) = b_3(F') - 4$. Next $D' = (F_y \cup F_z) \setminus F_0$ is a smooth divisor in F' and the projection

$$p : F' \setminus D' \to (\mathbb{C}^*)^2, \quad (x, y, z, w) \mapsto (y, z)$$

is a locally trivial fibration with contractible fibers and hence it is a homotopy equivalence. Moreover, it is easy to see that D' has five connected components, one homotopy equivalent to $\mathbb{C}$ minus 5 points, the other four homotopy equivalent to $\mathbb{C}^*$. It follows that $b_1(D') = 9$. The corresponding Gysin sequence of the pair (F', D') contains the sequence

$$0 = H_3(F' \setminus D') \to H_3(F') \to H_1(D') \to H_3(F' \setminus D') = \mathbb{Z},$$

which implies that $b_3(F') \in \{8, 9\}$. Therefore we get $b_3(F) \in \{4, 5\}$.

Running the algorithm described above, we get the following non-zero terms among those occurring in the inequality (3.4)

$$\dim E_2^{2,1}(f)_k = 1 \text{ for } k = 1, 2, 3 \text{ and } \dim E_2^{3,0}(f)_4 = 1. \tag{5.17}$$

The inequality (3.4) implies $b_3(F) = 4$ and hence $V : f = 0$ is $(k, 1)$-top-computable for all the integers k with $1 \leq k \leq 4$. In particular we have

$$Sp^0_P(f) = t^{\frac{4}{4}} + t^{\frac{5}{4}} + t^{\frac{6}{4}} + t^{\frac{7}{4}}. \tag{5.18}$$

Using [**17**, Proposition 4.4.3], we get as above $b_1(F) = 0$. Consider the partition of V given by $V = V_z \cup (V \setminus V_z)$, where $V_z = V \cap \{z = 0\}$. Then $V_z = \mathbb{P}^1$ and hence $\chi(V_z) = 2$. Using the theory of tame polynomials, see for instance [**5**], it follows that $\chi(V \setminus V_z) = 2$, and hence $\chi(V) = \chi(V_z) + \chi(V \setminus V_z) = 4$. It follows that $\chi(M) = \chi(\mathbb{P}^3) - \chi(V) = 0$. This implies that corresponding Alexander polynomials are $\Delta^3(V) = \Phi_1 \cdot \Phi_2 \cdot \Phi_4$, $\Delta^0(V) = \Phi_1$ and $\Delta^2(V) = \Phi_2 \cdot \Phi_4$. In particular $b_2(F) = 3$.

References

[1] E. Artal Bartolo, *Topology of arrangements and position of singularities* (English, with English and French summaries), Ann. Fac. Sci. Toulouse Math. (6) **23** (2014), no. 2, 223–265, DOI 10.5802/afst.1406. MR3205593

[2] P. Bailet, *On the monodromy of Milnor fibers of hyperplane arrangements*, Canad. Math. Bull. **57** (2014), no. 4, 697–707, DOI 10.4153/CMB-2014-032-4. MR3270792

[3] P. Bailet and S. Settepanella, *Homology graph of real arrangements and monodromy of Milnor fiber*, Adv. in Appl. Math. **90** (2017), 46–85, DOI 10.1016/j.aam.2017.04.006. MR3666711

[4] P. Bailet and M. Yoshinaga, *Degeneration of Orlik-Solomon algebras and Milnor fibers of complex line arrangements*, Geom. Dedicata **175** (2015), 49–56, DOI 10.1007/s10711-014-0027-7. MR3323628

[5] S. A. Broughton, *Milnor numbers and the topology of polynomial hypersurfaces*, Invent. Math. **92** (1988), no. 2, 217–241, DOI 10.1007/BF01404452. MR936081

[6] N. Budur and M. Saito, *Multiplier ideals, V-filtration, and spectrum*, J. Algebraic Geom. **14** (2005), no. 2, 269–282, DOI 10.1090/S1056-3911-04-00398-4. MR2123230

[7] N. Budur and M. Saito, *Jumping coefficients and spectrum of a hyperplane arrangement*, Math. Ann. **347** (2010), no. 3, 545–579, DOI 10.1007/s00208-009-0449-y. MR2640043

[8] N. Budur, A. Dimca, and M. Saito, *First Milnor cohomology of hyperplane arrangements*, Topology of algebraic varieties and singularities, Contemp. Math., vol. 538, Amer. Math. Soc., Providence, RI, 2011, pp. 279–292, DOI 10.1090/conm/538/10606. MR2777825

[9] F. Callegaro, *The homology of the Milnor fiber for classical braid groups*, Algebr. Geom. Topol. **6** (2006), 1903–1923, DOI 10.2140/agt.2006.6.1903. MR2263054

[10] F. Callegaro, *Salvetti complex, spectral sequences and cohomology of Artin groups* (English, with English and French summaries), Ann. Fac. Sci. Toulouse Math. (6) **23** (2014), no. 2, 267–296, DOI 10.5802/afst.1407. MR3205594

[11] F. J. Castro-Jiménez, L. Narváez-Macarro, and D. Mond, *Cohomology of the complement of a free divisor*, Trans. Amer. Math. Soc. **348** (1996), no. 8, 3037–3049, DOI 10.1090/S0002-9947-96-01690-X. MR1363009

[12] D. C. Cohen and A. I. Suciu, *On Milnor fibrations of arrangements*, J. London Math. Soc. (2) **51** (1995), no. 1, 105–119, DOI 10.1112/jlms/51.1.105. MR1310725

[13] CoCoA-5 (15 Sept 2014): a system for doing Computations in Commutative Algebra, available at http://cocoa.dima.unige.it

[14] W. Decker, G.-M. Greuel, G. Pfister and H. Schönemann. SINGULAR 4-0-1 — A computer algebra system for polynomial computations, available at http://www.singular.uni-kl.de (2014).

[15] P. Deligne, *Théorie de Hodge. II* (French), Inst. Hautes Études Sci. Publ. Math. **40** (1971), 5–57. MR0498551

[16] A. Dimca, *On the Milnor fibrations of weighted homogeneous polynomials*, Compositio Math. **76** (1990), no. 1-2, 19–47. Algebraic geometry (Berlin, 1988). MR1078856

[17] A. Dimca, *Singularities and topology of hypersurfaces*, Universitext, Springer-Verlag, New York, 1992. MR1194180

[18] A. Dimca, *Sheaves in topology*, Universitext, Springer-Verlag, Berlin, 2004. MR2050072

[19] A. Dimca, *Tate properties, polynomial-count varieties, and monodromy of hyperplane arrangements*, Nagoya Math. J. **206** (2012), 75–97, DOI 10.1215/00277630-1548502. MR2926485

[20] A. Dimca, *Jacobian syzygies, stable reflexive sheaves, and Torelli properties for projective hypersurfaces with isolated singularities*, Algebr. Geom. **4** (2017), no. 3, 290–303, DOI 10.14231/AG-2017-016. MR3652081

[21] A. Dimca, *Hyperplane arrangements*, Universitext, Springer, Cham, 2017. An introduction. MR3618796

[22] A. Dimca and G. Lehrer, *Cohomology of the Milnor fibre of a hyperplane arrangement with symmetry*, Configuration spaces, Springer INdAM Ser., vol. 14, Springer, [Cham], 2016, pp. 233–274. MR3615735

[23] A. Dimca and M. Saito, *Some consequences of perversity of vanishing cycles* (English, with English and French summaries), Ann. Inst. Fourier (Grenoble) **54** (2004), no. 6, 1769–1792 (2005). MR2134223

[24] A. Dimca, M. Saito, *Koszul complexes and spectra of projective hypersurfaces with isolated singularities*, arXiv:1212.1081.

[25] A. Dimca and M. Saito, *Generalization of theorems of Griffiths and Steenbrink to hypersurfaces with ordinary double points*, Bull. Math. Soc. Sci. Math. Roumanie (N.S.) **60(108)** (2017), no. 4, 351–371. MR3727394

[26] A. Dimca and G. Sticlaru, *Koszul complexes and pole order filtrations*, Proc. Edinb. Math. Soc. (2) **58** (2015), no. 2, 333–354, DOI 10.1017/S0013091514000182. MR3341443

[27] A. Dimca and G. Sticlaru, *A computational approach to Milnor fiber cohomology*, Forum Math. **29** (2017), no. 4, 831–846, DOI 10.1515/forum-2016-0044. MR3669005

[28] A. Dimca and G. Sticlaru, *Free divisors and rational cuspidal plane curves*, Math. Res. Lett. **24** (2017), no. 4, 1023–1042, DOI 10.4310/MRL.2017.v24.n4.a5. MR3723802

[29] A. Dimca and G. Sticlaru, *Free and nearly free surfaces in* $\mathbb{P}^3$, Asian J. Math. **22** (2018), no. 5, 787–810, DOI 10.4310/AJM.2018.v22.n5.a1. MR3878139

[30] A. Dimca and G. Sticlaru, *Computing the monodromy and pole order filtration on Milnor fiber cohomology of plane curves*, J. Symbolic Comput. **91** (2019), 98–115, DOI 10.1016/j.jsc.2018.06.015. MR3860886

[31] Lê Dũng Tráng, *La monodromie n'a pas de points fixes*, J. Fac. Sci. Univ. Tokyo Sect. IA Math. **22** (1975), no. 3, 409–427. MR0401756

[32] Lê Dũng Tráng, *The geometry of the monodromy theorem*, C. P. Ramanujam—a tribute, Tata Inst. Fund. Res. Studies in Math., vol. 8, Springer, Berlin-New York, 1978, pp. 157–173. MR541020

[33] Lê Dũng Tráng, *Le théorème de la monodromie singulier* (French, with English summary), C. R. Acad. Sci. Paris Sér. A-B **288** (1979), no. 21, A985–A988. MR540373

[34] A. Libgober, *Eigenvalues for the monodromy of the Milnor fibers of arrangements*, Trends in singularities, Trends Math., Birkhäuser, Basel, 2002, pp. 141–150. MR1900784

[35] L. Maxim, M.Saito, J. Schürmann, *Spectral Hirzebruch-Milnor classes of singular hypersurfaces*, arXiv:1606.02218.

[36] A. D. Măcinic and Ş. Papadima, *On the monodromy action on Milnor fibers of graphic arrangements*, Topology Appl. **156** (2009), no. 4, 761–774, DOI 10.1016/j.topol.2008.09.014. MR2492960

[37] D. A. Măcinic, Ş. Papadima, and C. R. Popescu, *Modular equalities for complex reflection arrangements*, Doc. Math. **22** (2017), 135–150. MR3609200

[38] L. Narváez Macarro, *A duality approach to the symmetry of Bernstein-Sato polynomials of free divisors*, Adv. Math. **281** (2015), 1242–1273, DOI 10.1016/j.aim.2015.06.012. MR3366865

[39] P. Orlik and H. Terao, *Arrangements of hyperplanes*, Grundlehren der Mathematischen Wissenschaften [Fundamental Principles of Mathematical Sciences], vol. 300, Springer-Verlag, Berlin, 1992. MR1217488

[40] S. Papadima and A. I. Suciu, *The Milnor fibration of a hyperplane arrangement: from modular resonance to algebraic monodromy*, Proc. Lond. Math. Soc. (3) **114** (2017), no. 6, 961–1004, DOI 10.1112/plms.12027. MR3661343

[41] C. A. M. Peters and J. H. M. Steenbrink, *Mixed Hodge structures*, Ergebnisse der Mathematik und ihrer Grenzgebiete. 3. Folge. A Series of Modern Surveys in Mathematics [Results in Mathematics and Related Areas. 3rd Series. A Series of Modern Surveys in Mathematics], vol. 52, Springer-Verlag, Berlin, 2008. MR2393625

[42] K. Saito, *Theory of logarithmic differential forms and logarithmic vector fields*, J. Fac. Sci. Univ. Tokyo Sect. IA Math. **27** (1980), no. 2, 265–291. MR586450

[43] M. Saito, *On b-function, spectrum and multiplier ideals*, Algebraic analysis and around, Adv. Stud. Pure Math., vol. 54, Math. Soc. Japan, Tokyo, 2009, pp. 355–379, DOI 10.2969/aspm/05410355. MR2499561

[44] M. Saito, *Multiplier ideals, b-function, and spectrum of a hypersurface singularity*, Compos. Math. **143** (2007), no. 4, 1050–1068, DOI 10.1112/S0010437X07002916. MR2339839

[45] M. Saito, *Bernstein-Sato polynomials of hyperplane arrangements*, Selecta Math. (N.S.) **22** (2016), no. 4, 2017–2057, DOI 10.1007/s00029-016-0268-4. MR3573952

[46] M. Saito, *Hilbert series of graded Milnor algebras and roots of Bernstein-Sato polynomials*, arXiv: 1509.06288.

[47] M. Saito, *Bernstein-Sato polynomials and graded Milnor algebras for projective hypersurfaces with weighted homogeneous isolated singularities*, arXiv:1609.04801.

[48] M. Saito, *Roots of Bernstein-Sato polynomials of homogeneous polynomials with two-dimensional singular loci*, arXiv:1703.05741v6.

[49] M. Salvetti and S. Settepanella, *Combinatorial Morse theory and minimality of hyperplane arrangements*, Geom. Topol. **11** (2007), 1733–1766, DOI 10.2140/gt.2007.11.1733. MR2350466

[50] E. Sernesi, *The local cohomology of the Jacobian ring*, Doc. Math. **19** (2014), 541–565. MR3218781

[51] S. Settepanella, *Cohomology of pure braid groups of exceptional cases*, Topology Appl. **156** (2009), no. 5, 1008–1012, DOI 10.1016/j.topol.2008.12.007. MR2498934

[52] A. Simis and Ş. O. Tohăneanu, *Homology of homogeneous divisors*, Israel J. Math. **200** (2014), no. 1, 449–487, DOI 10.1007/s11856-014-0025-3. MR3219587

[53] J. Steenbrink, *Intersection form for quasi-homogeneous singularities*, Compositio Math. **34** (1977), no. 2, 211–223. MR0453735

[54] A. I. Suciu, *Hyperplane arrangements and Milnor fibrations* (English, with English and French summaries), Ann. Fac. Sci. Toulouse Math. (6) **23** (2014), no. 2, 417–481, DOI 10.5802/afst.1412. MR3205599

[55] D. S. Tapp, *Picard-Lefschetz monodromy of products*, J. Pure Appl. Algebra **212** (2008), no. 10, 2314–2319, DOI 10.1016/j.jpaa.2008.03.017. MR2426511

[56] J. Wiens and S. Yuzvinsky, *De Rham cohomology of logarithmic forms on arrangements of hyperplanes*, Trans. Amer. Math. Soc. **349** (1997), no. 4, 1653–1662, DOI 10.1090/S0002-9947-97-01894-1. MR1407505

[57] H. Terao, *Forms with logarithmic pole and the filtration by the order of the pole*, Proceedings of the International Symposium on Algebraic Geometry (Kyoto Univ., Kyoto, 1977), Kinokuniya Book Store, Tokyo, 1978, pp. 673–685. MR578880

[58] H. Terao, *Generalized exponents of a free arrangement of hyperplanes and Shepherd-Todd-Brieskorn formula*, Invent. Math. **63** (1981), no. 1, 159–179, DOI 10.1007/BF01389197. MR608532

[59] M. Yoshinaga, *Milnor fibers of real line arrangements*, J. Singul. **7** (2013), 220–237. MR3090727

[60] M. Yoshinaga, *Freeness of hyperplane arrangements and related topics* (English, with English and French summaries), Ann. Fac. Sci. Toulouse Math. (6) **23** (2014), no. 2, 483–512, DOI 10.5802/afst.1413. MR3205600

Université Côte d'Azur, CNRS, LJAD, France
Email address: dimca@unice.fr

Faculty of Mathematics and Informatics, Ovidius University, Bd. Mamaia 124, 900527 Constanta, Romania
Email address: gabrielsticlaru@yahoo.com

Contemporary Mathematics
Volume **742**, 2020
https://doi.org/10.1090/conm/742/14938

Straight equisingular deformations and punctual Hilbert schemes

Gert-Martin Greuel

Dedicated to Lê Dũng Tráng on the occasion of his 70th Birthday

Abstract. We study "straight equisingular deformations", a linear subfunctor of all equisingular deformations and describe their seminuniversal deformation by an ideal containing the fixed Tjurina ideal. Moreover, we show that the base space of the seminuniversal straight equisingular deformation appears as the fibre of a morphism from the μ-constant stratum onto a punctual Hilbert scheme parametrizing certain zero-dimensional schemes concentrated in the singular point. Although equisingular deformations of plane curve singularities are very well understood, we believe that this aspect may give a new insight in their inner structure.

Introduction

Let z be a singular point of the reduced curve C contained in the smooth complex surface Σ. The topological type of the germ (C, z) can be described by the resolution graph of a good embedded resolution of (C, z), and an equisingular deformation of (C, z) is one with constant resolution graph. In the first section we fix the notations and recall the classical concepts of constellation, clusters and proximities, resulting in the notions of the essential tree $\mathscr{T}^*(C, z)$ and the cluster graph $\mathbb{G}(C, z)$, which is also a complete invariant of the toplogical type of (C, z). These notions go back to Enriques and Chisini, for a comprehensive and modern treatment of we refer to [**Cas00**].

A deformation of (C, z) is equisingular or an es-deformation, if there exist equimultiple sections through the points of the essential tree (i.e. through the non-nodes of a minimal good resolution) of (C, z) such that the family can be blown up successively along these sections. Those deformations for which these sections can be simultaneously trivialized are called straight equisingular or straight es. These deformations have been considered by J. Wahl in [**Wah74**] (not under this name) and they were further characterized in [**GLS07**], showing, e.g., that for semiquasi-homogeneous and Newton nondegenerate singularities every equisingular deformation is straight equisingular. In Section 2 we introduce, in addition, for any plane curve singularity the topological singularity ideal $I^s(C, z)$ such that

2010 *Mathematics Subject Classification.* 14B05, 14B07, 32S05, 32S15.

$I^s(C,z)/I^{ea}_{fix}(C,z)$, with $I^{ea}_{fix}(C,z)$ the fixed Tjurina ideal, is the tangent space to straight equisingular deformations. This provides an explicit description of the semiuniversal straight equisingular deformation as a linear subspace of all equisingular deformations. We illustrate this by an example using SINGULAR [**DGPS**].

In Section 3 we consider the punctual Hilbert scheme, parametrizing zero-dimensional schemes $Z \subset \Sigma$ with support in one point. We are interested in the *rooted Hilbert scheme*, a subscheme of the punctual Hilbert scheme consisting of the *topological singularity scheme* $Z^s(C,z)$ defined by the ideal $I^s(C,z)$, with fixed cluster graph $\mathbb{G} = \mathbb{G}(C,z)$. Mapping an equisingular family of plane curve singularities to the induced family of topological singularity schemes, provides a morphism from the functor of all es-deformations to the rooted Hilbert functor. We prove that the fibre of the induced morphism from the base space of the semiuniversal es-deformation of (C,z) to the rooted Hilbert scheme is exactly the base space of the semiuniversal straight equisingular deformations. This gives in addition a nice formula for the rooted Hilbert scheme in terms of the number of free vertices in the essential tree $\mathscr{T}^*$ of (C,z).

Let us finish the introduction by commenting on the positive characteristic case. Though we work here over the complex numbers, the results should hold also for algebraically closed ground fields of characteristic zero. Equisingular deformations of plane curve singularities in arbitrary characteristic have been studied in [**CGL07a**] and [**CGL07b**]. In positive characteristic one has to distinguish between "good" and "bad" characteristic and we conjecture that the results of this paper can be extended to good characteristic as well. However, in bad characteristic one has to distinguish between equisingular deformations and weakly equisingular deformations, i.e., those that become equisingular after a finite base change, with very different semiuniversal deformations. It would be interesting to study straight es-deformations in positive characteristic, in particular in bad characteristic.

Acknowledgement: The article is a revised excerpt of parts from Chapter I.1 of the forthcoming book [**GLS18**] by Chistoph Lossen, Eugenii Shustin and the author, where more details can be found. I would like to thank C. Lossen and E. Shustin for the many pleasant years of joint work on the book.

1. Equisingularity for Plane Curve Singularities

Before we define in the next section straight equisingular deformations, recall the definition of an equisingular deformation of a reduced plane complex curve singularity $(C,z) \subset (\Sigma,z)$.

It is well-known that we can resolve the singularity (C,z) by finitely many blowing ups of points. More precisely, there exists a *good embedded resolution* of the singularity of C at z, that is, a sequence of morphisms of smooth surfaces

$$\pi : \Sigma_{n+1} \xrightarrow{\pi_{n+1}} \Sigma_n \xrightarrow{\pi_n} \cdots \to \Sigma_1 \xrightarrow{\pi_1} \Sigma_0 = \Sigma,$$

such that π_i is the blowing up of a point $q_{i-1} \in \Sigma_{i-1}$ infinitely near to $q_0 = z$ and such that in a neighborhood of $E = \pi^{-1}(z)$, the *reduced total transform* of C

$$\pi^*(C)_{red} = \widetilde{C} + \sum_{i=1}^{n+1} E_i$$

is a divisor with normal crossings, that is, a hypersurface having only nodes as singularities (see [**GLS07**, Section I.3.3] for details). Here $\widetilde{C}$ is the (smooth) *strict transform* of C, $\pi^*(C)$ (with its scheme structure) is the *total transform* and $E_i = \pi_i^{-1}(q_{i-1}) \cong \mathbb{P}^1$, $i = 1, \ldots, n+1$, is the *exceptional divisor* of π_i in Σ_i (identified with its image in Σ_j, $j \geq i$). A good embedded resolution π is called a *minimal (good) embedded resolution*, if only non-nodal singular points of the reduced total transform of (C, z) are blown up in the resolution process (if (C, z) is smooth, we do not blow up anything, i.e. π is the empty map). It is well-known that such a minimal good resolution is unique up to isomorphism over Σ.

For $q \in \pi^{-1}(z)$ we denote by $C_{(q)}, \widehat{C}_{(q)}$ and $\widetilde{C}_{(q)}$ the respective germs at q of the strict, the total and the reduced total transform of C and and by $\operatorname{mt} C_{(q)}, \operatorname{mt} \widehat{C}_{(q)}$ and $\operatorname{mt} \widetilde{C}_{(q)}$ their multiplicities. If the germ $C_{(q)}$ is non-empty, we say that the curve C *goes through* the infinitely near point q (or that q *belongs to* C).

Recall that classically $E_1 \subset \Sigma_1$ and its strict transforms in Σ_i, $i > 1$, together with z, is called the *first infinitesimal neighbourhood* of $z \in \Sigma$. For $i > 1$ the i-th infinitesimal neighbourhood of z consists of points in the first infinitesimal neighbourhood of some point in the $i-1$-st infinitesimal neighbourhood of z. Any point belonging to some infinitesimal neighbourhood of $q_i, i \geq 0$, is called an *infinitely near point* of $q_i \in \Sigma_i$ or a point *infinitely near to* q_i.

DEFINITION 1.1. **(Constellation)** If $(q_0', \pi_1', q_1', \ldots, \pi_m', q_m')$ with $q_i' \in \Sigma_i'$ and $\pi_i' : \Sigma_i' \to \Sigma_{i-1}'$ blowing up q_{i-1}' is another sequence of a good embedded resolution of (C, z), then $(q_0, \pi_1, q_1, \ldots, \pi_n, q_n)$ and $(q_0', \pi_1', q_1', \ldots, \pi_m', q_m')$ are called equivalent if $m = n = -1$ or $m = n \geq 0$ and if there is a Σ-isomorphism $\Sigma_{n+1} \to \Sigma_{n+1}'$. An equivalence class of such sequences is called a *constellation* (of (C, z)) and denoted by $\mathscr{T}(C, z)$.

It is easy to see that the Σ-isomorphism preserves infinitely nearness and that successively blowing up two different points in the two possible orders gives rise to Σ-isomorphic surfaces, and hence the same constellation. The concept of a constellation has been introduced to describe isomorphism classes of good embedded resolutions.

DEFINITION 1.2. **(Constellation graph, essential tree, proximity)** Let $\mathscr{T} = \mathscr{T}(C, z) = (q_0, \pi_1, q_1, \ldots, \pi_n, q_n)$ be a constellation of (C, z).

(1) We introduce a partial ordering on the points $q_0, \ldots, q_n$ by

$$q_i \leq q_j \quad :\Longleftrightarrow \quad q_j \text{ is infinitely near to } q_i \,.$$

For each $0 \leq i \leq n$, we set the *level of* q_i as $\ell(q_i) := \#\{j \,|\, q_i \geq q_j\} - 1$. The point $z = q_0$, the *origin of* $\mathscr{T}$, is the only point of level 0.

(2) The *graph of the constellation* $\mathscr{T}$ is the oriented tree $\Gamma_{\mathscr{T}}$ (with root $z = q_0$) whose

- points are in 1-1 correspondence with $q_0, \ldots, q_n$ and
- edges with pairs (q_j, q_i) s.t. $\ell(q_j) = \ell(q_i) + 1$ and $q_j \geq q_i$.

Hence $p \geq q$ iff there is an oriented path in $\Gamma_{\mathscr{T}}$ from p to q.

(3) A point $z \neq q \in \mathscr{T}(C, z)$ is called *essential* for C iff the reduced total transform of C does not have a node at q. The origin z of $\mathscr{T}(C, z)$ is called essential for C, iff the germ (C, z) is not smooth. We call a point $q \in \mathscr{T}(C, z)$

a *singular essential point* for C if the strict transform of C at q is singular. The maximal finite sub-constellation of $\mathscr{T}(C,z)$ such that all of its points are essential is called the *essential constellation* of (C,z) and denoted by $\mathscr{T}^*(C,z)$.

By abuse of notation, a constellation $\mathscr{T}$ is also called a tree and $\mathscr{T}^*(C,z)$ is called the *essential tree* of (C,z). It describes the minimal embedded resolution of (C,z).

(4) Finally, we call q_j *proximate to* q_i (notation $q_j \dashrightarrow q_i$), if q_j is a point in E_{q_i}, the exceptional divisor of π_{i+1}, or in any of its strict transforms. q_j is a *satellite* point if it is proximate to (at least) two points q_i, $0 \le i \le j-1$, otherwise it is *free.*

It is clear that a point q_j cannot be proximate to more than two points since the exceptional divisors have normal crossings. Notice that each satellite point in $\mathscr{T}(C,z)$ is an essential point. It may well be a non-singular essential point for C.

It is classically known that the topological type of (C,z) can be characterized by several different sets of discrete data (see [**GLS07**, Section I.3.4] for a short overview). One characterization is by the graph $\Gamma_{\mathscr{T}}$, $\mathscr{T} = \mathscr{T}(C,z)$ a constellation of (C,z), together with the multiplicities $\operatorname{mt}\widehat{C}_{(q)}$, $q \in \mathscr{T}$, of the total transform (or, equivalently, the multiplicities $\operatorname{mt}\widetilde{C}_{(q)}$ of the reduced total transform). This weighted oriented graph is sometimes called a *resolution graph* of (C,z).

Remark 1.3. We have by [**GLS18**, Proposition I.1.11] the following *proximity equality*:

$$\operatorname{mt} C_{(p)} = \sum_{\substack{q\in\mathscr{T}\\ q\dashrightarrow p}} \operatorname{mt} C_{(q)} .$$

The difference between the multiplicities of the total and strict transform of C at a point $q \in \mathscr{T}$ is

$$\operatorname{mt}\widehat{C}_{(q)} - \operatorname{mt} C_{(q)} = \sum_{\substack{p\in\mathscr{T}\\ q\dashrightarrow p}} \operatorname{mt}\widehat{C}_{(p)}$$

(see [**GLS18**, Remark I.1.11.1]). Moreover $\operatorname{mt}\widetilde{C}_{(q)} - \operatorname{mt} C_{(q)} = 1$ or 2, depending if one or two points are proximate to q. In particular, the multiplicities of the strict transforms of C together with the proximities ($q \dashrightarrow p$) determine the multiplicities of the total transforms and hence can be used to describe the topological type of C. This is used in the following definition.

Definition 1.4. **(Cluster, Cluster graph)** Let $\mathscr{T}^* = \mathscr{T}^*(C,z) = (z = q_0, \pi_1, q_1, \ldots, \pi_n, q_n)$ be the essential tree of (C,z) and $\boldsymbol{m} = (m_0, \ldots, m_n)$ the vector of multiplicities with $m_0 := \operatorname{mt}(C,z)$ and $m_i := \operatorname{mt} C_{(q_i)}$, the multiplicity of the strict transform of C at q_i, $1 \le i \le n$. We call the tuple $(\mathscr{T}^*, \boldsymbol{m})$ the *cluster* of (C,z). The triple

$$\mathbb{G}(C,z) = \mathbb{G}(C,\mathscr{T}^*) := (\Gamma_{\mathscr{T}^*}, \dashrightarrow, \boldsymbol{m}),$$

consisting of $\Gamma_{\mathscr{T}^*}$, the graph of $\mathscr{T}^*$, the binary relation defined by $q \dashrightarrow p$ if q is proximate to p and the vector $\boldsymbol{m}$ of multiplicities of the strict transforms is called the *cluster graph* of (C,z).

Notice that the cluster graph $\mathbb{G}(C,z)$ determines (and is determined by) the topological type the curve singularity (C,z).

We consider now an *embedded deformation* (i, Φ, σ) of $(C, z) \subset (\Sigma, z)$, *with section* over a complex germ $(T, 0)$. This is a commutative diagram

$$\begin{array}{ccccc} (C,z) & \stackrel{i}{\hookrightarrow} & (\mathcal{C},z) & \hookrightarrow & (\mathcal{M},z) \\ \downarrow & & \sigma \uparrow \downarrow \Phi & \swarrow & \\ \{0\} & \hookrightarrow & (T,0) & & \end{array}$$

of morphisms of complex space germs with $(\mathcal{C}, z) \subset (\mathcal{M}, z)$ a hypersurface germ, σ a section of Φ. Φ is assumed to be flat as well as $(\mathcal{M}, z) \to (T, 0)$ which has (Σ, z) a special fibre. An embedded deformation (without section), denoted by (i, Φ), is given by a diagram as above but with σ deleted. We usually choose small representatives for the germs and denote them with the same letters, omitting the base points. Note that, for small representatives, we have $\mathcal{M} \simeq \Sigma \times T$ over T, and we have morphism $C \stackrel{i}{\hookrightarrow} \mathcal{C} \stackrel{\Phi}{\to} T$ with fibres the reduced curves $\mathcal{C}_t = \Phi^{-1}(t)$, where we identify C with $\mathcal{C}_0$ and write z instead of $(z, 0) \in \Sigma \times T$.

If $\mathcal{M} = \Sigma \times T$ with Φ the projection and if $\sigma(t) = (z, t)$, then σ is called the *trivial section*. By [**GLS07**, Proposition II.2.2] every section can be locally trivialized by an isomorphism $\mathcal{M} \simeq \Sigma \times T$ over T.

The following definition of equisingularity makes sense for arbitrary complex germs $(T, 0)$, even for Artinian ones. Let (C, z) be defined by $f \in \mathcal{O}_{\Sigma,z}$ with multiplicity $m = \mathrm{mt}(f)$.

DEFINITION 1.5. **(Equisingular deformation)** If (C, z) is smooth, any deformation of (C, z) is called equisingular. If (C, z) is singular let $\mathscr{T}^* = \mathscr{T}^*(C, z)$ be the essential tree of (C, z) and for each $q \in \mathscr{T}^*$, $q \neq z$, let $\widetilde{C}_{(q)}$ be the germ at q of the reduced total transform of (C, z). An embedded deformation (i, Φ) of (C, z) over $(T, 0)$ is then called *equisingular*, or an *es–deformation*, if the following three conditions hold:

(1) There exists a section σ of Φ, called *equimultiple section*, such that Φ is equimultiple along σ. If $(\mathcal{C}, z)$ is defined by $F \in \mathcal{O}_{\mathcal{M},z}$ this means that $F \in I_\sigma^m$, where I_σ denotes the ideal of $\sigma(T, 0) \subset (\mathcal{M}, z)$.
(2) For each $q \in \mathscr{T}^*$, $q \neq z$, there is a sequence of morphisms of germs (or of small representatives)
$$\Phi_{(q)} : \widetilde{\mathcal{C}}_{(q)} \hookrightarrow \mathcal{M}_{(q)} \stackrel{\pi_{(q)}}{\to} \mathcal{M} \to T,$$
where $\Phi_{(q)} : \widetilde{\mathcal{C}}_{(q)} \to T$ is an embedded deformation of $\widetilde{C}_{(q)} \subset \Sigma_{(q)}$ along an equimultiple section $\sigma_{(q)} : T \to \widetilde{\mathcal{C}}_{(q)}$ of $\Phi_{(q)}$.
(3) Each $\mathcal{M}_{(q)}$, $q \neq z$, is obtained by blowing up some $\mathcal{M}_{(p)}$ along the section $\sigma_{(p)}$, p of smaller level than q, with $\widetilde{\mathcal{C}}_{(q)}$ the reduced total transform of $\widetilde{\mathcal{C}}_{(p)}$ (for $p = z$ we take $\mathcal{M}_{(p)} = \mathcal{M}$, $\sigma_{(p)} = \sigma$ and $\widetilde{\mathcal{C}}_{(p)} = \mathcal{C}$).

See [**GLS07**, Definition II.2.6] for a more detailed description.

REMARK 1.6. (1) Condition (1) of Definition 1.5 implies that for $t \in T$ sufficiently close to 0 we have $F_t \in I_{\sigma(t)}^m$, where F_t defines the germ $(\mathcal{C}_t, \sigma(t))$ in $(\mathcal{M}_t, \sigma(t))$ and we have $I_{\sigma(t)} = \mathfrak{m}_{\mathcal{C}_t,\sigma(t)}$. Hence, the multiplicity of $(\mathcal{C}_t, \sigma(t))$ is constant for t near 0. For reduced base spaces this is equivalent to the given definition of equimultiplicity.

(2) Recall that, for an equisingular deformation, the equimultiple sections $\sigma_{(q)}$ through all essential points are unique (cf.[**GLS07**, Proposition II.2.8]). That is, after blowing up an equimultiple section, there is a unique section along which the blown up family is equimultiple.

(3) Let (i, Φ, σ) be an embedded deformation of $(C, z) \subset (\Sigma, z)$ along the section σ over a reduced complex germ $(T, 0)$. Then Φ is equisingular iff the cluster graph $\mathbb{G}(\mathcal{C}_t, \sigma(t))$ is constant for $t \in T$.

2. Straight Equisingular Deformations

We consider now a special class of equisingular deformations, originally introduced by Wahl, which we call *straight.*

Assume that the (unique) equimultiple sections $\sigma_{(q)} : T \to \widetilde{\mathcal{C}}_{(q)}$ from Definition 1.5 for an es-deformation of (C, z) over a complex germ $(T, 0)$ are all trivial. We know that they can be trivialized for each q by a local isomorphism of germs $\mathcal{M}_{(q)} \simeq \Sigma_{(q)} \times T$ over T at q, but in general not simultaneously for all q by an isomorphism of $(\mathcal{M}, z)$ over T (e.g. the cross-ratio of more than three sections through one exceptional component is an invariant). Those deformations for which these sections can be simultaneously trivialized are called straight equisingular.

Definition 2.1. **(Straight equisingular deformation)** A deformation with section of the reduced plane curve singularity (C, z) is called *straight equisingular* or a *straight es-deformation* if it is an equisingular deformation of (C, z) along the trivial section, such that the equimultiple sections $\sigma_{(q)}, q \in \mathscr{T}^*(C, z)$, through the non-nodes of the reduced total transform of (C, z) from Definition 1.5 are all trivial.

We denote by $\mathcal{D}ef^{s}_{(C,z)}$ the category of straight es-deformations, the full subcategory of the category $\mathcal{D}ef^{sec}_{(C,z)}$ of deformations with section, and by $\underline{\mathcal{D}ef}^{s}_{(C,z)}$ the functor of isomorphism classes of straight es-deformations.

Let us give a concrete description of straight equisingular deformations, using the notations from Definition 1.5:

Denote by $\widetilde{F}_{(q)} \in \mathcal{O}_{\mathcal{M}_{(q)}}$ a generator of the ideal of $\widetilde{\mathcal{C}}_{(q)} \subset \mathcal{M}_{(q)} = \Sigma_{(q)} \times T$ (the reduced total transform) and by $\mathfrak{m}_{(q)} \subset \mathcal{O}_{\Sigma_{(q)}}$ the maximal ideal. The condition that $\sigma_{(q)}$ is the trivial equimultiple section of $\Phi_{(q)}$ is equivalent to

$$\widetilde{F}_{(q)} \in \mathfrak{m}_{(q)}^{\widetilde{m}(q)} \cdot \mathcal{O}_{\Sigma_{(q)}} \times T \ ,$$

where $\widetilde{m}_{(q)}$ is the multiplicity of $\widetilde{C}_{(q)}$.

If $\widehat{F}_{(q)} \in \mathcal{O}_{\mathcal{M}_{(q)}}$ defines the total transform $\widehat{\mathcal{C}}_{(q)} \subset \Sigma_{(q)} \times T$ and if $\widehat{m}_{(q)}$ is the multiplicity of the total transform $\widehat{C}_{(q)}$ of the curve germ (C, z), then this is also equivalent to

$$\widehat{F}_{(q)} \in \mathfrak{m}_{(q)}^{\widehat{m}(g)} \cdot \mathcal{O}_{\Sigma_{(q)} \times T}.$$

This can be seen easily by induction on the number of blowing ups to resolve (C, z), using Remark 1.3. For $q = z$, we understand $\widetilde{\mathcal{C}}_{(q)} = \widehat{\mathcal{C}}_{(q)} = \mathcal{C}$, defined by $F \in \mathcal{O}_{\Sigma \times T}$, and both conditions mean $F \in \mathfrak{m}_z^m \cdot \mathcal{O}_{\Sigma \times T}$ with $m = \mathrm{mt}(C, z)$.

So far everything works for an arbitrary complex germ $(T, 0)$. If $(T, 0)$ is reduced then the equimultiplicity condition for the trivial sections $\sigma_{(q)}$, $q \in \mathcal{T}^*$, is equivalent to

$$\mathrm{mt}(C, z) = \mathrm{mt}(\mathcal{C}_t, z) \quad \text{for} \quad q = z$$

and for $q \neq z$ either to

$$\operatorname{mt} \widetilde{C}_{(q)} = \operatorname{mt}(\widetilde{\mathcal{C}}_{(q),t}, q)$$

or (equivalently) to

$$\operatorname{mt} \widehat{C}_{(q)} = \operatorname{mt}(\widehat{\mathcal{C}}_{(q),t}, q)$$

for all $t \in T$ sufficiently close to 0. $\widetilde{\mathcal{C}}_{(q),t}$ resp. $\widehat{\mathcal{C}}_{(q),t}$ denotes the reduced total, resp. the total transform of the fibre $\mathcal{C}_t$ of Φ over t.

For $(T,0) = (\mathbb{C},0)$ we can describe the straight equisingularity condition even more explicitly. We do this for the total transform $\widehat{F}_{(q)} \in \mathcal{O}_{\Sigma_{(q)}\times T}$ of $F \in \mathcal{O}_{\Sigma\times T} = \mathcal{O}_{\Sigma,z}\{t\}$. Then F can be written as

$$F = f + tg_1 + t^2 g_2 + \ldots$$

with $f, g_i \in \mathcal{O}_{\Sigma,z}$ and f defining (C,z). Then $\widehat{F}_{(q)}$ reads as

$$\widehat{F}_{(q)} = \widehat{f}_{(q)} + t\widehat{g}_{1,(q)} + t^2 \widehat{g}_{2,(q)} + \ldots$$

where $\widehat{f}_{(q)}, \widehat{g}_{i,(q)} \in \mathcal{O}_{\Sigma_{(q)}}$ denote the total transforms of f, g_i.

If we fix $t = t_0 \in T$ we write $F_{t_0} = F\big|_{t=t_0} \in \mathcal{O}_{\Sigma,z}$ and $\widehat{F}_{(q),t_0} = \widehat{F}_{(q)}\big|_{t=t_0} \in \mathcal{O}_{\Sigma(q),q}$. Then the equisingularity condition is equivalent to

$$\operatorname{mt}(f) = \operatorname{mt}(F_t)\ ,$$

$$\operatorname{mt}(\widehat{f}_{(q)}) = \operatorname{mt}(\widehat{F}_{(q),t})$$

for all $t \in T$ sufficiently close to 0 and for all $q \in \mathcal{T}^*, q \neq z$.

Thus we get:

LEMMA 2.2. *Let $f \in \mathbb{C}\{x,y\}$ define a reduced plane curve singularity $(C,0)$ with essential tree $\mathscr{T}^*$ and let $F(x,y,t) = f(x,y) + \sum_{i\geq 1} t^i g_i(x,y)$ define a one-parametric deformation of $(C,0)$. Then the following are equivalent:*

(i) *The deformation of $(C,0)$ defined by F is equisingular and the (unique) equimultiple sections through the infinitely near points $q \in \mathscr{T}^*$ are trivial, i.e., the deformation is straight equisingular.*

(ii) *For all $i \geq 1$,*

$$\operatorname{mt}(f) \leq \operatorname{mt}(g_i),$$

$$\operatorname{mt}(\widehat{f}_{(q)}) \leq \operatorname{mt}(\widehat{g}_{i,(q)}) \ \textit{for}\ q \in \mathscr{T}^*, q \neq 0.$$

We define now an ideal defining the topological singularity scheme, that will be identified in Corollary 2.8 with the tangent ideal to straight es-deformations.

DEFINITION 2.3. **(Topological singularity ideal and scheme)** Let $(C,z) \subset (\Sigma,z)$ be a reduced plane curve singularity and let $\mathscr{T}^* = \mathscr{T}^*(C,z)$ be the essential tree of (C,z). The ideal

$$\begin{aligned} I^s(C,z) &:= I(C, \mathscr{T}^*(C,z)) \subset \mathcal{O}_{\Sigma,z} \\ &:= \{g \in \mathcal{O}_{\Sigma,z} \mid \operatorname{mt} \widehat{g}_{(q)} \geq \operatorname{mt} \widehat{C}_{(q)} \text{ for each } q \in \mathscr{T}^*\} \end{aligned}$$

is called the *topological singularity ideal of* (C,z). It defines the *topological singularity scheme*

$$Z^s(C,z) := V(I^s(C,z)),$$

a zero-dimensional subscheme of Σ supported at $\{z\}$.

REMARK 2.4. Let $\mathscr{T} = \mathscr{T}(C, z)$ be any constellation of (C, z). Then we can define more generally the *cluster ideal* of (C, z) w.r.t. $\mathscr{T}$,

$$I(C, \mathscr{T}) = \{g \in \mathcal{O}_{\Sigma,z} \mid \operatorname{mt} \widehat{g}_{(q)} \geq \operatorname{mt} \widehat{C}_{(q)} \text{ for each } q \in \mathscr{T}\},$$

and $Z(C, \mathscr{T}) = V(I(C, \mathscr{T}))$ the *cluster scheme* of (C, z) w.r.t. $\mathscr{T}$, which is supported at $\{z\}$.

The following lemma can be proved by induction on the cardinality of $\mathscr{T}^*(C, z)$, see [**GLS18**, Lemma I.1.22].

LEMMA 2.5. *For $(C, z) \subset (\Sigma, z)$ a reduced plane curve singularity we have*

$$\deg Z^s(C, z) = \dim_{\mathbb{C}} \mathcal{O}_{\Sigma,z}/I^s(C, z) = \sum_{q \in \mathscr{T}^*(C,z)} \frac{m_q(m_q + 1)}{2},$$

with $m_q = \operatorname{mt} C_{(q)}$, the multiplicity of the strict transform of C at q.

EXAMPLE 2.6. (a) Let (C, z) be smooth. For the empty constellation $\mathscr{T}$ we obtain $I(C, \mathscr{T}) = \mathcal{O}_{\Sigma,z}$, that is, the scheme $Z(C, \mathscr{T})$ is the empty scheme. If (C, z) has the local equation $y = 0$ and if $\mathscr{T} = (z = q_0, q_1, \ldots, q_n)$ is the constellation obtained by blowing up (C, z) n times, then $I(C, \mathscr{T}) = \langle y, x^{n+1}\rangle \subset \mathbb{C}\{x, y\}$.

(b) If (C, z) is an *ordinary* m-fold *singularity* (i.e. m smooth branches with different tangents) then $\mathscr{T}^*(C, z) = (z)$ and $I^s(C, z) = \mathfrak{m}_{\Sigma,z}^m$.

The following lemma shows the relation of cluster schemes to equisingular deformations of curve germs:

LEMMA 2.7. *Let $Z^s(C, z)$ be the topological singularity scheme of (C, z) defined by $I^s(C, z)$.*

(a) *A generic element $g \in I^s(C, z)$ defines $Z^s(C, z)$, in the sense that $Z^s(C, z) = Z^s(C', z)$ for (C', z) the plane curve singulariy defined by g.*
(b) *The elements $g \in I^s(C, z)$ defining $Z^s(C, z)$ have no common (infinitely near) base point outside of $\mathscr{T}^*(C, z)$.*
(c) *Two generic elements $g, g' \in I^s(C, z)$ are topologically equivalent.*

Note that "generic" in (a) means: there exists a polynomial defining $Z^s(C, z)$, and if d_0 is the minimal degree of such a polynomial, then, for each $d \geq d_0$, the set of polynomials $g \in I^s(C, z)$ of degree at most d defining $Z^s(C, z)$ is a Zariski-open, dense subset in the vector space of all polynomials of degree at most d and contained in $I^s(C, z)$.

PROOF. (a) Let $f \in \mathcal{O}_{\Sigma,z}$ be a defining equation for (C, z). As f is (analytically) finitely determined, the essential tree $\mathscr{T}^*(C, z)$ depends only on a sufficiently high jet of f. We may therefore assume that f is a polynomial of degree d. Then the polynomials $g \in I^s(C, z)$ of degree $\leq d$ are parameterized by a finite dimensional vector space of positive dimension. The conditions $\operatorname{mt} \widehat{g}_{(q)} = \operatorname{mt} \widehat{C}'_{(q)}$, $q \in \mathscr{T}^*(C', z)$, define a Zariski-open subset of it. The density follows, since for almost all $t \in \mathbb{C}$ the germ $f + tg$ has precisely the multiplicities $\widehat{m}_q = \operatorname{mt} \widehat{C}'_{(q)}$ at each $q \in \mathscr{T}(C', z)$.

For the proof of (b) we refer to Proposition [**GLS18**, Proposition I.1.17]. (c) follows from (a), since Z^s determines the topological type. □

We relate below the topological singularity ideal $I^s(C,z)$ to the *equisingularity ideal*

$$I^{es}(C,z) = I^{es}(f) := \left\{ g \in \mathcal{O}_{\Sigma,z} \;\middle|\; \begin{array}{c} f+\varepsilon g \text{ is an } es\text{ -deformation of } (C,z) \\ \text{over } T_\varepsilon \text{ (along some section)} \end{array} \right\}$$

resp., the *fixed equisingularity ideal*

$$I^{es}_{\text{fix}}(C,z) = I^{es}_{\text{fix}}(f) := \left\{ g \in \mathcal{O}_{\Sigma,z} \;\middle|\; \begin{array}{c} f+\varepsilon g \text{ is an } es\text{ -deformation of } (C,z) \\ \text{over } T_\varepsilon \text{ along the trivial section} \end{array} \right\}.$$

They satisfy

$$I^{es}(C,z) = j(f) + I^{es}_{fix}(C,z).$$

Moreover, if $j(f)$ denote the Jacobian ideal we call

$$I^{ea}(C,z) := \langle f, j(f)\rangle \text{ resp. } I^{ea}_{fix}(C,z) := \langle f, \mathfrak{m}_z j(f)\rangle$$

the *Tjurina ideal* resp. the *fixed Tjurina ideal.* T_ε denotes the fat point with structure sheaf $\mathbb{C}[\varepsilon]$, $\varepsilon^2 = 0$, and $\underline{\mathcal{D}ef}^-_{(C,z)}(T_\varepsilon)$ is the tangent space to the deformation functor $\underline{\mathcal{D}ef}^-_{(C,z)}$, i.e.,

$$\underline{\mathcal{D}ef}^-_{(C,z)}(T_\varepsilon) = \{g \in \mathcal{O}_{\Sigma,z} | f+\varepsilon g \in \mathcal{D}ef^-_{(C,z)}(T_\varepsilon)\}/\text{isomorphism in } \mathcal{D}ef^-_{(C,z)}.$$

COROLLARY 2.8. *Let $(C,z) \subset (\Sigma,z)$ be a reduced plane curve singularity, defined by $f \in \mathcal{O}_{\Sigma,z}$.*

(1) If $g \in I^s(C,z)f$, then $f+tg$ are topologically equivalent for all but finitely many $t \in \mathbb{C}$ and $f+tg$ defines an equisingular deformation of (C,z) over $(\mathbb{C},0)$ along the trivial section.

(2) We get

$$I^s(C,z) = I^s(f) = \left\{ g \in \mathcal{O}_{\Sigma,z} \;\middle|\; \begin{array}{c} f+\varepsilon g \textit{ is a straight es-deformation of } (C,z) \\ \textit{over } T_\varepsilon \textit{ (along the trivial section)} \end{array} \right\}$$

and $I^s(C,z)/I^{ea}_{fix}(C,z)$ is the tangent space to the functor of (isomorphism classes of) straight es-deformations.

(3) We have the inclusions $I^{ea}_{fix}(C,z) \subset I^{ea}(C,z) \subset I^{es}(C,z)$ and

$$I^{ea}_{fix}(C,z) \subset I^s(C,z) \subset I^{es}_{fix}(C,z) \subset I^{es}(C,z).$$

$\underline{\mathcal{D}ef}^{es}_{(C,z)}(T_\varepsilon) = I^{es}(C,z)/I^{ea}(C,z)$ resp. $\underline{\mathcal{D}ef}^{es,fix}_{(C,z)}(T_\varepsilon) = I^{es}_{fix}(C,z)/I^{ea}_{fix}(C,z)$ is the tangent space to the functor of isomorphism classes of es-deformations resp. of es-deformations with (trivial) section. The dimensions satisfy

$$\dim_{\mathbb{C}} I^{es}(C,z)/I^{es}_{fix}(C,z) = \dim_{\mathbb{C}} I^{ea}(C,z)/I^{ea}_{fix}(C,z) = 2,$$

$$\dim_{\mathbb{C}} I^{es}_{fix}(C,z)/I^{ea}_{fix}(C,z) = \dim_{\mathbb{C}} I^{es}(C,z)/I^{ea}(C,z).$$

Moreover, the forgetful morphism

$$I^{es}_{fix}(C,z)/I^{ea}_{fix}(C,z) \to I^{es}(C,z)/I^{ea}(C,z)$$

is an isomorphism.

PROOF. (1) follows from the proof of Lemma 2.7, (2) from the proof of Lemma 2.2 applied to $f + tg, t^2 = 0$.

(3) The inclusion $I^{ea}_{fix}(C,z) \subset I^s(C,z)$ follows form the fact that for $g \in I^{ea}_{fix}(C,z)$ the deformation $f + \varepsilon g$ is trivial along the trivial section, hence straight, over T_ε. The other inclusions follow from the definitions. For the first dimension statements see [**GLS18**, Lemma I.2.13]. The second follows from the first and the following exact sequences,

$$0 \to I^{es}_{fix}(C,z)/I^{ea}_{fix}(C,z) \to I^{es}(C,z)/I^{ea}_{fix}(C,z) \to I^{es}(C,z)/I^{es}_{fix}(C,z) \to 0,$$

$$0 \to I^{ea}(C,z)/I^{ea}_{fix}(C,z) \to I^{es}(C,z)/I^{ea}_{fix}(C,z) \to I^{es}(C,z)/I^{ea}(C,z) \to 0.$$

The statement about the forgetful morphism follows, since both spaces have the same dimension and the cokernel is $I^{es}(C,z)/j(f) + I^{es}_{fix}(C,z) = 0$. □

REMARK 2.9. If a deformation functor has a semiuniversal object, then the tangent space to the functor (see [**GLS07**, App. C]) coincides with the Zariski tangent space of the semiuniversal deformation. We set

$$\begin{array}{rcl} \tau^s(C,z) & := & \dim_{\mathbb{C}} \mathcal{O}_{\Sigma,z}/I^s(C,z), \\ \tau^{es}(C,z) & := & \dim_{\mathbb{C}} \mathcal{O}_{\Sigma,z}/I^{es}(C,z), \\ \tau^{es}_{fix}(C,z) & := & \dim_{\mathbb{C}} \mathcal{O}_{\Sigma,z}/I^{es}_{\mathrm{fix}}(C,z), \\ \tau^{ea}(C,z) & := & \dim_{\mathbb{C}} \mathcal{O}_{\Sigma,z}/I^{ea}(C,z), \\ \tau^{ea}_{fix}(C,z) & := & \dim_{\mathbb{C}} \mathcal{O}_{\Sigma,z}/I^{ea}_{\mathrm{fix}}(C,z), \end{array}$$

with

$$\begin{array}{rcl} \tau^{ea}(C,z) & = & \dim_{\mathbb{C}} \underline{\mathcal{D}ef}_{(C,z)}(T_\varepsilon) \text{ (usual deformations)}, \\ \tau^{ea}(C,z)_{fix} & = & \dim_{\mathbb{C}} \underline{\mathcal{D}ef}^{sec}_{(C,z)}(T_\varepsilon) \text{ (deformations with section)}, \\ \tau^{ea}(C,z) - \tau^{es}(C,z) & = & \dim_{\mathbb{C}} \underline{\mathcal{D}ef}^{es}_{(C,z)}(T_\varepsilon) \text{ (es-deformations)}, \\ \tau^{ea}(C,z) - \tau^{es}_{fix}(C,z) & = & \dim_{\mathbb{C}} \underline{\mathcal{D}ef}^{es,fix}_{(C,z)}(T_\varepsilon) \text{ (es-deformations with section)}, \\ \tau^{ea}(C,z) - \tau^{s}(C,z) & = & \dim_{\mathbb{C}} \underline{\mathcal{D}ef}^{s}_{(C,z)}(T_\varepsilon) \text{ (straight es-deformations)}. \end{array}$$

By Corollary 2.8 we have

$$\tau^{es}(C,z) - \tau^{es}_{fix}(C,z) = \tau^{ea}(C,z) - \tau^{ea}_{fix}(C,z) = 2.$$

Moreover, by Lemma 2.5,

$$\tau^s(C,z) = \sum_{q \in \mathcal{T}^*(C,z)} \frac{m_q(m_q+1)}{2},$$

with m_q the multiplicity of the strict transform of (C,z) at q.

Let us recall the following result about straight es-deformations from [**GLS07**, Proposition II.2.69], basically due to Wahl [**Wah74**].

PROPOSITION 2.10. *Let $(C,z) \subset (\Sigma,z)$ be a reduced plane curve singularity defined by $f \in \mathcal{O}_{\Sigma,z}$. Then the following are equivalent:*

(a) *There are $\tau' = \tau(C,z) - \tau^{es}(C,z)$ elements $g_1, \dots, g_{\tau'} \in I^{es}(C,z)$ such that*

$$\varphi^{es} : V\left(f + \sum\nolimits_i t_i g_i\right) \subset (\Sigma \times \mathbb{C}^{\tau'}, (z,\mathbf{0})) \xrightarrow{\text{pr}} (\mathbb{C}^{\tau'}, \mathbf{0})$$

is a semiuniversal equisingular deformation of (C,z).

(b) *There exist* $g_1, \dots, g_{\tau'} \in I^{es}(C,z)$ *inducing a basis of* $I^{es}(C,z)/\langle f, j(f)\rangle$ *such that*

$$\varphi^{es} : V\left(f + \sum_i t_i g_i\right) \subset (\Sigma \times \mathbb{C}^{\tau'}, (z, \mathbf{0})) \xrightarrow{\text{pr}} (\mathbb{C}^{\tau'}, \mathbf{0})$$

is a semiuniversal equisingular deformation of (C,z).

(c) *Each locally trivial deformation of the reduced exceptional divisor* E *of a minimal embedded resolution of* $(C,z) \subset (\Sigma, z)$ *is trivial.*

(d) $I^{es}(C,z) = \langle f, j(f), I^s(C,z)\rangle$.

(e) *Each equisingular deformation of* (C,z) *is straight equisingular.*

Note that straight equisingular deformations are deformations with section while equisingular deformations are deformations without section. Forgetting the section we get a morphism from the category $\mathcal{D}ef^{s}_{(C,z)}$ of straight es-deformations to the category $\mathcal{D}ef^{es}_{(C,z)}$ of es-deformations. Saying that an equisingular deformation is straight equisingular means that it is isomorphic (as deformation without section) to the image in $\mathcal{D}ef^{es}_{(C,z)}$ of a straight equisingular deformation.

The image $\mathcal{D}ef^{ws}_{(C,z)}$ of $\mathcal{D}ef^{s}_{(C,z)}$ in $\mathcal{D}ef^{es}_{(C,z)}$ is called the category of *straight es-deformation without section.* It follows that

$$\underline{\mathcal{D}ef}^{ws}_{(C,z)}(T_\varepsilon) = \langle j(f), I^s(C,z)\rangle \,/\, \langle f, j(f)\rangle$$

is the tangent space to the functor of isomorphism classes of straight equisingular deformation of (C,z) without section.

We mention also the following consequence for semiquasi-homogeneous and Newton nondegenerate singularities from [**GLS07**, Proposition II.2.17 and Corollary II.2.71].

COROLLARY 2.11. *Let* $(C,z) \subset (\Sigma, z)$ *be a reduced plane curve singularity defined by* $f \in \mathcal{O}_{\Sigma,z}$, *and let* $\tau' = \tau(C,z) - \tau^{es}(C,z)$.

(a) *If* $f = f_0 + f'$ *is semiquasi-homogeneous with principal part* f_0 *being quasi-homogeneous of type* $(w_1, w_2; d)$, *then*

$$I^{es}(C,z) = \langle j(f), I^s(C,z)\rangle = \langle j(f),\, x^\alpha y^\beta \mid w_1\alpha + w_2\beta \geq d\rangle$$

and a semiuniversal equisingular deformation for (C,z) *is given by*

$$\varphi^{es} : V\left(f + \sum_{i=1}^{\tau'} t_i g_i\right) \subset (\Sigma \times \mathbb{C}^{\tau'}, (z, \mathbf{0})) \xrightarrow{\text{pr}} (\mathbb{C}^{\tau'}, \mathbf{0}),$$

for suitable $g_1, \dots, g_{\tau'}$ *representing a* $\mathbb{C}$*-basis for the quotient*

$$\langle j(f),\, x^\alpha y^\beta \mid w_1\alpha + w_2\beta \geq d\rangle / \,\langle f, j(f)\rangle.$$

(b) *If* f *is Newton non-degenerate with Newton diagram* $\Gamma(f)$ *at the origin, then*

$$I^{es}(C,z) = \langle j(f), I^s(C,z)\rangle = \langle j(f),\, x^\alpha y^\beta \mid x^\alpha y^\beta \text{ has Newton order} \geq 1\rangle$$

and a semiuniversal equisingular deformation for (C,z) *is given by*

$$\varphi^{es} : V\left(f + \sum_{i=1}^{\tau'} t_i g_i\right) \subset (\Sigma \times \mathbb{C}^{\tau'}, (z, \mathbf{0})) \xrightarrow{\text{pr}} (\mathbb{C}^{\tau'}, \mathbf{0}),$$

for suitable $g_1, \dots, g_{\tau'}$ *representing a monomial* $\mathbb{C}$*-basis for the quotient*

$$\langle j(f),\, x^\alpha y^\beta \mid x^\alpha y^\beta \text{ has Newton order } \geq 1\rangle / \langle f, j(f)\rangle.$$

Moreover, in both cases each equisingular deformation of $(C,0)$ is straight equisingular.

REMARK 2.12. (1) We can extend the chosen basis of $I^{es}(C,z)/\langle f, j(f)\rangle$ to a basis $g_1,\ldots,g_\tau$ of $\mathcal{O}_{\Sigma,z}/\langle f, j(f)\rangle$, showing that the base space of the semiuniversal es-deformation is the linear subspace $\{t_{\tau'+1} = \cdots = t_\tau = 0\}$ of the usual semiuniversal deformation of (C,z) given by $f + \sum_{i=1}^{\tau} t_i g_i$.

(2) Note that in Proposition 2.10(b) and Corollary 2.11 not every (monomial) basis of $I^{es}(C,z)/\langle f, j(f)\rangle$ has the claimed property (this was not clearly formulated in [**GLS07**]). We illustrate this by an example (due to Marco Mendes), using SINGULAR [**DGPS**].

EXAMPLE 2.13. Let $f = (y^3 + x^7)(y^3 + x^{10})$, which is Newton-nondegenerate. The quotient $Q := I^{es}(C,z)/\langle f, j(f)\rangle$ is the tangent space to the stratum of es-deformations (μ-constant stratum) of f, but in general not the stratum itself. We show that the monomial $x^{10}y^2$ is part of a monomial basis of Q. It is, however, below the Newton diagram (the monomials on or above the Newton diagram are those of Newton order[1] ≥ 1) and we will see that the deformation $f + tx^{10}y^2$ is not equisingular. At the end we construct a basis of monomials of Q of Newton order ≥ 1, explaining some of the peculiarities of SINGULAR.

```
LIB "all.lib";                  //loads all libraries
ring r  = 0,(x,y),ds;           //local ring with degree ordering
poly f = (y3+x7)*(y3+x10);
list L = esIdeal(f,1);          //computes all 3 es-ideals:
                                //L[1]: = I^{es} = es-ideal of Wahl
                                //L[2]: = I^{es}_fix
                                //L[3]: = I^s = top. singularity ideal
ideal Ifix = std(L[2]);         //we continue with I^{es}_{fix}
Ifix;
//-> Ifix[1]=2xy5+x8y2+x11y2
//-> Ifix[2]=y6
//-> Ifix[3]=x5y4
//-> Ifix[4]=x7y3
//-> Ifix[5]=x10y2
//-> Ifix[6]=x14y
//-> Ifix[7]=x17
ideal tj = std(jacob(f)+f);     //the Tjurina ideal
ideal Ies = std(L[1]);          //this is I^{es}
NF(Ies,tj)+0;                   //the non-zero elements not in tj
//-> _[1]=x5y4
//-> _[2]=x10y2
//-> _[3]=x14y
```

This shows that the monomial $x^{10}y^2$ is part of a monomial basis of Q. The deformation $f + tx^{10}y^2$ is not equisingular as we check by computing the Milnor numbers:

```
milnor(f);
```

---

[1] We say that a monomial has Newton order $d \in \mathbb{R}$ (w.r.t. f) iff it corresponds to a point on the hypersurface $d \cdot \Gamma(f) \subset \mathbb{R}^2$.

```
//-> 71
ring rt  = (0,t),(x,y),ds;
poly ft = (y3+x7)*(y3+x10) + t*x10y2;
milnor(ft);
//-> 70
```

To compute a monomial bases of Q with Newton order ≥ 1, we first construct a set of monomials containing the desired basis:

```
setring r;
int d = deg(highcorner(tj));  //determines the monomials inside tj
ideal k; int i;
for (i = 1; i <= d; i++)
{   k=k+maxideal(i);
}
```

The ideal k contains all monomials of degree $\leq d$. The monomials of degree $> d$ belong to the ideal tj (cf. [**GP07**, Lemma 1.7.17]).

To compute the monomials in k on or above the Newton diagram, let $n1, n2$ be the affine linear polynomials defining the faces of the Newton diagram. We compute the Newton order of the monomials by computing their exponents and then take the minimum value of $n1$ and $n2$ at these exponents.

```
poly n1 = y + 3/10x - 51/10;
poly n2 = y + 3/7x - 6;
int e1,e2;
poly mi ;
ideal J;
for (i = 1; i<=size(k); i++)
{ e1 = leadexp(k[i])[1];
  e2 = leadexp(k[i])[2];
  mi = min(subst(n1,x,e1,y,e2),subst(n2,x,e1,y,e2));
  if (mi >=0)
  {J = J + k[i];}
}
```

The elements of J generate Q, all have Newton order ≥ 1. We can now of course get a basis out of Q by homogenization and applying linear algebra methods. The SINGULAR command `reduce(J,tj)`; creates a basis of Q, however, the reduction process via standard bases produces $x^{10}y^2$, which is a monomial of the generator $2y^5 + x^7y^2 + x^{10}y^2$ of tj, as part of this basis. A good chance to get the right basis with elements of Newton order ≥ 1 is to use the reduction process in a ring with a *dp* ordering, since this ordering prefers higher degrees (and works in our example).

```
ring R = 0, (x,y),dp;
ideal J = imap(r,J);
ideal tj = std(imap(r,tj));
ideal N = reduce(J, tj);
N = normalize(N+0);   N;       //just to get a nice-looking basis
//->  N[1]=x3y5
//->  N[2]=x4y5
//->  N[3]=x5y4
//->  N[4]=x5y5
```

```
//-> N[5]=x6y4
//-> N[6]=x14y
//-> N[7]=x15y
```

The 7 elements of N have all Newton order ≥ 1. We check the dimension by computing $\dim_{\mathbb{C}} Q = \tau(f) - \tau^{es}(f)$.

```
setring r;
int tau = tjurina(f); tau;
//-> 59
int tes = vdim (Ies); tes;
//-> 52
tau - tes;
// -> 7
```

We show now that the functor of straight es-deformations is in general a linear subfunctor of all es-deformations (with and without section).

PROPOSITION 2.14. *Let $(C,z) \subset (\Sigma,z)$ be a reduced plane curve singularity defined by $f \in \mathcal{O}_{\Sigma,z}$.*

(1) There exist $g_1,\ldots,g_{\tau'} \in I^s(C,z)$ inducing a basis of $I^s(C,z)/\langle f, \mathfrak{m}_z j(f)\rangle$ such that

$$\varphi^s : V\left(f + \sum_i t_i g_i\right) \subset (\Sigma \times \mathbb{C}^{\tau'}, (z,\mathbf{0})) \xrightarrow{\text{pr}} (\mathbb{C}^{\tau'}, \mathbf{0}),$$

is a semiuniversal straight equisingular deformation of (C,z).

(2) There exist $g_1,\ldots,g_{\tau''} \in \langle j(f), I^s(C,z)\rangle$ inducing a basis of of the quotient $\langle j(f), I^s(C,z)\rangle \,/\, \langle f, j(f)\rangle$ such that

$$\varphi^s : V\left(f + \sum_i t_i g_i\right) \subset (\Sigma \times \mathbb{C}^{\tau''}, (z,\mathbf{0})) \xrightarrow{\text{pr}} (\mathbb{C}^{\tau'}, \mathbf{0}),$$

is a semiuniversal straight equisingular deformation of (C,z) without section.

PROOF. (1) That the deformation φ^s is straight es follows from Lemma 2.2 and its proof. Extending the chosen basis of $I^s(C,z)/\langle f, \mathfrak{m}_z j(f)\rangle$ to a basis of $\mathcal{O}_{\Sigma,z}/\langle f, \mathfrak{m}_z j(f)\rangle$ shows that the base space of φ^s is a linear subspace $B^s_{C,z}$ of the base space $B^{fix}_{C,z}$ of the semiuniversal deformation of (C,z) with (trivial) section. If a given deformation over an arbitrary base germ $(T,0)$ is straight es, it can be induced by a map $\varphi : (T,0) \to B^{fix}_{C,z}$. φ must factor through $B^s_{C,z}$ by definition of I^s because otherwise the deformation is not equimultiple along the trivial sections through all infinitely near points $q \in \mathcal{T}^*(C,z)$. This shows that φ^s is complete. Moreover, for extensions of Artinian base spaces $(T,0) \subset (T',0)$, the equimultiple deformations along the trivial sections over $(T,0)$ can can be clearly extended to equimultiple deformations along the trivial sections over $(T',0)$. Hence the versality follows. (For "complete" and "versal" see [**GLS07**, Definition II.1.8]).

(2) follows from (1), noting that $\langle j(f), I^s(C,z)\rangle \,/\, \langle f, j(f)\rangle$ is the tangent space to the functor of isomorphism classes of straight equisingular deformation of (C,z) without section. As in Remark 2.12 the base space of the semiuniversal straight equisingular deformation of (C,z) without section can be realized as a linear subspace of the usual semiuniversal deformation of (C,z). □

3. Relation to the Rooted Hilbert Scheme

In this section we show that straight es-deformations of a plane curve singularity $(C, z) \subset (\Sigma, z)$ appear as fibres of a morphism from all es-deformations to the "rooted Hilbert scheme", a certain punctual Hilbert scheme in Σ parametrizing topological singularity schemes of plane curve singularities with fixed cluster graph. Recall that the Hilbert scheme of points on Σ parametrizes families of zero-dimensional schemes, i.e., complex subspaces of Σ concentrated on finitely many points.

DEFINITION 3.1. Let T be an arbitrary complex space. A *family of zero-dimensional schemes (of degree n) on Σ over T* is a commutative diagram of complex spaces

$$\begin{array}{ccc} \mathcal{Z} & \overset{j}{\hookrightarrow} & \Sigma \times T \\ & {\scriptstyle\varphi}\searrow \quad \swarrow{\scriptstyle \mathrm{pr}} & \\ & T & \end{array}$$

with $j : \mathcal{Z} \hookrightarrow \Sigma \times T$ a closed embedding, pr the projection and $\varphi = \mathrm{pr} \circ j$ finite and flat. The fibre $\mathcal{Z}_t := \varphi^{-1}(t)$, $t \in T$, is mapped by j to a zero-dimensional subscheme of Σ and if its degree is n we say that $j : \mathcal{Z} \hookrightarrow \Sigma \times T$ is a *family of zero-dimensional schemes of degree n.*

We usually identify two families $j : \mathcal{Z} \hookrightarrow \Sigma \times T$ and $j' : \mathcal{Z}' \hookrightarrow \Sigma \times T$ if the images $j(\mathcal{Z})$ and $j'(\mathcal{Z}')$ coincide, and then we get subspaces $\mathcal{Z} \subset \Sigma \times T$ (of degree n) which are finite and flat over T. For Σ algebraic and algebraic families this is the Hilbert functor:

The *Hilbert functor* $\mathcal{H}ilb^n_\Sigma$ associates to an algebraic variety T the set

$$\mathcal{H}ilb^n_\Sigma(T) \quad := \quad \left\{ \begin{array}{c} \text{closed algebraic subvarieties } \mathcal{Z} \subset \Sigma \times T \\ \text{which are finite of degree } n \text{ and flat over } T \end{array} \right\}.$$

Hence $\mathcal{Z} \subset \Sigma \times T \to T$ is a family of zero-dimensional schemes of degree n. If Σ is a projective surface, the Hilbert functor is known to be representable by a smooth projective algebraic variety of dimension $2n$, denoted by Hilb^n_Σ, the *Hilbert scheme of points in* Σ. Representability means that there exists a closed algebraic subvariety $\mathscr{U}^n_\Sigma \subset \Sigma \times \mathrm{Hilb}^n_\Sigma$, finite of degree n and flat over Hilb^n_Σ, such that each element of $\mathcal{H}ilb^n_\Sigma(T)$ can be induced from the universal family $\mathscr{U}^n_\Sigma \to \mathrm{Hilb}^n_\Sigma$ via base change by a unique algebraic morphism $T \to \mathrm{Hilb}^n_\Sigma$.

There exists a birational morphism, the *Hilbert-Chow morphism*

$$\Phi : \mathrm{Hilb}^n_\Sigma \longrightarrow \mathrm{Sym}^n\Sigma\,,$$

$\mathrm{Sym}^n\Sigma$ denoting the n-th symmetric power of Σ. It assigns to a closed subscheme $Z \subset \Sigma$ of degree n the 0-cycle consisting of the points of Z with multiplicities given by the length of their local rings on Z. If Z consists of n distinct simple points $z_1, ..., z_n$ then Φ maps Z to the image of $(z_1, ..., z_n) \in \Sigma^n$ in $\mathrm{Sym}^n\,\Sigma = \Sigma^n/S_n$ where S_n is the symmetric group permuting the coordinates.

DEFINITION 3.2. **(Punctual Hilbert functor and scheme)** Fix $z \in \Sigma$ and let x, y be local coordinates at z. Then we have the *punctual Hilbert functor* $\mathcal{H}ilb^n_{\mathbb{C}\{x,y\}}$, which associates to a complex space T the set

$$\mathcal{H}ilb^n_{\mathbb{C}\{x,y\}}(T) := \big\{ (\mathcal{Z} \subset \Sigma \times T) \in \mathcal{H}ilb^n_\Sigma(T) \,\big|\, \mathrm{supp}(\mathcal{Z}) \subset \{z\} \times T \big\}.$$

Fogarty [**Fog68**] and Hartshorne [**Har66**] showed that the punctual Hilbert functor is representable (in the algebraic category) by a connected projective variety $\mathrm{Hilb}^n_{\mathbb{C}\{x,y\}}$, the *punctual Hilbert scheme*, which can be identified with the closed subvariety $\Phi^{-1}(n\cdot z)\subset \mathrm{Hilb}^n_\Sigma$. Briançon [**Bri77**] proved (in the analytic categroy) that $\mathrm{Hilb}^n_{\mathbb{C}\{x,y\}}$ is irreducible (in general not reduced) of dimension $n-1$.

REMARK 3.3. A functor analogous to the Hilbert functor can be defined for complex spaces T, associating to T the set of closed complex subspaces $\mathscr{Z}\subset\Sigma\times T$, finite of degree n and flat over T. This is the *Douady functor*, denoted by $\mathcal{D}ou^n_\Sigma$, which is representable by a complex space Dou^n_Σ, the *Douady space*. It has a universal family $\mathscr{V}^n_\Sigma\to\mathrm{Dou}^n_\Sigma$ with the same universal property as the Hilbert functor but for morphisms of complex spaces, i.e. each element of $\mathcal{D}ou^n_\Sigma(T)$ can be induced from $\mathscr{V}^n_\Sigma\to\mathrm{Dou}^n_\Sigma$ by a unique analytic morphism $T\to\mathrm{Dou}^n_\Sigma$.

The Hilbert scheme and the Douady space exist in much greater generality and their relation is discussed in detail in [**GLS18**, Section II.2.2.1]. We just mention that for families of zero-dimensional schemes on the smooth projective surface Σ, the analytification of Hilb^n_Σ and Dou^n_Σ are isomorphic as complex spaces and that $\mathscr{U}^n_\Sigma\subset\Sigma\times\mathrm{Hilb}^n_\Sigma$ has the universal property for complex spaces T and analytic morphisms $T\to\mathrm{Hilb}^n_\Sigma$ (proved in greater generality in [**GLS18**, Proposition II.2.28]).

We will consider the punctual Hilbert functor as above with z a point in an arbitrary smooth complex analytic surface Σ. Since we consider families ($\mathscr{Z}\subset\Sigma\times T$) $\in\mathcal{H}ilb^n_\Sigma(T)$ with $\mathrm{supp}(\mathscr{Z})\subset\{z\}\times T$, we may assume that Σ is projective and the above mentioned results hold in the algebraic as well as in the analytic category and we will always write Hilb, also in the analytic situation.

In order to define the "rooted Hilbert functor", a subfunctor of the punctual Hilbert functor, we have to extend the definition of a cluster graph to arbitrary zero-dimensional subschemes of $Z\subset\Sigma$. We just sketch the definition and refer to [**GLS18**] for details.

We set $\deg Z=\dim_{\mathbb{C}}H^0(Z,\mathcal{O}_Z)$, the degree of Z, and $\mathrm{mt}(Z,z)$ the minimum order at z of the elements contained in the ideal I_Z defining Z, the multiplicity of Z. Given a zero-dimensional scheme $Z_0:=Z\subset\Sigma=:\Sigma_0$, let $\pi_i:\Sigma_i\to\Sigma_{i-1}$ be the blowing up of $\mathrm{supp}(Z_{i-1})\subset\Sigma_{i-1}$, and let Z_i be the strict transform of Z_{i-1}, $i=1,\dots,r$. Note that $\mathrm{supp}(Z_i)$ consists of finitely many infinitely near points (of level i) and that $\deg Z_i$ is strictly decreasing (which can be seen, for instance, by using standard bases). We choose r minimal with the property $\mathrm{supp}(Z_r)=\emptyset$. Then the (isomorphism class of the) sequence

$$\mathscr{K}:=\big(\mathrm{supp}(Z_0),\pi_1,\mathrm{supp}(Z_1),\dots,\pi_{r-1},\mathrm{supp}(Z_{r-1})\big)$$

is called a constellation on Σ (generalizing constellations for plane curve singularities defined in Section 1). Setting

$$m_q:=\mathrm{mt}(Z_i,q)\quad\text{for each}\quad q\in\mathrm{supp}(Z_i)\ ,$$

and $\boldsymbol{m}=(m_q)_q$ then $\mathcal{C}\ell(Z):=(\mathscr{K},\boldsymbol{m})$ is called the *cluster defined by the zero-dimensional scheme* Z. Similar as in Definition 1.2 we define the graph $\Gamma_{\mathscr{K}}$ of $\mathscr{K}$ and a proximity relation ($q_j\dashrightarrow q_i$) on the points of $\mathscr{K}$ and call the triple

$$\mathbb{G}(Z):=\mathbb{G}(\mathcal{C}\ell(Z)):=(\Gamma_{\mathscr{K}},\dashrightarrow,\boldsymbol{m})$$

the *cluster graph of the zero-dimensional scheme* Z. We set $n := \sum_{q\in\mathscr{K}} \frac{m_q(m_q+1)}{2}$. The support of Z is called the *root of* $\mathbb{G}$.

Let $\mathbb{G}$ be the cluster graph of a zero-dimensional scheme on Σ. We define the *Hilbert functor with fixed cluster graph* $\mathbb{G}$ on the category of reduced complex spaces as

$$\mathcal{H}ilb^{\mathbb{G}}_{\Sigma}(T) = \big\{(\mathcal{Z} \subset \Sigma \times T) \in \mathcal{H}ilb^{n}_{\Sigma}(T) \,\big|\, \mathbb{G}(\mathcal{Z}_t) = \mathbb{G} \text{ for all } t \in T\big\}.$$

It is proved in [**GLS18**, Proposition I.1.52] that for $(\mathcal{Z} \subset \Sigma \times T) \in \mathcal{H}ilb^{\mathbb{G}}_{\Sigma}(T)$ there exists a complex space T' and a finite surjective morphism $\alpha : T' \to T$ such that the induced family $\alpha^*\mathcal{Z} \to T'$ is resolvable by a sequence of blowing ups $\pi_i : \mathscr{X}^{(i+1)} \to \mathscr{X}^{(i)}$ along equimultiple sections $\sigma^{(i)}_j : T' \to \mathscr{X}^{(i)}$ $(i = 0, \cdots, N, j = 0, \cdots k_i)$ with $\mathscr{X}^{(0)} = \Sigma \times T, \mathcal{Z}^{(0)} = \mathcal{Z}$ and $\mathcal{Z}^{(i+1)}$ the strict transform of $\mathcal{Z}^{(i)}$. Moreover, $\operatorname{supp}(\mathcal{Z}^{(i)}) = \bigcup_{j=1}^{k_i} \sigma^{(i)}_j(T)$ and $\operatorname{supp}(\mathcal{Z}^{(N+1)}) = \emptyset$. The initial sections $\sigma^{(0)}_j$ satisfy $\{\sigma^{(0)}_1(t), \cdots, \sigma^{(0)}_{k_1}(t)\} = \operatorname{root}(\mathbb{G}(\mathcal{Z}_t))$. The rooted Hilbert functor is the subfunctor of $\mathcal{H}ilb^{\mathbb{G}}_{\Sigma}$ where the initial sections are trivial.

Consider now a reduced plane curve singularity $(C, z) \subset (\Sigma, z)$ with essential tree $\mathscr{T}^*$ and cluster graph $\mathbb{G}(C, z) = (\Gamma_{\mathscr{T}^*}, \dashrightarrow, \boldsymbol{m})$, defined in Section 1. The definition of the cluster graph of a zero-dimensional scheme is modeled in such a way that we have

$$\mathbb{G}(C, z) = \mathbb{G}(Z^s(C, z)),$$

where $Z^s(C, z)$ is the topological singularity scheme from Definition 2.3. The root of $\mathbb{G}(C, z)$ is $\{\mathrm{z}\}$.

DEFINITION 3.4. **(Rooted Hilbert functor)** Let $\mathbb{G} = \mathbb{G}(C, z)$ be the cluster graph of a reduced plane curve singularity. The *rooted Hilbert functor* or the *punctual Hilbert functor fixing* $\mathbb{G}$ is the subfunctor $\mathcal{H}ilb^{\mathbb{G}}_{\mathbb{C}\{x,y\}}$ of $\mathcal{H}ilb^{\mathbb{G}}_{\Sigma}$ associating to a reduced complex space T the subspaces $(\mathcal{Z} \subset \Sigma \times T) \in \mathcal{H}ilb^{\mathbb{G}}_{\Sigma}(T)$ such that the initial section $\sigma := \sigma^{(0)}_1$ passing through z is trivial.

The main result about the rooted Hilbert functor is the following theorem. It is used in [**GLS18**, Section IV.6] to prove asymptotically sufficient conditions for the irreducibility of the variety of plane projective curves of given degree with a fixed number of singularities of given topological type.

THEOREM 3.5. ([**GLS18**, Theorem I.1.55, I.1.57]) *The rooted functor* $\mathcal{H}ilb^{\mathbb{G}}_{\mathbb{C}\{x,y\}}$ *is representable by* $\mathrm{Hilb}^{\mathbb{G}}_{\mathbb{C}\{x,y\}}$*, the* rooted Hilbert scheme, *an irreducible, quasi-projective subvariety of the projective variety* $\mathrm{Hilb}^{n}_{\mathbb{C}\{x,y\}}$*, of dimension equal to the number of free vertices in* $\mathscr{T}^*(C, z) \setminus \{z\}$.

Let S denote the topological type of (C, z). Since $\mathbb{G} = \mathbb{G}(C, z)$ is a complete invariant of S, $\mathcal{H}ilb^{\mathbb{G}}_{\mathbb{C}\{x,y\}}$ depends only on S and we can introduce the *punctual Hilbert scheme associated to a topological type* S,

$$\mathcal{H}^s_0(S) := \mathrm{Hilb}^{\mathbb{G}}_{\mathbb{C}\{x,y\}}\ .$$

It is shown in [**GLS18**, Proposition I.1.17] that each point in $\mathcal{H}^s_0(S)$ corresponds to a topological singularity scheme $Z^s(C, w)$ of a plane curve singularity (C, w) of topological type S and with $\deg Z^s(C, w) = \dim_{\mathbb{C}} \mathcal{O}_{\Sigma,z}/I^s(C, w) = n$, with

$n = \sum_{q\in\mathscr{T}^*(C,w)} \frac{m_q(m_q+1)}{2}$, m_q the multiplicity of the strict transform of (C,w) at q, and with fixed cluster graph $\mathbb{G} = \mathbb{G}(Z^s(C,w))$.

Before we relate $\mathcal{H}_0^s(S)$ to straight equisingular deformations, we consider several related semiuniversal base spaces.

LEMMA 3.6. *For a reduced plane curve singularity (C,z) the following base spaces of semiuniversal deformations of (C,z) are smooth of the given dimension.*

(1) *$B_{C,z}$ the base space of the (usual) semiuniversal deformation of dimension $\tau^{ea}(C,z) = \tau(C,z)$,*
(2) *$B^{fix}_{C,z}$ the base space of the semiuniversal deformation with section of dimension $\tau^{ea}_{fix}(C,z) = \tau_{fix}(C,z)$,*
(3) *$B^{es}_{C,z}$ the base space of the seminuniversal equisingular deformation of dimension $\tau^{ea}(C,z) - \tau^{es}(C,z)$,*
(4) *$B^{es,fix}_{C,z}$ the base space of the seminuniversal es-deformation with section of dimension $\tau^{ea}_{fix}(C,z) - \tau^{es}_{fix}(C,z) = \tau^{ea}(C,z) - \tau^{es}(C,z)$,*
(5) *$B^{s}_{C,z}$ the base space of the seminuniversal straight es-deformation of dimension $\tau^{ea}_{fix}(C,z) - \tau^{s}(C,z)$.*

Note that the definitions of τ^{ea} and τ^{es} are consistent in the following sense. $\tau^{es}(C,z)$ is the codimension in $B_{C,z}$ of the base space of the semiuniversal *equisingular* deformation of (C,z) (which coincides with the μ-constant stratum in $B_{C,z}$), while $\tau^{ea}(C,z)$ is the codimension in $B_{C,z}$ of the base space of the semiuniversal *equianalytic* deformation of (C,z) (which is the reduced point).

PROOF. For $B_{C,z}$ and $B^{fix}_{C,z}$ the smoothness and the dimension statements are well known. By [**GLS07**, Theorem II.2.61, Corollary II.2.67 and Proposition II.2.63] we know that $B^{es}_{C,z}$ is isomorphic to the μ-constant stratum in $B_{C,z}$ and smooth, while $B^{es,fix}_{C,z}$ is isomorphic to the μ-constant stratum in $B^{fix}_{C,z}$. The smoothness of $B^{s}_{C,z}$ follows from Proposition 2.14. The dimension statements are a consequence of the smoothness and of Corollary 2.8, since then the dimension coincides with the dimension of the Zariski tangent space.

To see the smoothness of $B^{es,fix}_{C,z}$ and the second equality in (4) note that in general deformations with section are isomorphic to deformations with trivial section (cf. [**GLS07**, Proposition II.2.2]) and we will identify the corresponding semiuniversal base spaces $B^{sec}_{(C,z)} = B^{fix}_{(C,z)}$. Moreover, the forgetful morphism of functors from deformations with section to deformations without section is smooth by [**GLS07**, Corollary II.1.6]. It follows that the induced morphism of base spaces $B^{es,fix}_{C,z} \to B^{es}_{C,z}$ is smooth (flat with smooth fibre). Hence $B^{es,fix}_{C,z}$ is smooth and of the same dimension as its tangent space, namely $\dim_{\mathbb{C}} T^{1,es,fix}_{C,z} = \tau_{fix}(C,z) - \tau^{es}_{fix}(C,z) = \tau(C,z) - \tau^{es}(C,z)$ by Corollary 2.8. As this is the dimension of $B^{es}_{C,z}$ by Proposition 2.14, we get that the morphism $B^{fix,es}_{C,z} \to B^{es}_{C,z}$ is an isomorphism (reflecting the fact that the equisingular section is unique). □

We state now the main result about the relation between equisingular deformations of (C,z) along the trivial section, the induced deformation of the topological singularity scheme $Z^s(C,z)$, and straight equisingular deformations of (C,z).

THEOREM 3.7. *With the above notations there exists a surjective morphism of germs*

$$\psi : B^{es,fix}_{C,z} \to \mathcal{H}^s_0(S),$$

with smooth fibre $\psi^{-1}(Z^s(C,z)) \cong B^s_{C,z}$. *Moreover, we have*

$$\begin{aligned}\dim \mathcal{H}^s_0(S) &= \#\{\text{free vertices } q \in \mathscr{T}^*(C,z) \setminus \{z\}\} \\ &= \dim_{\mathbb{C}} I^{es}_{fix}(C,z)/I^s(C,z) \\ &= \deg Z^s(C,z) - \tau^{es}((C,z) - 2 \\ &= \tau^s(C,z) - \tau^{es}_{fix}(C,z) \\ &= \textstyle\sum_{q\in\mathscr{T}^*(C,z)} \frac{m_q(m_q+1)}{2} - \tau^{es}_{fix}(C,z)\end{aligned}$$

with m_q *the multiplicity of the strict transform of* (C,z) *at* q.

PROOF. Equisingular deformations $\mathscr{C} \subset \Sigma \times T \to T$ of (C,z) along the trivial section over a reduced germ T are exactly those, which are equimultiple along the trivial section $\sigma(t) = z$ and equimultiple along the (not necessarily trivial) sections $\sigma_{(q)}, q \in \mathscr{T}^*(C,z)$, through the non-nodes of the reduced total transform of (C,z) in a resolution of (C,z). Equisingularity implies that the cluster graph $\mathbb{G}(\mathscr{C}_t,z) = \mathbb{G}(Z^s(\mathscr{C}_t,z))$ is constant and equal to $\mathbb{G} = \mathbb{G}(Z^s(C,z))$. Hence $\deg Z^s(\mathscr{C}_t,z)$ is constant and $\mathcal{Z} = \{Z^s(\mathscr{C}_t,z)\}_{t\in T}$ is a flat family of zero-dimensional schemes in $\mathcal{H}ilb^{\mathbb{G}}_{\mathbb{C}\{x,y\}}(T)$, a deformation of $Z^s(C,z)$.

Since the association $(\mathscr{C} \to T) \to (\mathcal{Z} \to T)$ is functorial in T and respects isomorphism classes, we get a morphism of functors

$$\underline{\mathcal{D}ef}^{es,fix}_{C,z} \to \mathcal{H}ilb^{\mathbb{G}}_{\mathbb{C}\{x,y\}},$$

where $\underline{\mathcal{D}ef}^{es,fix}_{C,z}$ is the functor of isomorphism classes of equisingular deformations of (C,z) with trivial section. A family in $\mathcal{H}ilb^{\mathbb{G}}_{\mathbb{C}\{x,y\}}$ is trivial if it is given by the subspace $Z^s(C,z) \times T \subset \Sigma \times T$, i.e., iff the sections through the infinitely near points $q \in \mathscr{T}^*(C,z)$ are trivial. Hence the fibres of the morphism of functors are exactly the straight equisingular deformations of (C,z).

The morphism of functors induces a morphism of germs $\psi : B^{es,fix}_{C,z} \to \mathcal{H}^s_0(S)$, which is surjective (by [**GLS18**, Proposition I.1.17]) with fibre over $Z^s(C,z)$ being isomorphic to $B^s_{C,z}$, since every straight es-deformation of (C,z) over T can be induced via a map $T \to B^{es,fix}_{C,z}$ which must factor through $\psi^{-1}(Z^s(C,z))$.

By Theorem 2.9 we get the first formula for the dimension of $\mathcal{H}^s_0(S)$. For the others we use the theorems of Frisch and Sard to see that there are points $Z^s(C',z')$ in $\mathcal{H}^s_0(S)$ arbitrary close to $Z^s(C,z)$, over which ψ is flat and hence satisfies

$$\begin{aligned}\dim(\mathcal{H}^s_0(S), Z^s(C',z')) &= \tau(C',z') - \tau^{es}(C',z') - \dim_{\mathbb{C}} I^s(C',z')/I^{ea}(C',z') \\ &= \deg Z^s(C',z') - \tau^{es}_{fix}(C',z').\end{aligned}$$

Since $\deg Z^s(C',z') = \deg Z^s(C,z)$ and since $\mathcal{H}^s_0(S)$ is irreducible (by Theorem 3.5) its dimension is constant and thus we get $\tau^{es}_{fix}(C',z') = \tau^{es}_{fix}(C,z)$. Using Remark 2.9 we get the last four dimension formulas. □

REMARK 3.8. For any es-deformations there exists (unique) sections $\sigma_{(q)}$ through $q \in \mathscr{T}^* = \mathscr{T}^*(C,z)$ along which the deformation of the reduced total transform is equimultiple. For straight es-deformations $f + tg$ over $(\mathbb{C},0)$ with $g \in I^s(f)$ these sections are all trivial. For arbitrary es-deformations with tangent directions in $I^{es}_{\text{fix}}(f)$ the sections through satellite points $q \in \mathscr{T}^*$ have to stay at satellite points. But the sections through free points in $\mathscr{T}^* \setminus \{z\}$ may move along the exceptional divisor, giving one degree of freedom for every free point. Since $\mathcal{H}_0^s(S)$ represents es-deformations with trivial initial section, we get a geometric interpretation of the formula $\dim \mathcal{H}_0^s(S) = \#(\text{free vertices } q \in \mathscr{T}^* \setminus \{z\})$.

References

[Bri77] Briançon, J.: *Description de* $Hilb^n\mathbb{C}\{x, y\}$. Invent. Math. **41**, 45–89 (1977). Zbl 0353.14004

[CGL07a] Campillo, A.; Greuel, G.-M.; Lossen, C.: *Equisingular Deformations of Plane Curves in Arbitrary Characteristic.* Compos. Math. **143**, 829–882 (2007). Zbl 1121.14003

[CGL07b] Campillo, A.; Greuel, G.-M.; Lossen, C.: *Equisingular calculations for plane curve singularities.* J. Symb. Comput. **42**, No. 1-2, 89-114 (2007). Zbl 1128.14003

[Cas00] Casas-Alvero, E.: *Singularities of Plane Curves.* LMS 276, Cambridge Univ. Press (2000). Zbl 0967.14018

[DGPS] Decker, W.; Greuel, G.-M.; Pfister, G.; Schönemann, H.: Singular 4-1-1 — A computer algebra system for polynomial computations. `http://www.singular.uni-kl.de` (2018).

[Fog68] Fogarty, J.: *Algebraic families on an algebraic surface.* Amer. J. Math. **90**, 511–521 (1968). Zbl 0176.18401

[GLS07] Greuel, G.-M.; Lossen, C.; Shustin, E.: *Introduction to Singularities and Deformations.* Springer (2007). Zbl 1125.32013

[GLS18] Greuel, G.-M.; Lossen, C.; Shustin, E.: *Singular Algebraic Curves.* Springer (2018). Zbl 06982463

[GP07] Greuel, G.-M.; Pfister, G.: *A* Singular *Introduction to Commutative Algebra. With contributions by Olaf Bachmann, Christoph Lossen and Hans Schönemann.* 2nd extended ed., Springer, Berlin, Heidelberg, New York (2007). Zbl 1133.13001

[Har66] Hartshorne, R.: *Connectedness of the Hilbert scheme.* Publ. Math. IHES **29**, 261–304 (1966). Zbl 0171.41502

[Wah74] Wahl, J.: *Equisingular deformations of plane algebroid curves.* Trans. Amer. Math. Soc. **193**, 143–170 (1974). Zbl 0294.14007

Department of Mathematics, University of Kaiserslautern
Email address: `greuel@mathematik.uni-kl.de`

Contemporary Mathematics
Volume **742**, 2020
https://doi.org/10.1090/conm/742/14939

Divisor class groups of singular varieties

Helmut A. Hamm

ABSTRACT. In this paper we look at Weil divisor class groups of complex algebraic varieties which are not supposed to be smooth in codimension 1. We prove a Lefschetz theorem in this context. The hypothesis is of geometric nature (transversality condition).

1. Introduction

This paper relies on results which were obtained during a long collaboration with Lê Dũng Tráng about Lefschetz theorems which started more than 45 years ago. We looked first at geometric objects: (co)homology or homotopy groups. Later on we passed to the Picard group of quasi-projective varieties and realized that we could use our geometric results in this context. The object of the paper is to prove a Lefschetz theorem for the Weil divisor class group (which coincides with the Picard group in the smooth case). It is based on results obtained with Lê Dũng Tráng concerning the homotopy type resp. the Picard group.

So let X be a purely n-dimensional complex algebraic variety (by convention, reduced but not necessarily irreducible) and $Cl\,X$ its Weil divisor class group. Let X^{an} be the corresponding complex analytic space.

The main result of the present paper is Theorem 3.1, which is a Lefschetz theorem for the Weil divisor class group Cl. The essential point is that we do not suppose - as often done, e.g. in [**3**] - that X is smooth in codimension 1, i.e. that $Sing\,X$ is of codimension ≥ 2!

We will use topological methods to study $Cl\,X$.

First look at the case where X is non-singular. Then $Cl\,X$ is isomorphic to the Picard group $Pic\,X = H^1(X, \mathcal{O}_X^*)$ of X, i.e. the group of isomorphism classes of line bundles on X. So we have the following Lefschetz type theorem which is well-known:

2010 *Mathematics Subject Classification.* Primary 14C15, 14C20, 14C30, 32S50.

THEOREM 1.1. *Suppose that X is a smooth subvariety of $\mathbb{P}_N$ of pure dimension n, H a hyperplane which is transversal to X, $Y := X \cap H$. Then $Cl(X) \to Cl(Y)$ is bijective (resp. injective) if $n \geq 4$ (resp. $n \geq 3$).*

N.B.: Subvarieties are supposed to be closed unless stated otherwise.

PROOF. Note that Y is smooth, too, so we may work with Pic instead of Cl. This explains, by the way, how to get the mapping in question. The classical Lefschetz theorem on hyperplane sections implies that $H^j(X^{an};\mathbb{Z}) \to H^j(Y^{an};\mathbb{Z})$ is bijective if $j \leq 2$ (resp. bijective if $j \leq 1$ and injective if $j = 2$). By Hodge theory, we deduce a corresponding statement for $H^j(X^{an}, \mathcal{O}_{X^{an}}) \to H^j(Y^{an}, \mathcal{O}_{Y^{an}})$. Now look at the exponential sequence: we obtain that $Pic(X^{an}) \to Pic(Y^{an})$ is bijective (resp. injective). Finally we can pass to the algebraic case by GAGA. □

When we pass to the singular case, we lose the connection between $Cl(X)$ and $Pic(X)$. A manageable case is when $Sing\, X$ is of codimension ≥ 2, e.g. X normal: Then $Cl(X) = Cl(X \setminus Sing\, X) \simeq Pic(X \setminus Sing\, X)$.

THEOREM 1.2. *(cf. [**7**] Theorem 1.5): Suppose that X is a subvariety of $\mathbb{P}_N$ which is of pure dimension n, H a hyperplane which is transversal to some Whitney statification of X, $Y := X \cap H$. Suppose that $Sing(X)$ has codimension ≥ 2. Then $Cl(X) \to Cl(Y)$ is bijective (resp. injective) if $n \geq 4$ (resp. $n \geq 3$).*

We will explain in the proof ad-hoc how to obtain the mapping $Cl\, X \to Cl\, Y$. This will be justified later on (Remark 2.3).

PROOF. Note that $Sing(Y) = Y \cap Sing(X)$ has codimension ≥ 2 in Y, too, and $Y \setminus Sing\, Y \subset X \setminus Sing\, X$. By Zariski-Lefschetz (see [**5**] Theorem 1.1.1), $H^j(X^{an} \setminus Sing(X^{an});\mathbb{Z}) \to H^j(Y^{an} \setminus Sing(Y^{an});\mathbb{Z})$ is bijective if $j \leq 2$ (resp. bijective if $j \leq 1$ and injective if $j = 2$). So we get a mapping $Cl\, X \simeq Pic(X \setminus Sing\, X) \to Pic(Y \setminus Sing\, Y) \simeq Cl(Y)$. Now use the following, with f being the inclusion of $Y \setminus Sing\, Y$ in $X \setminus Sing\, X$: □

PROPOSITION 1.3. *(cf. [**7**] Theorem 5.1): Suppose that $f : Y \to X$ is a morphism between smooth complex algebraic varieties such that $H^j(X^{an};\mathbb{Z}) \to H^j(Y^{an};\mathbb{Z})$ is bijective for $j \leq 2$ (resp. bijective for $j \leq 1$, injective for $j = 2$). Then $Pic(X) \to Pic(Y)$ is bijective (resp. injective).*

It is mainly because of this condition on singular cohomology that we imposed the transversality hypothesis.

The condition that $codim\, Sing(X) \geq 2$ in Theorem 1.2 seems acceptable because e.g. Hartshorne [**3**] looks at Weil divisor class groups only under this additional assumption, as said before. Also, when looking at factorial rings we may work with this condition: for an affine complex variety X the coordinate ring is factorial if and only if X is normal and $Cl(X) = 0$, see [**3**] II Prop. 6.2; for examples see [**10**].

But if one looks at the framework of Chow groups this assumption is no longer natural, see [**1**]. Assume that X has pure dimension n. Then we define

$$A^k(X) := A_{n-k}(X)$$

where $A_l(X)$ is the Chow group defined in [**1**] 1.3, p. 10. This definition agrees with the one in [**1**] 8.3 in the case when X is smooth. The elements of $A^k(X)$

are represented by formal integral linear combinations of irreducible subvarieties of codimension k. There is no condition on the codimension of the singular set.

Recall that $A^0(X)$ is the group of formal integral linear combinations of irreducible components of X.

By definition, $A^1(X) = Cl(X)$ with a suitable definition of $Cl(X)$: One must be careful when extending the usual definition of $Cl(X) := \{$Weil divisors on $X\}/\{$principal divisors$\}$ to this case, see [**1**] 1.2: Principal divisors are of the form $div\, f := \sum_D (ord_D f)[D]$, where f is an invertible rational function on X and D runs through the set of irreducible (reduced) hypersurfaces in X. The order $ord_D f$ of f along D is defined as follows: Let us look at the local ring $\mathcal{O}_{X,D}$ of X at D with maximal ideal $\mathfrak{m}_D$. The image of f in the quotient ring of $K_{X,D}$ of $\mathcal{O}_{X,D}$ can be written as g/h with $g, h \in \mathcal{O}_{X,D}$. Then $ord_D f := ord_D g - ord_D h$, so we have to define $ord_D f$ for $f \in \mathcal{O}_{X,D}$. If X is non-singular at D we know that the ideal $(f) \subset \mathcal{O}_{X,D}$ is of the form $\mathfrak{m}_D^k$, then $ord_D f := k$. In general we use the length: $ord_D f := l_{\mathcal{O}_{X,D}}(\mathcal{O}_{X,D}/(f))$. We prefer to avoid the use of length by passing to the normalization: f can be regarded as a rational function $\widehat{f}$ on $\widehat{X}$, where $\pi : \widehat{X} \to X$ is the normalization; if $D_1, \dots, D_k$ are the irreducible components of $\pi^{-1}(D)$, $ord_D f := \sum (ord_{\widehat{D}_j} \widehat{f}) deg(\widehat{D}_j \to D)$. Note that $\widehat{X}$ is non-singular at D_j. See [**1**] Example 1.2.3.
In general we can have that $deg(\widehat{D}_j \to D) > 1$:

EXAMPLE 1.4. $X := \{z^2 - yx^2 = 0\} \subset \mathbb{C}^3$. We may take $\widehat{X} := \{z - yx = 0\} \subset \mathbb{C}^3$, $\pi(x, y, z) := (x, y^2, z)$. Then $D := Sing\, X = \{z = x = 0\}$, $\widehat{D} := \pi^{-1}(D) = \{z = x = 0\}$ is irreducible, and $deg(\widehat{D} \to D) = 2$. So we see that $Cl(X) \simeq \mathbb{Z}/2\mathbb{Z}$.

This phenomenon is possible because X^{an} is not locally irreducible, then π is not a homeomorphism.

Here are two further illustrative examples to illustrate the concept of order:

EXAMPLE 1.5. $X := \{x^2 - y^3 = 0\} \subset \mathbb{C}^2$. Then $\widehat{X} = \mathbb{C}$, and $\pi : \widehat{X} \to X : t \mapsto (t^3, t^2)$. There is a rational function f on X whose divisor is $[D]$ with $D := \{0\}$: $\frac{x}{y}$, because $ord_D x = 3$, $ord_D y = 2$. But there is no such function which is regular on X. Anyhow, $f^{an} \to 0$ when approaching $D = \{0\}$.

EXAMPLE 1.6. $X := \{y^2 = x^2(1 + x)\} \subset \mathbb{C}^2$, cf. also [**3**] Expl. 4.9.1, p. 29. Of course, X is irreducible and reduced. Look at $D := \{0\}$. Then $\pi^{-1}(\{0\}) = \widehat{D}_1 \cup \widehat{D}_2$ where $\widehat{D}_1$ corresponds to $y = x\sqrt{1 + x}$ and $\widehat{D}_2$ to $y = -x\sqrt{1 + x}$, where $\sqrt{1 + x} = 1 + \frac{1}{2}x + \dots$ Then $ord_{\widehat{D}_1}(y - x)^3 = 6$, $ord_{\widehat{D}_1}(y + x)^2 = 2$, $ord_{\widehat{D}_2}(y - x)^3 = 3$, $ord_{\widehat{D}_2}(y + x)^2 = 4$. Put $f := \frac{(y-x)^3}{(y+x)^2}$: then $ord_{\widehat{D}_1} f = 4$, $ord_{\widehat{D}_2} f = -1$. This implies that $ord_D f = 4 - 1 = 3$.
This does not mean that $f^{an} \to 0$ when approaching D: if one approaches D along the local analytic irreducible component which corresponds to $\widehat{D}_1$ we have $f^{an} \to 0$, whereas $f^{an} \to \infty$ if we approach D along the irreducible component which corresponds to $\widehat{D}_2$.

2. The Gysin homomorphism

There are difficulties to define a pull-back for Chow groups. As explained in [**1**] 1.7 one can define it for a flat morphism. We look at another case: l.c.i. (local complete intersection) morphism where one prefers to speak of a Gysin homomorphism. See [**1**] 6.6. Recall that a l.c.i. morphism is a regular embedding (i.e. the subspace being locally defined by a regular system of parameters) followed by a smooth morphism.

In certain cases one can avoid to go back to the explicit definition of the Gysin homomorphism by use of its properties instead, as we will see in this section. We restrict to A^0 and A^1.

Let $f : Y \to X$ be a l.c.i. morphism between pure-dimensional complex algebraic varieties.

The case of A^0 is easy:

LEMMA 2.1. *$f^*([C]) = [f^{-1}(C)] = [D_1] + \ldots + [D_r]$ if C is an irreducible component of X and $D_1, \ldots, D_r$ the irreducible components of $f^{-1}(C)$.*
In particular, $f^* : A^0(X) \to A^0(Y)$ can be identified with $f^* : H^0(X^{an}; \mathbb{Z}) \to H^0(Y^{an}; \mathbb{Z})$ if X, Y are smooth.

PROOF. Let $i : C \to X$ be the inclusion. Note that i is proper, hence we have a push-forward $i_* : A^0(C) \to A^0(X)$, see [**1**] 1.4. We have a commutative diagram

$$\begin{array}{ccc} f^{-1}(C) & \to & Y \\ g\downarrow & & \downarrow f \\ C & \stackrel{i}{\to} & X \end{array}$$

with $g := f|f^{-1}(C) : f^{-1}(C) \to C$. It leads to a commutative diagram

$$\begin{array}{ccc} A^0(C) & \stackrel{i_*}{\to} & A^0(X) \\ g^*\downarrow & & \downarrow f^* \\ A^0(f^{-1}(C)) & \to & A^0(Y) \end{array}$$

For the commutativity it is sufficient to look at the case where f is a regular imbedding resp. a smooth morphism, then apply [**1**] Theorem 6.2a) resp. Prop. 1.7.
Note that $g^*([C]) = [f^{-1}(C)]$ by [**1**] Example 6.2.1. Altogether,

$$\begin{array}{ccc} [C] & \mapsto & [C] \\ g^*\downarrow & & \downarrow f^* \\ [f^{-1}(C)] & \mapsto & [f^{-1}(C)] = f^*([C]) \end{array}$$

So $f^*([C]) = [f^{-1}(C)]$. Now $[X] = [C_1] + \ldots + [C_s]$ if $C_1, \ldots, C_s$ are the irreducible components of X because X is reduced. Then $[Y] = [f^{-1}(C_1)] + \ldots + [f^{-1}(C_s)]$ and Y is reduced, so we must have $[f^{-1}(C)] = [D_1] + \ldots + [D_r]$. □

Now we turn to $A^1 = Cl$. First we have the following compatibility. Note that the usual mapping $Pic\, X \to Cl\, X$ can be described as $[\mathcal{L}] \to c^1(\mathcal{L}) \cap [X]$, see [**1**], beginning of 2.5.

LEMMA 2.2. *There is a commutative diagram*

$$\begin{array}{ccc} Pic\,X & \to & Cl\,X \\ f^*\downarrow & & \downarrow f^* \\ Pic\,Y & \to & Cl\,Y \end{array}$$

PROOF. (which was communicated to me by Jörg Schürmann): We use the framework of Fulton [**1**].
Recall that $Cl\,X = A^1(X)$, and the upper horizontal mapping is of the form $[\mathcal{L}] \mapsto c^1(\mathcal{L}) \cap [X]$, see above.
Since f is a l.c.i. morphism, we have

$$f^*(c^1(\mathcal{L}) \cap [X]) = c^1(f^*\mathcal{L}) \cap f^*[X]$$

In fact, if f is smooth, apply [**1**] Prop. 2.5(d),
if f is a regular immersion, apply [**1**] Prop. 6.3.
Moreover, $f^*[X] = [Y]$, see [**1**] Example 6.2.1. So we obtain

$$f^*(c^1(\mathcal{L}) \cap [X]) = c^1(f^*\mathcal{L}) \cap [Y]$$

But this means that our diagram is commutative:

$$\begin{array}{ccc} [\mathcal{L}] & \mapsto & c^1(\mathcal{L}) \cap [X] \\ \downarrow & & \downarrow \\ [f^*\mathcal{L}] & \mapsto & c^1((f^*\mathcal{L}) \cap [Y]) = f^*(c^1(\mathcal{L}) \cap [X]) \end{array}$$

□

REMARK 2.3. a) Assume that X is smooth. Then the upper horizontal map in Lemma 2.2 is an isomorphism, so we get an explicit description of $f^* : Cl\,X \to Cl\,Y$.
b) Assume that X_2 is a closed subspace of X of codimension ≥ 2 such that $X \setminus X_2$ is smooth and $Y_2 := f^{-1}(X_2)$ has codimension ≥ 2 in Y, too. Then we have a commutative diagram

$$\begin{array}{ccccc} Pic(X \setminus X_2) & \overset{\cong}{\to} & Cl(X \setminus X_2) & \overset{\cong}{\leftarrow} & Cl(X) \\ \downarrow & & \downarrow & & \downarrow \\ Pic(Y \setminus Y_2) & \to & Cl(Y \setminus Y_2) & \overset{\cong}{\leftarrow} & Cl(Y) \end{array}$$

which leads to an explicit description of $f^* : Cl\,X \to Cl\,Y$.
Note that Gysin homomorphisms are compatible with flat morphisms, in particular open imbeddings: if f is a regular immersion, see [**1**] Theorem 6.2b), if f is smooth, hence flat, see [**1**] Lemma 1.7.1 and the subsequent remark.

3. A Lefschetz theorem for Weil divisor class groups

The main result of this paper is the following:

THEOREM 3.1. *Let X be a locally closed subvariety of $\mathbb{P}_m$ of pure dimension, $\pi : \widehat{\overline{X}} \to \overline{X}$ the normalization of $\overline{X}$. Let us fix some stratification of $\overline{X}$ which comes from a stratification of π such that X is a union of strata. Let L be a linear subspace of $\mathbb{P}_m$ which is transversal to the stratification of $\overline{X}$ and $Y := X \cap L$. Suppose* $\dim Y \geq 3$ *(resp. ≥ 2). Then $Cl\,X \to Cl\,Y$ is bijective (resp. injective).*

Note that the inclusion $Y \to X$ is a regular embedding, hence we have a mapping $Cl\,X \to Cl\,Y$.

The proof consists of two steps: to prove that the hypothesis of Proposition 3.3 below is fulfilled and to prove the Proposition itself.

First:

LEMMA 3.2. *Let $f : Y \to X$ be a l.c.i. morphism of complex algebraic varieties of pure dimension, X_1 a purely 1-codimensional subvariety of X, $Y_1 := f^{-1}(X_1)$ purely 1-codimensional in Y. Let $\pi : \widehat{X} \to X$ be the normalization of X, $\widehat{Y} := Y \times_X \widehat{X}$, $\widehat{X}_1 := \pi^{-1}(X_1)$, $\widehat{Y}_1 := Y_1 \times_{X_1} \widehat{X}_1$. Then $f_1 := f|Y_1 : Y_1 \to X_1$ is a l.c.i. morphism, too, and we have a commutative diagram with exact rows*

$$\begin{array}{ccccccc} A^0(X_1) & \longrightarrow & Cl\,X & \longrightarrow & Cl(X \setminus X_1) & \to & 0 \\ \downarrow & & \downarrow & & \downarrow & & \\ A^0(Y_1) & \longrightarrow & Cl\,Y & \longrightarrow & Cl(Y \setminus Y_1) & \to & 0 \end{array}$$

PROOF. In order to show that f_1 is a l.c.i. morphism, too, it is sufficient to look at the case where f is a regular embedding. Locally on X, we have a regular system of parameters for Y, and we see that this yields a regular system of parameters on X_1 for Y_1.
The exact sequences follow from [**1**] Prop. 1.8. The commutativity follows from functoriality statements: [**1**] Th. 6.2, Prop. 1.7. □

PROPOSITION 3.3. *Beyond the hypothesis of Lemma 3.2, assume:*
a) $Cl(X \setminus X_1) \to Cl(Y \setminus Y_1)$ is bijective (resp. injective),
b) $Cl(\widehat{X}) \to Cl(\widehat{Y})$ is injective,
c) $A^0(X_1) \overset{\cong}{\to} A^0(Y_1)$,
d) $A^0(\widehat{X}_1) \overset{\cong}{\to} A^0(\widehat{Y}_1)$.
Then $Cl\,X \to Cl\,Y$ is bijective (resp. injective).

PROOF. By Lemma 3.2, we have a commutative diagram

$$\begin{array}{ccccccc} A^0(X_1) & \overset{\phi}{\longrightarrow} & Cl\,X & \longrightarrow & Cl(X \setminus X_1) & \to & 0 \\ \downarrow & & \downarrow & & \downarrow & & \\ A^0(Y_1) & \overset{\psi}{\longrightarrow} & Cl\,Y & \longrightarrow & Cl(Y \setminus Y_1) & \to & 0 \end{array}$$

By c), the first vertical arrow is bijective, by a), the third one is bijective resp. injective.
So it is sufficient to show that the mapping $Im\,\phi \to Im\,\psi$ is bijective. It is surjective, of course, so we must show that it is injective, too.
Let $D_1, \ldots, D_r$ be the irreducible components of X_1. Because of c), the inverse images $D_1^*, \ldots, D_r^*$ with respect to f_1 are the irreducible components of $\widehat{Y}_1$:
Note that $D_i^* \to D_i$ is a l.c.i. morphism, and $(f_1)^*([D_i]) = [D_i^*]$ by Lemma 2.1. Then use assumption c).
Now suppose that $\sum n_j[D_j]$ represents an element of the kernel of $Im\,\phi \to Im\,\psi$, we must show that this divisor is principal. By assumption, $\sum n_j[D_j^*]$ is principal, i.e. there is a rational function f on Y such that $\sum n_j[D_j^*] = div\,f$. This means that $n_j = ord_{D_j^*} f$.
Let $\widehat{D}_{j1}, \ldots$ be the irreducible components of $\pi^{-1}(D_j)$. Because of d), the inverse images $\widehat{D}_{jk}^*$ with respect to the induced mapping $\widehat{f}_1 : \widehat{Y}_1 \to \widehat{X}_1$ are the irreducible components of $\widehat{Y}_1$.

By [**1**] Example 1.2.3, $n_j = ord_{D_j^*} f = \sum_k (ord_{\widehat{D}_{jk}^*} \widehat{f}) deg(\widehat{D}_{jk}^* \to D_j^*)$, where f is considered a a rational function $\widehat{f}$ on $\widehat{Y}$, too. Now $\sum_{j,k}(ord_{\widehat{D}_{jk}^*} \widehat{f})[\widehat{D}_{jk}]$ represents an element of $Cl(\widehat{X})$ whose image in $Cl(\widehat{Y})$ is 0, so it is a principal divisor because $Cl(\widehat{X}) \to Cl(\widehat{Y})$ is injective, by b). So there is a rational function $\widehat{g}$ on $\widehat{X}$ such that the divisor of $\widehat{g}$ is $\sum_{j,k}(ord_{\widehat{D}_{jk}^*} \widehat{f})[\widehat{D}_{jk}]$. This means that $ord_{\widehat{D}_{jk}^*} \widehat{f} = ord_{\widehat{D}_{jk}} \widehat{g}$.
Also, $deg(\widehat{D}_{jk}^* \to D_j^*) = deg(\widehat{D}_{jk} \to D_j)$.
Altogether, $n_j = \sum_k (ord_{\widehat{D}_{jk}} \widehat{g}) deg(\widehat{D}_{jk} \to D_j) = ord_{D_j} g$: here we view $\widehat{g}$ as a rational function g on X, too.
So $\sum n_j [D_j] = \sum (ord_{D_j} g)[D_j]$ is indeed a principal divisor. □

How to verify the hypotheses of Proposition 3.3? Here the following lemma is useful:

LEMMA 3.4. *Beyond the hypothesis of Lemma 3.2, assume that $X \setminus X_1$, $Y \setminus Y_1$, $\widehat{X}$, $\widehat{Y}$, X_1, Y_1, $\widehat{X}_1$, $\widehat{Y}_1$ are smooth, and:*
a) $H^j(X^{an} \setminus X_1^{an}; \mathbb{Z}) \to H^j(Y^{an} \setminus Y_1^{an}; \mathbb{Z})$ is bijective, $j \leq 2$ (resp. bijective, $j \leq 1$, injective, $j = 2$),
b) $H^j(\widehat{X}^{an}; \mathbb{Z}) \to H^j(\widehat{Y}^{an}; \mathbb{Z})$ is bijective, $j \leq 1$, injective, $j = 2$,
c) $H^0(X_1^{an}; \mathbb{Z}) \simeq H^0(Y_1^{an}; \mathbb{Z})$,
d) $H^0(\widehat{X}_1^{an}; \mathbb{Z}) \simeq H^0(\widehat{Y}_1^{an}; \mathbb{Z})$.
Then the hypothesis of Proposition 3.3 is fulfilled.

PROOF. Use Proposition 1.3 and Remark 2.3a), Lemma 2.1. □

Finally we will use the following Lefschetz theorem for homotopy groups, see [**5**] Theorem 1.1.1:

PROPOSITION 3.5. *Let X be a pure-dimensional complex algebraic variety which is nonsingular, $f : X \to \mathbb{P}_N$ an algebraic morphism with finite fibres. Let us fix a Whitney stratification of $\mathbb{P}_N$ adapted to f. Suppose that L is a linear subspace of $\mathbb{P}_N$ which is transversal to all strata, $\dim(X \cap L) \geq k$. Then $(X^{an}, (f^{-1}(L))^{an})$ is k-connected.*

PROOF OF THEOREM 3.1: Let X_j be the union of strata of X of codimension $\geq j$. Then we may replace X by $X \setminus X_2$. Then $X \setminus X_1$, X_1, $\widehat{X}$, $\widehat{X}_1$ are smooth, similarly for $Y := X \cap L$. By Proposition 3.5 we have:
$H^j(X^{an} \setminus X_1^{an}; \mathbb{Z}) \to H^j(Y^{an} \setminus Y_1^{an}; \mathbb{Z})$ is bijective, $j \leq 2$ (resp. bijective, $j \leq 1$, injective, $k = 2$) (put $k = 3$ resp. 2),
$H^j(\widehat{X}^{an}; \mathbb{Z}) \to H^j(\widehat{Y}^{an}; \mathbb{Z})$ is bijective, $j \leq 1$, injective, $j = 2$ (put $k = 2$),
$H^0(X_1^{an}; \mathbb{Z}) \simeq H^0(Y_1^{an}; \mathbb{Z})$,
$H^0(\widehat{X}_1^{an}; \mathbb{Z}) \simeq H^0(\widehat{Y}_1^{an}; \mathbb{Z})$.
Note that the normalization of $\overline{X}$ is projective. So we have the hypothesis of Lemma 3.4, hence of Proposition 3.3.

We may weaken the transversality hypothesis of Theorem 3.1. For simplicity we look at the case where L is a hyperplane:

COROLLARY 3.6. *Let X be a locally closed subvariety of $\mathbb{P}_m$ of pure dimension, $\pi : \widehat{X} \to \overline{X}$ the normalization of $\overline{X}$. Let us fix some stratification of $\overline{X}$ which comes* *from a stratification of* π *such that* X *is a union of strata. Let* H *be a hyperplane of* $\mathbb{P}_m$ *which is transversal to the stratification of* $\overline{X}$ *outside some subset of* $\overline{X} \cap H$ *of codimension* ≥ 4 *(resp.* ≥ 3*),* $Y := X \cap H$*. Then* $Cl\,X \to Cl\,Y$ *is bijective (resp. injective).*

PROOF. Let L be a general linear subspace of H such that $dim\, X \cap L = 3$ (resp. $= 2$). Then L is transversal to the stratification of $\overline{X}$, so we can apply Theorem 3.1 and obtain that $Cl(X) \to Cl(X \cap L)$ and $Cl(X \cap H) \to Cl(X \cap L)$ are bijective (resp. injective). □

One cannot simply drop the transversality condition at all:

EXAMPLE 3.7. $X := \mathbb{P}_4$, $Y := \{f = 0\}$, where $f := z_1 z_2 - z_3 z_4$. Note that Y is the projective closure of the affine cone over the quadric $V := \{z_1 z_2 - z_3 z_4 = 0\}$ in $\mathbb{P}_3$. We have $Cl(Y) = Cl(V) \simeq \mathbb{Z}^2$, see [**3**] p. 146 below, resp. [**3**] Ex. 6.5c), p. 147. So $Cl(Y) \not\simeq Cl(X) \simeq \mathbb{Z}$. Note that we can represent Y by a hyperplane section if we replace X by its image under a Veronese embedding. By Corollary 3.6 we have at least injectivity.

If we look at the Picard group instead we must deal with different conditions. Recall that $Pic(X) \simeq Pic(Y)$ as soon as X is a local complete intersection in $\mathbb{P}_N$ of dimension ≥ 4 and the codimension of $Sing(Y)$ in Y is at least 3, see [**8**] Theorem 1.1.
In our example above we obtain $\mathbb{Z} \simeq Pic(X) \simeq Pic(Y)$.
So $Pic(Y) \not\simeq Cl(Y)$.
This is not unexpected: In general, we have that $Pic(Y) \simeq Cl(Y)$ if the local rings of Y are factorial, see [**3**] II Prop. 6.11. This is known to be the case if Y is a local complete intersection which is smooth outside a set of codimension ≥ 4, see [**2**] XI Cor. 3.14, and this condition fails in our case.

Finally let us treat a situation which is very different from Theorem 3.1 and which shows that it is worth while to work in a general framework as long as possible:

PROPOSITION 3.8. *Suppose that* $f : Y \to X$ *is smooth and defines a locally trivial bundle with fibre* F*. Assume that* F *is smooth in codimension* 1, $\tilde{H}^j(F^{an} \setminus Sing\, F^{an}; \mathbb{Z}) = 0, j \leq 3$*. Then* f *induces an isomorphism* $Cl(X) \to Cl(Y)$*.*

PROOF. Note that our assumption implies $R^0 f_*^{an} \mathbb{Z}_{Y^{an}} \simeq \mathbb{Z}_{X^{an}}$, $R^j f_*^{an} \mathbb{Z}_{Y^{an}} = 0, j = 1,2,3$, similarly for $f_1, \widehat{f}, \widehat{f_1}$, where $X_1 := Sing\, X$, $Y_1 := f^{-1}(X_1)$, $\widehat{X} :=$ normalization of X, $\widehat{Y} := Y \times_X \widehat{X}$, similarly $\widehat{Y}_1$, $\widehat{f} : \widehat{Y} \to \widehat{X}$ etc. the obvious mappings. We may assume that X_1 and $\widehat{X}$ are smooth, X_1 of pure codimension 1. So we obtain the hypothesis of Lemma 3.4, hence of Proposition 3.3. □

EXAMPLE 3.9. a) $F = \mathbb{C}^m$: then we get $Cl(X) \simeq Cl(Y)$. The surjectivity is proved by [**1**] Prop. 1.9.
b) $F = \bar{F} \setminus Sing(\bar{F})$, where $\bar{F}$ is a singular weighted homogeneous complete intersection in $\mathbb{C}^m$, $codim_{\overline{F}} Sing(\bar{F}) \geq 5$.
In fact, let S_ϵ be the sphere of radius $\epsilon > 0$ in $\mathbb{C}^m$, $\Sigma := S_\epsilon \cap \overline{F}^{an}$, then F^{an} has the same homotopy type as $\Sigma \cap F^{an}$. Cutting down by a general linear subspace of codimension $dim\, Sing\, \overline{F}$ we see that $\Sigma \cap F^{an}$ is 3-connected: use [**9**] Korollar 1.3 and [**6**] Theorem I.1.1.

For instance, let $\mathcal{L}$ be a line bundle on X, $\overline{F}$ defined by polynomials $g_1, \ldots, g_k$ with respect to weights $w_1, \ldots, w_m$, then $g_1, \ldots, g_k$ define a subspace $\overline{Y}$ of $\mathcal{L}^{\otimes w_1} \oplus \ldots \oplus \mathcal{L}^{\otimes w_m}$, and we have a mapping $\bar{Y} \to X$ which is a locally trivial fibre bundle with fibre $\overline{F}$.

4. Remarks on Chow groups

Let us assume that X is a purely n-dimensional smooth subvariety of $\mathbb{P}_N$ and $Y := X \cap H$ is a smooth hypersurface.
First note that $Y \to X$ is a regular embedding in the sense of [**1**] App. B.7.1, so there are Gysin maps $A^k(X) \to A^k(Y)$, see [**1**] 6.2, p. 98.
For $k = 0$ we have $A^0(X) = H^0(X^{an}; \mathbb{Z})$, so indeed $A^0(X) \to A^0(Y)$ is bijective for $n \geq 2$ and injective for $n \geq 1$, by Lefschetz, cf. Lemma 2.1.
For $k = 1$ we have that $A^1(X) \to A^1(Y)$ is bijective for $n \geq 4$ and injective for $n \geq 3$, by Theorem 1.1.

This may suggest the keen question whether $A^k(X) \to A^k(Y)$ is bijective for $2k \leq n-2$ and injective for $2k \leq n-1$.

There is another motivation for this question: We have a mapping $A^k(X) \to H_{2n-2k}(X^{an}; \mathbb{Z})$ to Borel-Moore homology: the cycle map, see [**1**] 19.1. By Poincaré duality, $H_{2n-2k}(X^{an}; \mathbb{Z}) \simeq H^{2k}(X^{an}; \mathbb{Z})$. So altogether we have a mapping $A^k(X) \to H^{2k}(X^{an}; \mathbb{Z})$. In our case, $A^k(X) \to A^k(Y)$ should (hopefully) correspond to the mapping $H^{2k}(X^{an}; \mathbb{Z}) \to H^{2k}(Y^{an}; \mathbb{Z})$ induced by the inclusion.
If X, Y are smooth we know by Lefschetz that $H^{2k}(X^{an}; \mathbb{Z}) \to H^{2k}(Y^{an}; \mathbb{Z})$ is bijective for $2k \leq n-2$ and injective for $2k \leq n-1$, so one may expect the same for $A^k(X) \to A^k(Y)$.

But for $k > 1$ almost nothing seems to be known, even in the special case where X, Y are complete intersections in $\mathbb{P}_N$: in this case we deal with a question asked by Hartshorne, see [**4**], [**1**] Expl. 19.3.10.

5. A different approach to the Weil divisor class group

Here we develop a different approach to the Weil divisor class group which avoids the restriction to l.c.i. morphisms in section 3.

Let $f : Y \to X$ be a morphism between pure-dimensional complex algebraic varieties which is not supposed to be l.c.i.

First note that f still induces a homomorphism $f^* : A^0(X) \to A^0(Y)$ defined by $[C] \mapsto [f^{-1}(C)]$, C being an irreducible component of X. This definition is obviously compatible with the one for l.c.i. morphisms, see Lemma 2.1.

THEOREM 5.1. *Let $X_2 \subset X_1$ be subvarieties of X of codimension ≥ 2 resp. of pure codimension 1, $X \setminus X_1$ and $X_1 \setminus X_2$ smooth. Let $\pi : \widehat{X} \to X$ be the normalization, $\widehat{X} \setminus \widehat{X}_2$ smooth, where $\widehat{X}_j := \pi^{-1}(X_j)$,*

$$\mathrm{Cl}(X) := A^0(X_1 \setminus X_2) +_{A^0(\widehat{X}_1 \setminus \widehat{X}_2)} Cl(\widehat{X} \setminus \widehat{X}_2)$$

the push-out. Then the induced homomorphism $\mathrm{Cl}(X) \to Cl(X \setminus X_2) \simeq Cl(X)$ is an isomorphism.

PROOF. We may assume $X_2 = \emptyset$, replacing X by $X \setminus X_2$. Then the induced homomorphism $\mathrm{Cl}\, X \to Cl\, X$ comes from the commutative diagram

$$\begin{array}{ccc} A^0(\widehat{X}_1) & \to & Cl(\widehat{X}) \\ \downarrow & & \downarrow \\ A^0(X_1) & \to & Cl(X) \end{array}$$

We have a commutative diagram with exact rows:

$$\begin{array}{ccccccc} A^0(X_1) & \stackrel{\phi}{\to} & \mathrm{Cl}(X) & \to & Cl(\widehat{X} \setminus \widehat{X}_1) & \to & 0 \\ \downarrow= & & \downarrow & & \downarrow\simeq & & \\ A^0(X_1) & \stackrel{\psi}{\to} & Cl(X) & \to & Cl(X \setminus X_1) & \to & 0 \end{array}$$

It is sufficient to show that $im\, \phi \to im\, \psi$ is bijective. The surjectivity is obvious. Injectivity: An element $x \in \mathrm{Cl}(X)$ is represented by an element x_1 of

$$\mathrm{Div}(X) := A^0(X_1) +_{A^0(\widehat{X}_1)} Div(\widehat{X})$$

Now we have a commutative diagram

$$\begin{array}{ccc} A^0(X_1) & \to & \mathrm{Div}(X) \\ \downarrow= & & \downarrow \\ A^0(X_1) & \to & Div(X) \end{array}$$

If $x \in im\, \phi$, the element x_1 above can be assumed to come from an element $\sum n_i[D_i] \in A^0(X_1)$.
Suppose that $x \mapsto 0 \in im\, \psi$:
This means that the image of $\sum n_i[D_i]$ in $Div\, X$ is a principal divisor, i.e. $\sum n_i[D_i] = div\, f$ for some invertible rational function f on X.
Now π is birational, so f comes from an invertible rational function $\widehat{f}$ on $\widehat{X}$.
This implies that the image of x_1 in $\mathrm{Cl}(X)$ is represented by $(0, div\, \widehat{f}) \in A^0(X_1) \oplus Div(\widehat{X})$, so $x = 0$ as we claimed:
We must show that $(\sum n_i[D_i], 0)$ and $(0, div\, \widehat{f})$ represent the same element of Div(X), i.e. $(\sum n_i[D_i], -div\, \widehat{f})$ represents 0.
But this is because $(\sum n_i[D_i], -div\, \widehat{f}) \mapsto 0 \in Div\, X$ and there are bijective mappings

$$\mathrm{Div}(X) \stackrel{\cong}{\to} A^0(X_1) \oplus Div(\widehat{X} \setminus \widehat{X}_1) \stackrel{\cong}{\to} Div(X)$$

□

Note that the assumptions can be fulfilled as follows: Take a Whitney stratification of the mapping π and let X_1, X_2 be the union of strata of codimension ≥ 1 resp. ≥ 2.

Theorem 5.1 allows to prove the following modification of Lemma 3.2:

LEMMA 5.2. *Let $f : Y \to X$ be a morphism of complex algebraic varieties of pure dimension, X_1 a purely 1-codimensional subvariety of X, $Y_1 := f^{-1}(X_1)$ purely 1-codimensional in Y, $X \setminus X_1$, $Y \setminus Y_1$ smooth. Let $\pi : \widehat{X} \to X$ be the normalization of X, $\widehat{Y} := Y \times_X \widehat{X}$, $\widehat{X}_1 := \pi^{-1}(X_1)$, $\widehat{Y}_1 := Y_1 \times_{X_1} \widehat{X}_1$. Then we* *have a commutative diagram with exact rows*

$$\begin{array}{ccccccc} A^0(X_1) & \longrightarrow & Cl\, X & \longrightarrow & Cl(X \setminus X_1) & \to & 0 \\ \downarrow & & \downarrow & & \downarrow & & \\ A^0(Y_1) & \longrightarrow & Cl\, Y & \longrightarrow & Cl(Y \setminus Y_1) & \to & 0 \end{array}$$

PROOF. We work with Cl instead of Cl. Note that f induces a homomorphism $\mathrm{Cl}(X) \to \mathrm{Cl}(Y)$. The rest is clear. □

Similarly in case of Proposition 3.3:

PROPOSITION 5.3. *Proposition* 3.3 *holds if we drop the assumption that* f *is a l.c.i. morphism provided that* $X \setminus X_1$, $Y \setminus Y_1$ *are smooth.*

Finally we obtain:

LEMMA 5.4. *Lemma* 3.4 *holds without assuming that* f *is a l.c.i. morphism.*

This yields a new proof of Theorem 3.1.

References

[1] W. Fulton, *Intersection theory*, Ergebnisse der Mathematik und ihrer Grenzgebiete (3) [Results in Mathematics and Related Areas (3)], vol. 2, Springer-Verlag, Berlin, 1984. MR732620

[2] A. Grothendieck, *Cohomologie locale des faisceaux cohérents et théorèmes de Lefschetz locaux et globaux* (*SGA* 2) (French), North-Holland Publishing Co., Amsterdam; Masson & Cie, Éditeur, Paris, 1968. Augmenté d'un exposé par Michèle Raynaud; Séminaire de Géométrie Algébrique du Bois-Marie, 1962; Advanced Studies in Pure Mathematics, Vol. 2. MR0476737

[3] R. Hartshorne, *Algebraic geometry*, Springer-Verlag, New York-Heidelberg, 1977. Graduate Texts in Mathematics, No. 52. MR0463157

[4] R. Hartshorne, *Equivalence relations on algebraic cycles and subvarieties of small codimension*, Algebraic geometry (Proc. Sympos. Pure Math., Vol. 29, Humboldt State Univ., Arcata, Calif., 1974), Amer. Math. Soc., Providence, R.I., 1973, pp. 129–164. MR0369359

[5] H. A. Hamm and Lê Dũng Tráng, *Lefschetz theorems on quasiprojective varieties* (English, with French summary), Bull. Soc. Math. France **113** (1985), no. 2, 123–142. MR820315

[6] H. A. Hamm and Lê Dũng Tráng, *Local generalizations of Lefschetz-Zariski theorems*, J. Reine Angew. Math. **389** (1988), 157–189, DOI 10.1515/crll.1988.389.157. MR953670

[7] H. A. Hamm and D. T. Lê, *On the Picard group for non-complete algebraic varieties* (English, with English and French summaries), Singularités Franco-Japonaises, Sémin. Congr., vol. 10, Soc. Math. France, Paris, 2005, pp. 71–86. MR2145949

[8] H. A. Hamm and D. T. Lê, *A Lefschetz theorem on the Picard group of complex projective varieties*, Singularities in geometry and topology, World Sci. Publ., Hackensack, NJ, 2007, pp. 640–660, DOI 10.1142/9789812706812_0021. MR2311503

[9] H. Hamm, *Lokale topologische Eigenschaften komplexer Räume* (German), Math. Ann. **191** (1971), 235–252, DOI 10.1007/BF01578709. MR0286143

[10] H. A. Hamm, *Divisor class groups of affine complete intersections*, Singularities and computer algebra, Springer, Cham, 2017, pp. 147–160. MR3675725

MATHEMATISCHES INSTITUT, UNIVERSITÄT MÜNSTER, EINSTEINSTRASSE 62, 48149 MÜNSTER, GERMANY

Email address: hamm@uni-muenster.de

Contemporary Mathematics
Volume **742**, 2020
https://doi.org/10.1090/conm/742/14940

The augmented deformation space of rational maps

Eriko Hironaka

ABSTRACT. It was recently shown that the Epstein deformation space of marked rational maps with prescribed combinatorial and dynamical structure can be disconnected. For example, the family of quadratic rational maps with a periodic critical cycle of order 4 and an extra critical value not lying in this cycle has a deformation space with infinitely many components. We study the structure of the augmented deformation space for this example, and show, in particular, that the closure of deformation space in augmented deformation space is also disconnected in this case.

In celebration of Lê Dũng Tráng's 70th birthday

1. Introduction

Let A and B be two finite subsets of the 2-dimensional sphere S^2 such that $|A|, |B| \geq 3$, and let $f, \iota : (S^2, A) \to (S^2, B)$ be two maps of pairs:

$$f : (S^2, A) \to (S^2, B)$$

a branched covering with branch values contained in B, and

$$\iota : (S^2, A) \hookrightarrow (S^2, B)$$

a homeomorphism identifying domain and range, and A with a subset of B.

The *Epstein deformation space* $\mathcal{D}_{f,\iota}$ is defined as the equalizer of the induced maps on Teichmüller spaces

$$f^*, \iota^* : \mathcal{T}_{(S^2,B)} \to \mathcal{T}_{(S^2,A)}.$$

In unpublished work from the 1990s, A. Epstein showed that if f is not Lattès[1], then $\mathcal{D}_{f,\iota}$ is either empty or a complex $|B| - |A|$ dimensional submanifold of $\mathcal{T}_{(S^2,B)}$ [**7**]. This generalizes a seminal result of W. Thurston who showed that in the *postcritically finite* case, when $A = B$, either $\mathcal{D}_{f,\iota}$ is empty, f is a Lattès example, or $\mathcal{D}_{f,\iota}$ contains exactly one point [**6**]. In particular, $\mathcal{D}_{f,\iota}$ is always connected in the postcritically finite case. When f is not post-critically finite, then $\mathcal{D}_{f,\iota}$ need not be connected [**11**], but so far there is only one known class of examples.

PROBLEM 1.1. What are necessary and sufficient conditions for $\mathcal{D}_{f,\iota}$ to be connected?

This work was partially supported by a grant from the Simons Foundation #426722.
[1] See [**15**] for definitions.

The counter-example to connectedness in [**11**] is part of a class of quadratic rational maps studied by Milnor in [**14**]. These are denoted $\mathrm{Per}_n(0)$ and have one periodic n-cycle containing a single critical point, and the behavior of the other critical value under iteration is unspecified. If in addition the extra critical value does not lie in the first critical n-cycle, the rational map is said to be in $\mathrm{Per}_n(0)^*$. Given an element $F \in \mathrm{Per}_n(0)^*$ let $f, \iota : (S^2, A) \to (S^2, B)$ be defined so that f is the topological covering underlying F, ι is a identification of domain and range of f so that $A \subset B$, A is the critical n-cycle, and B is the union of A and the extra critical point. For $n = 3$, $\mathcal{T}_{(S^2,A)}$ is a singleton set, and hence $\mathcal{D}_{f,\iota}$ is the entire space $\mathcal{T}_{(S^2,B)}$. For $n = 4$, the work in [**11**] shows that $\mathcal{D}_{f,\iota}$ is not connected, and in fact has infinitely many connected components. This leads to the following question.

QUESTION 1.2. Does $\mathcal{D}_{f,\iota}$ have a natural connected closure?

The points in $\mathcal{D}_{f,\iota}$ are sometimes called the *dynamical points* in the space of complex structures on $\mathcal{T}_{(S^2,B)}$. The *augmented deformation space* $\mathcal{AD}_{f,\iota}$, or *ideal dynamical points*, is the subset of the Bers augmented deformation space $\mathcal{AT}_{(S^2,B)}$ (see [**2**]) defined as the equalizer of extensions

$$\widetilde{f}^*, \widetilde{\iota}^* : \mathcal{AT}_{(S^2,B)} \to \mathcal{AT}_{(S^2,A)}$$

of f^* and ι^*.

At the time of this writing, it is not known whether $\mathcal{AD}_{f,\iota}$ is connected in the $\mathrm{Per}_4(0)^*$ case. Our goal in this paper is to extend the techniques of [**11**], and apply them to give a partial description of the structure of $\mathcal{AD}_{f,\iota}$ and the closure of $\mathcal{D}_{f,\iota}$ within it. We prove the following theorem.

THEOREM 1.3. *For (f, ι) associated to an element of $\mathrm{Per}_4(0)^*$, the closure of $\mathcal{D}_{f,\iota}$ in augmented deformation space $\mathcal{AD}_{f,\iota}$ is not connected.*

1.1. Some background and ideas behind the proofs. The question of whether and when $\mathcal{D}_{f,\iota}$ is connected has roots in work of Thurston from the 1980s, in which he showed that if

$$F : \mathbb{P}^1 \to \mathbb{P}^1$$

is a non-Lattès rational map from the complex projective line to itself with a finite post-critical set

$$\bigcup_{n=1}^{\infty} F^{(n)}(\mathrm{Crit}_F),$$

and $f : (S^2, P) \to (S^2, P)$ is the corresponding branched covering of pointed spheres with domain and range identified, and postcritical set P, then the lifting map on holomorphic markings defines a contracting map on Teichmüller space

$$f^* : \mathcal{T}_{(S^2,P)} \to \mathcal{T}_{(S^2,P)}.$$

Thus, f^* has a unique fixed point, and hence $\mathcal{D}_{f,\iota}$ is a singleton set.

This classical result suggests that there could be a contracting flow in the general case when the identification map $\iota : (S^2, A) \to (S^2, B)$ is a strict inclusion on A. If such a flow exists, there are two possibilities: one is that $\mathcal{D}_{f,\iota}$ is connected, and the other is that some points may be pushed out to the boundary of $\mathcal{T}_{(S^2,B)}$. This suggests looking at dynamical elements of the augmented Teichmüller space to find a natural connected completion of $\mathcal{D}_{f,\iota}$.

The proof in [**11**], that $\mathcal{D}_{f,\iota}$ can be disconnected translates the question about flows on Teichmüller spaces to one about the topology of algebraic varieties. A key ingredient is to define an intermediate covering $\mathcal{M}_f$ of $\mathcal{T}_{(S^2,B)} \to \mathcal{M}_{(S^2,B)}$ that is a natural quotient of the space of marked rational maps topologically equal to f. The projection $\mathcal{M}_f \to \mathcal{M}_{(S^2,B)}$ is a finite covering, and hence has the structure of a quasi-projective variety.

In the case when $F \in \mathrm{Per}_4(0)^*$, the space $\mathcal{M}_f$ is isomorphic to a Zariski dense subset of $\mathbb{P}^1 \times \mathbb{P}^1$:

$$\mathcal{M}_f = \mathbb{P}^1 \times \mathbb{P}^1 \setminus \mathcal{Z} \cup \mathcal{L}$$

where $\mathbb{P}^1 \times \mathbb{P}^1 \setminus \mathcal{L} = \mathcal{M}_{(S^2,A)} \times \mathcal{M}_{(S^2,A)}$, and $\mathcal{Z}$ indicates the locus where the both critical points of f lie in the same periodic cycle. With respect to this parameterization, the image of $\mathcal{D}_{f,\iota}$ in $\mathcal{M}_f$ is the diagonal

$$\mathcal{V}_{f,\iota} = \{(x,x) \in \mathbb{P}^1 \times \mathbb{P}^1 \mid (x,x) \notin \mathcal{Z} \cup \mathcal{L}\}.$$

Let L_f be the group of covering automorphisms of $\mathcal{T}_{(S^2,B)}$ over $\mathcal{M}_{f,\iota}$. It was shown in [**11**] that the projection of $\mathcal{D}_{f,\iota}$ on $\mathcal{V}_{f,\iota}$ is a regular covering with covering automorphism group a proper subgroup $S_{f,\iota} \subset L_f$. In particular, $S_{f,\iota}$ acts transitively on fibers of $\mathcal{D}_{f,\iota} \to \mathcal{V}_{f,\iota}$.

Choose a basepoint $d_0 \in \mathcal{D}_{f,\iota}$ and let v_0 be its image in $\mathcal{V}_{f,\iota}$. Then we can identify L_f with $\pi_1(\mathcal{M}_f, v_0)$ and $S_{f,\iota}$ with a subgroup of the fundamental group. Let $E_{f,\iota}$ be the image of $\pi_1(\mathcal{V}_{f,\iota}, v_0) \to \pi_1(\mathcal{M}_f, v_0)$ induced by the inclusion map. Then $E_{f,\iota}$ is the subgroup of $S_{f,\iota}$ that fixes the component of $\mathcal{D}_{f,\iota}$ containing d_0; $E_{f,\iota}$ has infinite index in $S_{f,\iota}$; and hence $\mathcal{D}_{f,\iota}$ has infinitely many components. In other words, the image of $\pi_1(\mathcal{V}_{f,\iota}, v_0)$ is not sufficiently large in $\pi_1(\mathcal{M}_f, v_0)$. One way to try to rectify this is to put both $\mathcal{M}_f$ and $\mathcal{V}_{f,\iota}$ into a larger ambient space.

Let $\mathcal{AT}_{(S^2,B)}$ be the augmented Teichmüller space of (S^2, B). Then L_f extends to an action on $\mathcal{AT}_{(S^2,B)}$ giving a quotient $\mathcal{AM}_f$ and a commutative diagram

$$\begin{array}{ccc} \mathcal{T}_{(S^2,B)} & \hookrightarrow & \mathcal{AT}_{(S^2,B)} \\ \downarrow {\scriptstyle /L_f} & & \downarrow {\scriptstyle /L_f} \\ \mathcal{M}_f & \hookrightarrow & \mathcal{AM}_f. \end{array}$$

Similarly, $S_{f,\iota}$ acts on $\mathcal{AD}_{f,\iota}$ with quotient denoted $\mathcal{AV}_{f,\iota}$. The spaces $\mathcal{AV}_{f,\iota}$ and $\mathcal{AM}_f$ have the advantage of being algebraic geometric sets, and can be studied via singularity theory.

We define a connected pure 1-dimensional algebraic subset X of a blowup $\widetilde{\mathbb{P}^1 \times \mathbb{P}^1}$ of $\mathbb{P}^1 \times \mathbb{P}^1$ and an embedding $X \to \mathcal{AV}_{f,\iota}$ that is surjective except possibly a finite set of points (Proposition 4.2). By studying properties of X and its embedding in $\widetilde{\mathbb{P}^1 \times \mathbb{P}^1}$ we prove Theorem 1.3.

1.2. Organization. In Section 2.1, we give necessary definitions of Teichmüller and moduli spaces for rational maps, and their augmented versions. In Section 3, we prove some general properties of complements of plane algebraic curves and their blowups. In Section 4, we apply these ideas to the $\mathrm{Per}_4(0)^*$ case, and prove Theorem 1.3.

1.3. Further questions. There are still many open questions along the lines of this investigation. So far, the $\mathrm{Per}_4(0)^*$ example is the only known case when $\mathcal{D}_{f,\iota}$ is disconnected. Are there others? Is $\mathcal{AD}_{f,\iota}$ connected in the $\mathrm{Per}_4(0)^*$ case, and is $\mathcal{AD}_{f,\iota}$ connected in general? The analysis in this paper suggest general approaches to these questions, which we leave for further investigation.

1.4. Some comments on notation. This paper grew out of the ideas in [**11**], but some of the notation has changed. Most notably, we refer to the Teichmüller space $\mathcal{T}_f$ of rational maps topologically equivalent to f, and its corresponding Moduli space $\mathcal{M}_f$. Thus, Epstein's deformation space $\mathcal{D}_{f,\iota}$ will be considered as a subspace of $\mathcal{T}_f$ rather than of the isomorphic space $\mathcal{T}_{(S^2,B)}$ (also known as the *parameter space*). The space $\mathcal{M}_f$ may be more familiarly known as a connected component of a Hurwitz space of rational maps [**16**] and was denoted by $\mathcal{W}_f$ in [**11**]. A smaller change is the use of $\mathcal{D}_{f,\iota}$ for the deformation space $\mathcal{D}_f$ to emphasize the dependence on both the topological branched covering of pairs f and the identification ι. Similarly, we changed the notation of S_f to $S_{f,\iota}$. These ease our transition from a discussion of deformation space to augmented deformation space.

1.5. Acknowledgments. I would like to thank X. Buff, S. Koch, C. McMullen, R. Ramadas, and L. D. Tráng for helpful references and discussions, and the anonymous referee for their careful reading and useful comments.

2. Background definitions

In this section, we recall definitions and properties of Teichmüller space, moduli space and deformation space for rational maps and their augmented versions. For more details about the general theory see, for example, [**5**], [**2**], [**12**], [**16**] and [**17**]. We also recall definitions of liftables L_f and special liftables $S_{f,\iota}$ from [**11**], and examine their extensions to the augmented spaces.

2.1. Teichmüller spaces for rational maps. Let A be a finite subset of 3 or more points of the topological 2-sphere S^2. The *Teichmüller space* $\mathcal{T}_{(S^2,A)}$ of *holomorphic markings* on (S^2, A) is the collection of orientation preserving homeomorphisms

$$\phi : (S^2, A) \to (\mathbb{P}^1, \phi(A))$$

defined up to post-composition by automorphisms of $\mathbb{P}^1$ (i.e., Möbius transformations) and pre-composition by self-homeomorphisms of S^2 that are isotopic rel A to the identity.

Similarly, given a finite branched covering of pairs $f : (S^2, A) \to (S^2, B)$, where $\infty > |A|, |B| \geq 3$ and B contains the branch values (or critical values) of f, we define the *Teichmüller space* $\mathcal{T}_f$ of *holomorphic markings* on (f, A, B) as the set of commutative diagrams

$$\begin{array}{ccc} (S^2, A) & \xrightarrow{\psi} & (\mathbb{P}^1, \psi(A)) \\ {\scriptstyle f}\downarrow & & \downarrow{\scriptstyle F} \\ (S^2, B) & \xrightarrow{\phi} & (\mathbb{P}^1, \phi(B)) \end{array}$$

also denoted by (ϕ, ψ, F), where F is a rational map, ϕ and ψ are homeomorphisms. Two triples (ϕ, ψ, F) and (ϕ_1, ψ_1, F_1) in $\mathcal{T}_f$ are *equivalent* if there are homeomorphisms $\alpha : (S^2, A) \to (S^2, A)$ isotopic to the identity map rel A and

$\beta : (S^2, B) \to (S^2, B)$ isotopic to the identity rel B, and biholomorphic maps $\mu, \nu : \mathbb{P}^1 \to \mathbb{P}^1$ so that the diagram

$$\begin{array}{ccccccc}
(\mathbb{P}^1, \psi_1(A)) & \xleftarrow{\psi_1} & (S^2, A) & \xrightarrow{\alpha} & (S^2, A) & \xrightarrow{\psi} & (\mathbb{P}^1, \psi(A)) \\
\downarrow {\scriptstyle F_1} & & \downarrow {\scriptstyle f} & & \downarrow {\scriptstyle f} & & \downarrow {\scriptstyle F} \\
(\mathbb{P}^1, \phi_1(B) & \xleftarrow{\phi_1} & (S^2, B) & \xrightarrow{\beta} & (S^2, B) & \xrightarrow{\phi} & (\mathbb{P}^1, \phi(B))
\end{array}$$

commutes, and $\mu = \psi \circ \alpha \circ \psi_1^{-1}$ and $\nu = \phi \circ \beta \circ \phi_1^{-1}$.

By these definitions, $\mathcal{T}_f$ comes with natural surjections

$$\begin{array}{ccccc}
 & & \mathcal{T}_f & & \\
 & \swarrow {\scriptstyle q} & & {\scriptstyle p_U} \searrow & \\
\mathcal{T}_{(S^2,B)} & & & & \mathcal{T}_{(S^2,A)}
\end{array}$$

recording the holomorphic markings of the domain space (p_U) and the range space (q). Furthermore, $\mathcal{T}_f$ can be thought of as the graph of a *lifting map*

$$f^* : \mathcal{T}_{(S^2,B)} \to \mathcal{T}_{(S^2 A)}$$

defined by f. Thus, q is an isomorphism of holomorphic spaces.

We create an iteration scheme from f by partially identifying the domain (S^2, A) and range (S^2, B) of f. That is, we fix a homeomorphism $\iota : S^2 \to S^2$ that restricts to an inclusion $\iota|_A : A \hookrightarrow B$. Then we have a map of pairs

$$\iota : (S^2, A) \to (S^2, B).$$

Let $p_L = \iota^* \circ q$, where $\iota^* : \mathcal{T}_{(S^2,B)} \to \mathcal{T}_{(S^2,A)}$ is the map that takes $\phi \in \mathcal{T}_{(S^2,B)}$ to the class in $\mathcal{T}_{(S^2,A)}$ defined by

$$\phi \circ \iota : (S^2, A) \to (\mathbb{P}^1, (\phi \circ \iota)(A)) = (\mathbb{P}^1, \phi(A)).$$

Then the elements of $\mathcal{T}_f$ can be thought of as the holomorphic markings of the branched covering f, and p_U and p_L record the induced marked holomorphic structures (S^2, A) on the domain and range.

The Epstein's *deformation space* $\mathcal{D}_{f,\iota}$ is the subspace of $\mathcal{T}_f$ consisting of holomorphic structures on the covering and base of f that are equivalent relative to A. That is, $\mathcal{D}_{f,\iota}$ consists of the triples (ϕ, ψ, F) so that $\phi^{-1} \circ \psi$ is isotopic to the identity relative to A, or equivalently

$$f^*\phi = \iota^*\phi.$$

Another way to say this is that $\mathcal{D}_{f,\iota}$ is the equalizer in $\mathcal{T}_f$ of the maps p_U and p_L, that is

$$\mathcal{D}_{f,\iota} = \{(\phi, \psi, F) \in \mathcal{T}_f \mid p_U(\phi, \psi, F) = p_L(\phi, \psi, F)\}.$$

We think of $\mathcal{D}_{f,\iota}$ as the *dynamical Teichmüller space* for (f, ι).

2.2. Moduli spaces. Let $\mathcal{M}_{(S^2,A)}$ be the space of embeddings

$$A \hookrightarrow \mathbb{P}^1$$

up to post-composition by a Möbius transformation. Then the restriction map $[\phi] \mapsto [\phi|_A]$ defines a regular covering map

$$\mathcal{T}_{(S^2,A)} \to \mathcal{M}_{(S^2,A)}$$

with covering automorphism group equal to the *mapping class group* $\mathrm{Mod}(S^2, A)$ of orientation preserving homeomorphisms $h : S^2 \to S^2$ that fix the points of A up to isotopy rel A. (Unlike the case for Teichmüller spaces and moduli space of general surfaces, $\mathrm{Mod}(S^2, A)$ on $\mathcal{T}_{(S^2,A)}$ acts without fixed points.)

Similarly let $\mathcal{M}_f$ be the space of commutative diagrams

$$\begin{array}{ccc} A & \stackrel{j}{\hookrightarrow} & \mathbb{P}^1 \\ {\scriptstyle f|_A}\downarrow & & \downarrow{\scriptstyle F} \\ B & \stackrel{i}{\hookrightarrow} & \mathbb{P}^1. \end{array}$$

where (i, j, F) is defined up to modifications of the domain and range of F by Möbius transformations. Then the map $(\phi, \psi, F) \mapsto (\phi|_B, \psi|_A, F)$ defines a covering map

$$\mathcal{T}_f \to \mathcal{M}_f,$$

with covering automorphism group $L_f \subset \mathrm{Mod}(S^2, B)$ called the subgroup of *liftables* consisting of elements $h \in \mathrm{Mod}(S^2, B)$ such that for some $h' \in \mathrm{Mod}(S^2, A)$ the diagram

$$\begin{array}{ccc} (S^2, A) & \xrightarrow{h'} & (S^2, A) \\ {\scriptstyle f}\downarrow & & \downarrow{\scriptstyle f} \\ (S^2, B) & \xrightarrow{h} & (S^2, B) \end{array}$$

commutes.

REMARK 2.1. Since A and B are assumed to contain at least three points, $f : (S^2, A) \to (S^2, B)$ can have no non-trivial covering automorphisms. In the degree 2 case, this follows from the fact that f must have exactly two branch points and two branch values. Any other marked point would lie in the unbranched part of the covering. Even in the case when the degree of f is greater than 2, the fact that f may be realized as a rational map implies that all covering automorphisms must be conjugate to a Möbius transformation. If there are at least three points in A, then the Möbius transformation must be the identity.

By the definition of L_f and since f has no non-trivial covering automorphisms, f defines a unique lifting map

$$f^\sharp : L_f \to \mathrm{Mod}(S^2, A)$$

where $f^\sharp h = h'$.

The following Proposition was shown using slightly different language in [**13**] (cf. [**11**] Proposition 2.2). For the convenience of the reader, we include a proof below.

PROPOSITION 2.2. *The map $\mathcal{T}_f \to \mathcal{M}_f$ is a regular covering with automorphism group L_f.*

PROOF. Let $h \in L_f$. Take any $(\phi, \psi, F) \in \mathcal{T}_f$. Then we have the commutative diagram

$$
\begin{array}{ccccc}
(S^2, A) & \xleftarrow{f^\sharp h} & (S^2, A) & \xrightarrow{\psi} & (\mathbb{P}^1, \psi(A)) \\
\Big\downarrow f & & \Big\downarrow f & & \Big\downarrow F \\
(S^2, B) & \xleftarrow{h} & (S^2, B) & \xrightarrow{\phi} & (\mathbb{P}^1, \mathcal{B}).
\end{array} \tag{2.1}
$$

Since h and $f^\sharp h$ fix B and A, respectively, (ϕ_B, ψ_A, F) and $(\phi \circ h, \psi \circ f^\sharp h, F)$ map to the same element in $\mathcal{M}_f$.

Conversely, suppose (ϕ, ψ, F) and (ϕ_1, ψ_1, F_1) both map to equivalent elements in $\mathcal{M}_f$. Then by definition, there are embeddings $i : B \to \mathbb{P}^1$ and $j : A \to \mathbb{P}^1$ so that $i = \phi|_B = \phi_1|_B$, $j = \psi_1|_A = \psi|_A$ and there are Möbius transformations μ and ν such that $F = \mu^{-1} \circ F_1 \circ \nu$. We can assume for simplicity that $F = F_1$ by replacing (ϕ_1, ψ_1, F_1) by the equivalent element $(\mu^{-1} \circ \phi_1, \nu^{-1} \circ \psi_1, F)$ in $\mathcal{T}_f$. Then the common image of (ϕ, ψ, F) and (ϕ_1, ψ_1, F) is some (i, j, F) in $\mathcal{M}_f$. Thus the images of ϕ and ϕ_1 are the same in $\mathcal{M}_{(S^2,B)}$, the images of ψ and ψ_1 are the same in $\mathcal{M}_{(S^2,A)}$, and hence there are mapping classes $h \in \mathrm{Mod}(S^2, B)$ and $f^\sharp h \in \mathrm{Mod}(S^2, A)$ such that $\phi = h \circ \phi_1$ and $\psi = f^\sharp h \circ \psi_1$. The maps h and $f^\sharp h$ complete a diagram of the form (2.1), and thus $h \in L_f$. $\square$

Let $\overline{q} : \mathcal{M}_f \to \mathcal{M}_{(S^2,B)}$ be the map sending (i, j, F) to the equivalence class containing i in $\mathcal{M}_{(S^2,B)}$; let $\overline{p}_U : \mathcal{M}_f \to \mathcal{M}_{(S^2,A)}$ be the map sending (i, j, F) to the equivalence class containing j in $\mathcal{M}_{(S^2,A)}$; let $\overline{\iota}^* : \mathcal{M}_{(S^2,B)} \to \mathcal{M}_{(S^2,A)}$ be the forgetful map sending an inclusion $i : B \to \mathbb{P}^1$ to $i \circ \iota|_A$; and let $\overline{p}_L = \overline{\iota}^* \circ \overline{q}$. Then we have the commutative diagram

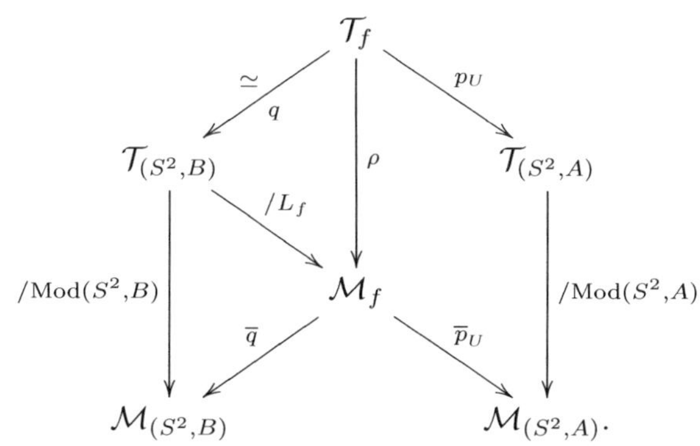

All vertical arrows and the left three diagonal arrows are unbranched covering maps. The right diagonal arrows may not be surjective (see [**3**]). Let $\mathcal{V}_{f,\iota}$ be the image of the deformation space $\mathcal{D}_{f,\iota}$ in $\mathcal{M}_f$. Then we have

$$\mathcal{V}_{f,\iota} = \mathrm{Eq}(\overline{p}_L, \overline{p}_U).$$

2.3. Stabilizer of deformation space. Fix a basepoint $d_0 \in \mathcal{D}_{f,\iota}$, and let $m_0 \in \mathcal{M}_{f,\iota}$ be the image of d_0 under the map ρ. Then we have an identification

$$\ell : \pi_1(\mathcal{M}_{f,\iota}, m_0) \to L_f,$$

defined by the path-lifting theorem for coverings. That is, for each $\gamma \in \pi_1(\mathcal{M}_{f,\iota}, m_0)$, we lift γ to a path γ' on $\mathcal{T}_{(S^2,B)}$ based at d_0, and let $\ell(\gamma)$ be the mapping class that takes d_0 to the end point of γ'.

PROPOSITION 2.3. *The restriction of $\rho : \mathcal{T}_f \to \mathcal{M}_f$ to $\mathcal{D}_{f,\iota}$ defines a covering map*

$$\rho_{\mathcal{D}} : \mathcal{D}_{f,\iota} \to \mathcal{V}_{f,\iota},$$

and the image E_{f,ι,d_0} of

$$\pi_1(\mathcal{V}_{f,\iota}, m_0) \to \pi_1(\mathcal{M}_{f,\iota}, m_0) \xrightarrow{\ell} L_f$$

is the stabilizer of the connected component of $\mathcal{D}_{f,\iota}$ that contains d_0.

PROOF. We show that the projection $\rho_{\mathcal{D}_{f,\iota}}$ satisfies the path-lifting theorem. Let $d_0 = (\phi_0, \psi_0, F_0)$, and let (i_t, j_t, F_t) be a path in $\mathcal{V}_{f,\iota}$ with $m_0 = (j_0, i_0, F_0)$. Let ξ and η be representatives of the class of ϕ_0 and ψ_0 so that the diagram commutes

$$\begin{array}{ccc} S^2 & \xrightarrow{\eta} & \mathbb{P}^1 \\ {\scriptstyle f}\downarrow & & \downarrow{\scriptstyle F} \\ S^2 & \xrightarrow{\xi} & \mathbb{P}^1, \end{array}$$

$\eta|_A = \psi_0|_A$ and $\xi|_B = \psi|_B$. Let $p_t : (S^2, B) \to (S^2, B_t)$ be any continuous family of homeomorphisms. Then

$$(\xi \circ p_t|_{B_t}, \eta \circ p_t|_{A_t}, F_t) \sim (i_t, j_t, F_t)$$

and $(\xi \circ p_t, \eta \circ p_t, F_t)$ is a lift of (i_t, j_t, F_t) and lies in $\mathcal{D}_{f,\iota}$.

Thus, as a restriction of an unbranched covering map, $\rho_{\mathcal{D}}$ is itself is a covering map. The rest follows from basic covering space theory. □

The lifting map $f^\sharp : L_f \to \mathrm{Mod}(S^2, A)$ is uniquely defined and satisfies the commutative diagram

$$\begin{array}{ccc} (S^2, A) & \xrightarrow{f^\sharp h} & (S^2, A) \\ {\scriptstyle f}\downarrow & & \downarrow{\scriptstyle f} \\ (S^2, B) & \xrightarrow{h} & (S^2, B). \end{array}$$

For a homeomorphism $h : (S^2, B) \to (S^2, B)$, let h_A be the element of $\mathrm{Mod}(S^2, A)$ defined by ignoring the points of $B \setminus A$, that is h_A is the isotopy class of h defined up to homeomorphisms isotopic to the identity rel A. Define

$$\begin{array}{rcl} \iota^\sharp : \mathrm{Mod}(S^2, B) & \to & \mathrm{Mod}(S^2, A) \\ h & \mapsto & h_A. \end{array}$$

Let $S_{f,\iota} \subset L_f$ be the equalizer

$$S_{f,\iota} = \mathrm{Eq}(f^\sharp, \iota^\sharp) \subset L_f.$$

By the identification of L_f with $\pi_1(\mathcal{M}_{f,\iota}, m_0)$, $S_{f,\iota}$ can equivalently be defined as the equalizer of the homomorphisms

$$(\overline{p}_L)_*, (\overline{p}_U)_* : \pi_1(\mathcal{M}_{f,\iota}, m_0) \to \pi_1(\mathcal{M}_{(S^2,A)}, a_0)$$

where $a_0 = \overline{p}_L(m_0) = \overline{p}_U(m_0)$, where $m_0 \in \mathcal{V}_{f,\iota} = \mathrm{Eq}(\overline{p}_L, \overline{p}_U)$.

PROPOSITION 2.4 ([**11**] Proposition 2.5). *The stabilizer in L_f of $\mathcal{D}_{f,\iota}$ equals $S_{f,\iota}$, and $S_{f,\iota}$ acts transitively on the fibers of the covering*

$$\mathcal{D}_{f,\iota} \to \mathcal{V}_{f,\iota}.$$

Thus the covering is regular and $S_{f,\iota}$ is the group of covering automorphisms.

COROLLARY 2.5. *The deformation space $\mathcal{D}_{f,\iota}$ is connected if and only if $\mathcal{V}_{f,\iota}$ is connected and $S_{f,\iota} = E_{f,\iota,d_0}$.*

In the case when (f,ι) is associated to an element of $\mathrm{Per}_4(0)^*$, E_{f,ι,d_0} has infinite index in $S_{f,\iota}$ ([**11**], Proposition 2.11).

2.4. Augmented spaces. By a *rational curve* $\mathcal{C}$ we mean a connected nodal curve with the following properties

(a) the irreducible components of $\mathcal{C}$ are isomorphic to $\mathbb{P}^1$, and
(b) the fundamental group of $\mathcal{C}$ is trivial.

A *pre-stable rational curve* $(\mathcal{C},\mathcal{A})$ is a rational curve $\mathcal{C}$ together with a finite set $\mathcal{A}$ contained in the complement of the nodes of $\mathcal{C}$. The set of nodes of $\mathcal{C}$ union the points of $\mathcal{A}$ form the *distinguished points* of $\mathcal{C}$. A *stable rational curve* is a pre-stable rational curve with the following additional property:

(c) the number of distinguished points on each irreducible component of $\mathcal{C}$ is greater than or equal to 3.

For each component C of a pre-stable rational curve $(\mathcal{C},\mathcal{A})$ there are three possibilities:

(1) C is stable, i.e., it contains at least 3 distinguished points;
(2) C is unstable, and contains two nodes; or
(3) C is unstable, and contains one node and zero or one point in $\mathcal{A}$.

Let $\Sigma^{\mathrm{pre}}_{(S^2,A)}$ be the set of pre-stable rational curves $(\mathcal{C},\mathcal{A})$ with a bijection $A \to \mathcal{A}$, and let $\Sigma_{(S^2,A)} \subset \Sigma^{\mathrm{pre}}_{(S^2,A)}$ be the space of stable rational curves. Define a map

$$\mathfrak{s} : \Sigma^{\mathrm{pre}}_{(S^2,A)} \to \Sigma_{(S^2,A)}$$

sending $(\mathcal{C},\mathcal{A})$ to the result of contracting components of $\mathcal{C}$ using the following rule: in case (1) leave C alone; and in case (2) and (3) contract C to a point. If C contains a point of $\mathcal{A}$, then mark the image of the contraction by that point. This map is well-defined since in case (3) there is at most one point of $\mathcal{A}$ in C. We call $\mathfrak{s}$ the *stabilization map.*

A *marking* of a pre-stable rational curve $(\mathcal{C},\mathcal{A}) \in \Sigma^{\mathrm{pre}}_{(S^2,A)}$ is a quotient map

$$\phi : (S^2, A) \to (\mathcal{C},\mathcal{A}),$$

such that ϕ is a homeomorphism when restricted to $S^2 \setminus \gamma$ for some multi-curve $\gamma \subset S^2 \setminus A$, ϕ restricts to a bijection $A \to \mathcal{A}$, and the components of γ are in one-to-one correspondence with the nodes of $\mathcal{C}$. The curve γ is called the *contracting curve* for ϕ. We consider two markings *equivalent* if they are the same up to post-composition by automorphisms of $\mathcal{C}$ and pre-composition by homeomorphisms $(S^2,A) \to (S^2,A)$ that are isotopic to the identity rel A.

The collection of markings of pre-stable and stable rational curves by (S^2,A) is denoted by $\mathcal{AT}^{\mathrm{pre}}_{(S^2,A)}$ and $\mathcal{AT}_{(S^2,A)}$, respectively. Post-composition by $\mathfrak{s}$ defines a surjection

$$\mathcal{AT}^{\mathrm{pre}}_{(S^2,A)} \to \mathcal{AT}_{(S^2,A)}.$$

The space $\mathcal{AT}_{(S^2,A)}$ is called the *augmented Teichmüller space* of (S^2, A). There is a natural topology on $\mathcal{AT}_{(S^2,A)}$ such that points on $\mathcal{AT}_{(S^2,A)} \setminus \mathcal{T}_{(S^2,A)}$ are the limits of sequences points on $\mathcal{T}_{(S^2,A)}$ for which the length of γ tends to zero (see [**2**] for more precise definitions).

REMARK 2.6. The mapping class group $\mathrm{Mod}(S^2, A)$ extends to actions on $\mathcal{AT}^{\mathrm{pre}}_{(S^2,A)}$ and $\mathcal{AT}_{(S^2,A)}$. Given a point in $\mathcal{AT}_{(S^2,A)}$, the stabilitzer is the free abelian group of mapping classes generated by Dehn twists along the components of the contracting curve.

The action of $\mathrm{Mod}(S^2, A)$ on $\mathcal{AT}_{(S^2,A)}$ defines a branched covering

$$\mathcal{AT}_{(S^2,A)} \to \mathcal{AM}_{(S^2,A)}$$

where $\mathcal{AM}_{(S^2,A)}$ is the space of inclusions

$$A \hookrightarrow \mathcal{C}$$

of A into a rational curve $\mathcal{C}$ up to holomorphic automorphism of $\mathcal{C}$ that do not permute components (cf. [**5**]).

We now define the augmented Teichmüller and moduli spaces of f. A rational map $F : (\mathcal{C}_U, \mathcal{A}) \to (\mathcal{C}_L, \mathcal{B})$ is *pre-admissible* if

(a) F defines a surjective map from $\mathcal{C}_U$ to $\mathcal{C}_L$ of generically constant degree that maps nodes to nodes;
(b) locally near each node of $\mathcal{C}_U$, F has generically constant degree; and
(c) $(\mathcal{C}_L, \mathcal{B})$ is a stable rational curve and $(\mathcal{C}_U, \mathcal{A})$ is pre-stable.

We say that F is *admissible* if in addition

(b) $(\mathcal{C}_U, \mathcal{A})$ is stable.

The *augmented Teichmüller space* $\mathcal{AT}_f$ for f is the collection of holomorphic markings

$$\begin{array}{ccc} (S^2, A) & \xrightarrow{\ \psi\ } & (\mathcal{C}_U, \mathcal{A}) \\ {\scriptstyle f}\downarrow & & \downarrow{\scriptstyle F} \\ (S^2, B) & \xrightarrow{\ \phi\ } & (\mathcal{C}_L, \mathcal{B}) \end{array}$$

where the horizontal maps are markings in $\mathcal{AT}_{(S^2,A)}$ and $\mathcal{AT}_{(S^2,B)}$ respectively, and F is a pre-admissible covering. Here, as in the definition of $\mathcal{T}_f$, we take (ϕ, ψ, F) up to the natural equivalences.

With this definition, the projection $\widetilde{q} : \mathcal{AT}_f \to \mathcal{AT}_{(S^2,B)}$ defines an isomorphism. Let $\widetilde{p}_U, \widetilde{p}_L : \mathcal{AT}_f \to \mathcal{AT}_{(S^2,A)}$ be defined by

$$\begin{aligned} \widetilde{p}_L(\phi, \psi, F) &= \mathfrak{s}(\phi) \\ \widetilde{p}_U(\phi, \psi, F) &= \mathfrak{s}(\psi) \end{aligned}$$

PROPOSITION 2.7. *The subgroup of liftables* $L_f \subset \mathrm{Mod}(S^2, B)$ *extends to an action on* $\mathcal{AT}_f$.

PROOF. Let $\phi : (S^2, B) \to (\mathcal{C}_L, \mathcal{B})$ be an element of $\mathcal{AT}_{(S^2,B)}$, and let γ be the contracting multi-curve. Since f is unbranched outside of B, it follows that $f^{-1}(\gamma)$ is a multi-curve on $S^2 \setminus A$, and since h is liftable

$$f^{\sharp}h(f^{-1}(\gamma)) = f^{-1}(h(\gamma)).$$

Thus $h \in L_f$ takes (ϕ, ψ, F) to $(\phi \circ h, \psi \circ f^\sharp h, F)$ where $\phi \circ h$ and $\psi \circ f^\sharp h$ contract the curves $h(\gamma)$ and $f^{-1}(h(\gamma))$ respectively. $\square$

Let $\mathcal{AM}_f = \mathcal{AT}_f / L_f$. Then the points of $\mathcal{AM}_f$ are defined by diagrams

$$\begin{array}{ccc} A & \xhookrightarrow{\ i\ } & \mathcal{C}_U \\ {\scriptstyle f|_A}\big\downarrow & & \big\downarrow{\scriptstyle F} \\ B & \xhookrightarrow{\ j\ } & \mathcal{C}_L \end{array}$$

where $(\mathcal{C}_L, \mathcal{B})$ is stable, $(\mathcal{C}_U, \mathcal{A})$ is pre-stable, and the inclusions i, j are defined up to holomorphic automorphisms of $\mathcal{C}_U$ and $\mathcal{C}_L$. We denote an element by (i, j, F). Let $\overline{p}_U, \overline{p}_L : \mathcal{AM}_f \to \mathcal{AM}_{(S^2,A)}$ be defined by

$$\begin{aligned} \overline{p}_L([\phi, \psi, F]) &= \mathfrak{s} \circ j \\ \overline{p}_U([\phi, \psi, F]) &= \mathfrak{s} \circ i \end{aligned}$$

While the action of L_f on $\mathcal{T}_f$ has no fixed points and the quotient map $\mathcal{T}_f \to \mathcal{M}_f$ is a covering, the quotient map $\mathcal{AT}_f \to \mathcal{AM}_f$ can have branch points.

Summarizing, we have a commutative diagram of augmented spaces:

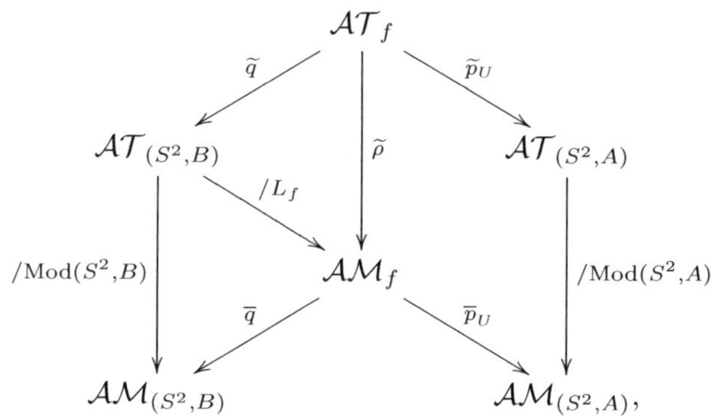

The augmented deformation space is defined to be the equalizer

$$\mathcal{AD}_{f,\iota} = \mathrm{Eq}(\widetilde{p}_U, \widetilde{p}_L)$$

and contains the deformation space $\mathcal{D}_{f,\iota}$. Let $\mathcal{AV}_{f,\iota} = \widetilde{p}(\mathcal{AD}_{f,\iota})$. Then we have

$$\mathcal{AV}_{f,\iota} = \mathrm{Eq}(\overline{p}_U, \overline{p}_L),$$

2.5. Stabilizer of augmented deformation space. In this section, we give a necessary and sufficient condition for $\mathcal{AD}_f$ to be connected. Unlike in the case for Corollary 2.5, $\mathcal{AD}_{f,\iota} \to \mathcal{AV}_{f,\iota}$ is not an unbranched covering. To get around this we use the notion of regular neighborhoods. Recall that, for a simplicial subcomplex V embedded in a manifold X, a *regular neighborhood* $N(V)$ of V is an open subset of X containing V that has a deformation retract to V.

PROPOSITION 2.8. *The stabilizer in L_f of $\mathcal{AD}_{f,\iota}$ equals $S_{f,\iota}$, that is, if $g \in L_f$ is such that $g(\alpha) = \alpha$ for all $\alpha \in \mathcal{AD}_{f,\iota}$, then $g \in S_{f,\iota}$.*

PROOF. The stabilizer in L_f of $\mathcal{AD}_{f,\iota}$ must be contained in $S_{f,\iota}$, since $\mathcal{D}_{f,\iota} \subset \mathcal{AD}_{f,\iota}$. Let $h \in S_{f,\iota}$ and let $(\phi, \psi, F) \in \mathcal{AD}_{f,\iota}$. Then by definition $f^\sharp h = i^\sharp h$. Thus, we have a commutative diagram

$$\begin{array}{ccccc} (S^2, A) & \xrightarrow{f^\sharp h} & (S^2, A) & \xrightarrow{\psi} & (\mathcal{C}_U, \mathcal{A}) \\ {\scriptstyle f}\downarrow & & \downarrow{\scriptstyle f} & & \downarrow{\scriptstyle F} \\ (S^2, B) & \xrightarrow{h} & (S^2, B) & \xrightarrow{\phi} & (\mathcal{C}_L, \mathcal{B}) \end{array}$$

and we have

$$\widetilde{f}^*(h([\phi])) = [\psi \circ f^\sharp h] = [\psi \circ \iota^\sharp h] = \widetilde{\iota}^*(h([\phi])).$$

Thus h stabilizes $\mathcal{AD}_{f,\iota}$. □

REMARK 2.9. Conversely, one can ask whether if $g \in L_f$ and $g(\alpha) = \alpha$ for some $\alpha \in \mathcal{AD}_{f,\iota}$, then does it follow that $g \in S_{f,\iota}$? This is not true in general, since points in the boundary of $\mathcal{AD}_{f,\iota}$ have extra automorphisms that need not be in $S_{f,\iota}$.

PROPOSITION 2.10. *Suppose there is a connected quasi-projective variety X with $\mathcal{V}_{f,\iota} \subset X \subset \mathcal{AV}_{f,\iota}$ such that X has a regular neighborhood $N(X) \subset \mathcal{AM}_f$ with the properties*

(1) *$N(X) \cap \mathcal{M}_f$ is connected, and*
(2) *the image of the homomorphism induced by inclusion*

$$\pi_1(N(X) \cap \mathcal{M}_f, m_0) \to \pi_1(\mathcal{M}_f, m_0)$$

contains $S_{f,\iota}$.

Then $\mathcal{D}_{f,\iota}$ is contained in a connected component of $\mathcal{AD}_{f,\iota}$.

PROOF. Let $Y_0 \subset \mathcal{AT}_f$ be the connected component of the preimage of X in $\mathcal{AT}_f$ containing d_0. Let U be the connected component of the preimage of $N(X)$ in $\mathcal{AT}_f$ containing d_0. Then since $N(X)$ has a retract to X, the set U has a corresponding retraction to a component of $\widetilde{\rho}^{-1}(X)$. This component is necessarily Y_0, since $Y_0 \cap U \neq \emptyset$. Since the image of $\pi_1(N(X) \cap \mathcal{M}_f, m_0)$ in $L_f = \pi_1(\mathcal{M}_f, m_0)$ contains $S_{f,\iota}$, it follows that the action of $S_{f,\iota}$ on $\mathcal{AT}_f$ preserves U and hence Y_0. Thus, we have

$$\mathcal{D}_{f,\iota} \subset Y_0 \subset \mathcal{AD}_{f,\iota}.$$

□

3. Blowups and topology of curve complements

In this section we study the topology of surface/curve pairs and the effect of blowups (see for example [**8**] or [**9**] for a review of elementary blowup theory for surfaces).

3.1. Regular neighborhoods of algebraic curves on surfaces. Let X be a smooth complex projective surface, and let $V \subset X$ be a pure codimensional one subvariety. We first observe that we may assume that V is a finite union of smooth curves with normal crossings. Let Q be the set of singular points on V. Then $V \setminus Q$ is a finite union of smoothly embedded punctured Riemann surfaces in $X \setminus Q$. By successively blowing up X at the points of Q (and at points of the preimages of Q), it is possible to obtain a new projective surface $\widetilde{X}$ and a surjective morphism $\sigma : \widetilde{X} \to X$ such that the preimage (or *total transform*) $\widetilde{V} = \sigma^{-1}(V)$ is a union of smoooth curves with normal crossings. That is $(\widetilde{X}, \widetilde{V})$ is locally isomorphic near a point of intersection on $\widetilde{V}$ to $(\mathbb{C}^2, \{x=0\} \cup \{y=0\})$.

Hereafter in this section we assume that V has smooth components intersecting in normal crossing. As before, let Q be the set of intersections of V, and let $V_1, \dots, V_k$ be the irreducible components of V. Since each V_i is smooth there is an embedded tubular neighborhood $T(V_i) \subset X$ so that for $i \neq j$, $T(V_i)$ only intersects $T(V_j)$ near points in Q where V_i and V_j intersect, and if V_i and V_j intersect at q, then

$$N(q) = T(V_i) \cap T(V_j)$$

is a neighborhood of q so that $(N(q), V, q)$ is homeomorphic to

$$(\{|x| < 1\} \times \{|y| < 1\}, \{x = 0\} \cup \{y = 0\}, (0, 0)).$$

For $i = 1, \dots, k$, let

$$V_i^c = V_i \setminus \bigcup_{q \in Q} N(q)$$

and let $T(V_i^c)$ be the tubular neighborhood given by

$$T(V_i^c) = T(V_i) \setminus \bigcup_{q \in Q} N(q).$$

Let $T(V) = \bigcup_{i=1}^k T(V_i^c)$, called the *tubular neighborhood* of V.

Let $S(V_i^c)$ be the circle bundle over V_i^c contained in the boundary of $T(V_i^c)$. Then $S(V_i^c)$ is an oriented 3-manifold with torus boundary components corresponding to the intersections of V_i with other components of V. In particular, if V_i is isomorphic to $\mathbb{P}^1$, then $S(V_i^c)$ is the complement of thickened Hopf links in the 3-sphere. The *boundary manifold* of V is given by

$$S(V) = \bigcup_{i=1}^{k} S(V_i^c).$$

This manifold and its embedding in $X \setminus V$ is uniquely determined up to homeomorphisms of X that are isotopic to the identity rel V.

LEMMA 3.1. *The punctured tubular neighborhood $T(V) \setminus V$ has a deformation retraction to $S(V)$.*

PROOF. For each i, $T(V_i^c) \setminus V_i^c$ has a deformation retract to $S(V_i^c)$ corresponding from the retraction of a punctured disk to its boundary circle. Thus, we have only to consider what happens near the intersection points $q \in Q$. In $N(q)$ it is enough to show that

$$\{|x| < 1\} \times \{|y| < 1\} \setminus \{x = 0\} \cup \{y = 0\}$$

has a deformation retract to $\{|x| = 1\} \times \{|y| = 1\}$. Such a retraction is defined by the map

$$((x, y), t) \mapsto \left(\frac{x}{1 + t(|x| - 1)}, \frac{y}{1 + t(|y| - 1)} \right).$$

□

REMARK 3.2. By its construction, $S(V)$ is naturally homeomorphic to a graph manifold (see, for example, [**10**] for definitions) over the incidence graph Γ of the components of V; this is the bipartite graph with vertices

$$\{v_i \mid i = 1, \dots k\}$$

corresponding to the components of V, and edges between v_i and v_j for each $q \in Q$ such that V_i and V_j intersect at q. To each vertex v_i associate the manifold $S(V_i^c)$ and to each edge of Γ between v_i and v_j associate the common torus boundary component of their associated vertex manifolds.

3.2. Regular neighborhoods and fundamental groups. In this section we prove an easy variation of the Lefschetz hyperplane theorem [**1**] and a useful corollary.

LEMMA 3.3. *Let $\mathfrak{p} : X \to \mathbb{P}^1$ be a smooth projective surface fibered over the complex projective line, and let V be a fiber. Let $C \subset X$ be a pure codimension one algebraic subset none of whose components are fibers of $\mathfrak{p}$. Then V has a regular neighborhood $N(V)$ so that*

$$\pi_1(N(V) \setminus C) \to (X \setminus C)$$

induced by inclusion is surjective.

PROOF. Let $P \subset \mathbb{P}^1$ be the set of points p where $\mathfrak{p}$ restricted to C drops in cardinality, i.e., at least one intersection point of C and the fiber above p has higher multiplicity. Let $\mathfrak{p}^o$ be the restriction of $\mathfrak{p}$ to $X^o = \mathfrak{p}^{-1}(\mathbb{P}^1 \setminus P) \setminus C$. Then

$$\mathfrak{p}^o : X^o \to \mathbb{P}^1 \setminus P$$

is a fiber bundle, and the Zariski-Van Kampen theorem [**18**] [**4**] implies that for a general fiber V' of $\mathfrak{p}^o$

$$\pi_1(X \setminus C) \simeq \pi_1(V')/K$$

where K is the subgroup of $\pi_1(V')$ generated by the relations $\gamma^{-1}\beta(\gamma)$, where β ranges over automorphisms of $\pi_1(V')$ determined by the action of $\pi_1(\mathbb{P}^1 \setminus P)$ on V'. In particular, the homomorphism

$$\pi_1(V') \to \pi_1(X \setminus C)$$

is surjective. Let $N(V) = (\mathfrak{p}^o)^{-1}(U)$ where U is a small neighborhood of $\mathfrak{p}(V)$. Then $N(\widetilde{V}) \setminus C$ contains a general fiber of $\mathfrak{p}^o$, and the claim follows. $\square$

COROLLARY 3.4. *Let $C \subset \mathbb{C}^2$ be an algebraic curve, and let $V \subset \mathbb{C}^2$ be a line not contained in C. Then we can include $\mathbb{C}^2$ as a Zariski open subset of a smooth projective surface X, and find a sequence of blowups $\sigma : \widetilde{X} \to X$ such that*

(1) *σ is an isomorphism over $\mathbb{C}^2 \setminus C$; and*
(2) *the total transform $\widetilde{V}$ of the closure of V in X has a regular neighborhood $N(\widetilde{V})$ in $\widetilde{X}$ such that the map on fundamental groups induced by inclusion*

$$\pi_1(N(\widetilde{V}) \cap \mathbb{C}^2 \setminus C) \to \pi_1(\mathbb{C}^2 \setminus C)$$

is surjective.

PROOF. Let $\mathfrak{p} : \mathbb{C}^2 \to \mathbb{C}$ be a linear projection so that V is a fiber. Then we can define completions $\mathbb{C}^2 \subset \mathbb{P}^1 \times \mathbb{P}^1$ and $\mathbb{C} \subset \mathbb{P}^1$, so that $\mathfrak{p}$ extends to a projection

$$\overline{\mathfrak{p}} : \mathbb{P}^1 \times \mathbb{P}^1 \to \mathbb{P}^1$$

so that the closure $\overline{V}$ of V in $\mathbb{P}^1 \times \mathbb{P}^1$ is a fiber. Let $\overline{C}$ be the union of the closure of C in $\mathbb{P}^1 \times \mathbb{P}^1$, and the two lines in $\mathbb{P}^1 \times \mathbb{P}^1 \setminus \mathbb{C}^2$.

Let $\sigma : \widetilde{\mathbb{P}^1 \times \mathbb{P}^1} \to \mathbb{P}^1$ be sequence of blowups over points of $\overline{C} \cap \overline{V}$ such that the total transform $\widetilde{V} = \sigma^{-1}(\overline{V})$ and the proper transform $\widehat{C}$ over $\overline{C}$ meet in normal crossings. Let $N(\widetilde{V})$ be a regular neighborhood of $\widetilde{V}$ as in Lemma 3.3. Then the homomorphism

$$\pi_1(N(\widetilde{V}) \setminus \widehat{C}) \to \pi_1(\widetilde{\mathbb{P}^1 \times \mathbb{P}^1} \setminus \widehat{C})$$

induced by inclusion is surjective.

Since σ is an isomorphism outside the preimage of $\overline{C} \cap \overline{V}$, we have a commutative diagram

$$\begin{array}{ccc} \widetilde{V} & \longrightarrow & N(\widetilde{V}) \\ \downarrow & & \downarrow \\ \overline{V} & \longrightarrow & N(\overline{V}) \end{array}$$

where the vertical arrows are quotient maps that contract the exceptional curves (or 1-dimensional fibers) over points on $\overline{V} \cap \overline{C}$. It follows that the retraction of $N(\widetilde{V}) \setminus \widehat{C}$ to $\widetilde{V} \setminus \widehat{C}$ descends to a retraction of $N(\overline{V}) \setminus C$ to $\overline{V} \setminus C$, and hence $N(\overline{V}) \setminus \overline{C}$ is a regular neighborhood of $\overline{V} \setminus \overline{C}$. Finally, $P^1 \times \mathbb{P}^1 \setminus \overline{C} = \mathbb{C}^2 \setminus C$, so setting $N(V) = N(\overline{V}) \cap \mathbb{C}^2$, it follows that $N(V)$ is a regular neighborhood of V and the homomorphism

$$\pi_1(N(V) \setminus C) \to \pi_1(\mathbb{C}^2 \setminus C)$$

defined by inclusion is surjective. $\square$

Corollary 3.4 is used in our discussion of connectivity of augmented deformation space in Section 4.4.

REMARK 3.5. The boundary manifold $S(\widetilde{V})$ associated to $N(\widetilde{V})$ in the previous proof has the structure of a boundary manifold over the incidence graph of the irreducible components of $\widetilde{V}$ (cf. Remark 3.2). Furthermore, each component of $\widetilde{V}$ is isomorphic to a line, each vertex manifold is an S^1 fiber bundle over S^2 with a finite set of thickened fibers removed.

4. Application to the Main Example

Let $F \in \mathrm{Per}_4(0)^*$, and let $f, \iota : (S^2, A) \to (S^2, B)$ be the underlying branched covering and identification of domain and range so that A is the periodic 4 cycle, and $B = A \cup \{v\}$ where v is the extra critical point. We study the inclusion $\mathcal{D}_{f,\iota} \subset \mathcal{AD}_{f,\iota}$ by looking at their images $\mathcal{V}_{f,\iota}$ and $\mathcal{AV}_{f,\iota}$ in $\mathcal{M}_f$ and $\mathcal{AM}_f$.

4.1. Parameterization of moduli space. We begin by embedding $\mathcal{M}_f$ in $\mathbb{P}^1 \times \mathbb{P}^1$ as follows. Consider an element of $\mathcal{M}_f$ represented as a commutative diagram

$$\begin{array}{ccc} A & \overset{i}{\hookrightarrow} & \mathbb{P}^1 \\ {\scriptstyle f|_A}\downarrow & & \downarrow{\scriptstyle F} \\ B & \underset{j}{\hookrightarrow} & \mathbb{P}^1. \end{array}$$

By applying automorphisms of $\mathbb{P}^1$ on the right side, we can assume that

$$i(B) = \{0, 1, \infty, y, z\} \quad j(A) = \{0, 1, \infty, x\},$$

where

(i) ∞ and z are the critical values of f;
(ii) 0 is a critical point in A with $f(0) = \infty$; and

(iii) $f(\infty) = 1$, $f(x) = 0$.

The above data completely determines $F : \mathbb{P}^1 \times \mathbb{P}^1$ as a rational function in the variable t:

$$F(t) = \frac{(t-x)(t-r)}{t^2} \qquad r = \frac{x+y-1}{x-1}.$$

It follows that in this example z is determined by x and y:

$$z = -\frac{(1-2x+x^2-y)^2}{4x(x-1)(x+y-1)}. \tag{4.1}$$

We have the following (see also, [**11**]).

LEMMA 4.1. *There is an identification*

$$\mathcal{M}_f = \mathbb{P}^1 \times \mathbb{P}^1 \setminus \mathcal{L} \cup \mathcal{Z},$$

assigning (i, j, F) *to* (x, y)*, where*

$$\mathcal{L} = \{x = 0\} \cup \{y = 0\} \cup \{x = 1\} \cup \{y = 1\} \cup \{x = \infty\} \cup \{y = \infty\}$$

and $\mathcal{Z}$ *is the closure in* $\mathbb{P}^1 \times \mathbb{P}^1$ *of the affine union of curves*

$$\{1 - 2x + x^2 - y = 0\} \cup \{x^2 + y = 1\} \cup \{x + y = 1\} \cup \{2xy + x^2 - y - 2x + 1 = 0\}$$

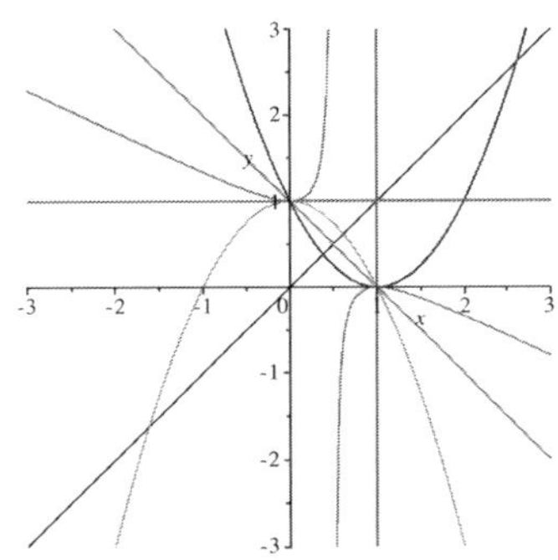

FIGURE 1. Picture of $\mathcal{L} \cup \mathcal{Z}$ in the affine plane. The lines in $\mathcal{L}$ are shown here as the two horizontal lines $y = 0$ and $y = 1$, and two vertical lines $x = 0$ and $x = 1$.

By this parameterization, the image $\mathcal{V}$ of $\mathcal{D}_{f,\iota}$ in $\mathcal{M}_f$ equals the diagonal

$$\mathcal{V} = \{(x, y) \in \mathbb{P}^1 \times \mathbb{P}^1 \setminus \mathcal{L} \cup \mathcal{Z} \mid x = y\}.$$

Figure 1 gives a picture[2] of the real part of $\mathcal{L} \cup \mathcal{Z}$ in the affine open subset $\mathbb{C}^2 \subset \mathbb{P}^1 \times \mathbb{P}^1$, and Figure 2 gives a picture near (∞, ∞).

[2]This figure was provided courtesy of Sarah Koch.

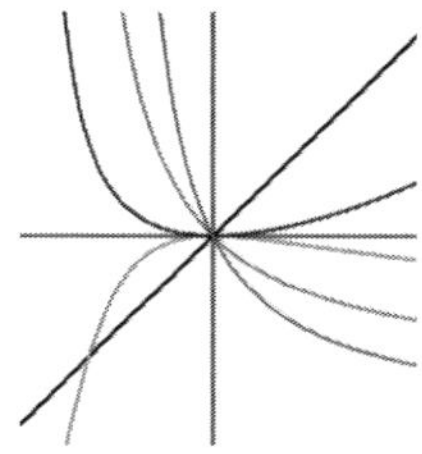

FIGURE 2. Picture of $\mathcal{L} \cup \mathcal{Z}$ in $\mathbb{P}^1 \times \mathbb{P}^1$ near (∞, ∞). The lines $x = \infty$ and $y = \infty$ are the horizontal and vertical lines, respectively, and $\mathcal{V}$ is the diagonal line $x = y$.

4.2. The quotient of augmented deformation space. Let $\mathcal{AM}_f$ be the quotient of $\mathcal{AT}_f$ by the action of L_f, and let $\mathcal{AV}_{f,\iota}$ be the image of $\mathcal{AD}_{f,\iota}$ in $\mathcal{AM}_f$. Our goal in this section is to concretely describe a subspace $X \subset \mathcal{AV}_{f,\iota}$ satisfying the properties in Proposition 2.10.

First we recall that the elements of $\mathcal{AV}_{f,\iota}$ are the elements of $\mathcal{AM}_f$ that equalize the two maps

$$p_L, p_U : \mathcal{AM}_f \to \mathcal{AM}_{(S^2,A)}.$$

Each stratum of $\mathcal{AM}_f$ is described by a partition of $\{0, 1, \infty, y, z\}$ into two or three sets by an admissible multi-curve γ, as in Figure 3. Here, the empty multi-curve corresponds to the principal stratum $\mathcal{M}_f \subset \mathcal{AM}_f$. The corresponding stable curves

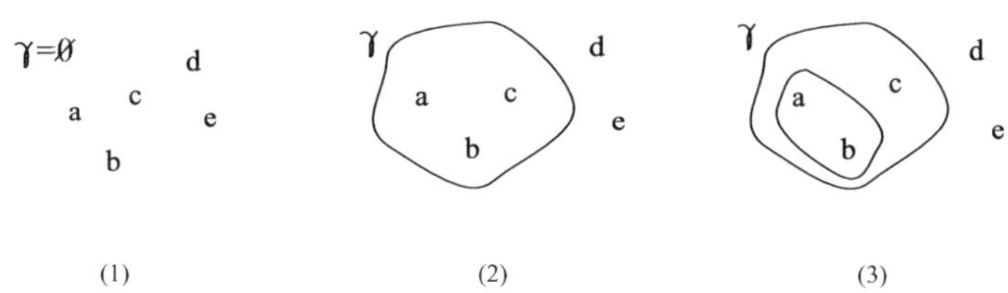

FIGURE 3. Possible partitions of five points by an admissible multi-curve.

are shown in Figure 4, with each $\mathbb{P}^1$, homeomorphic to S^2, is drawn as a line.

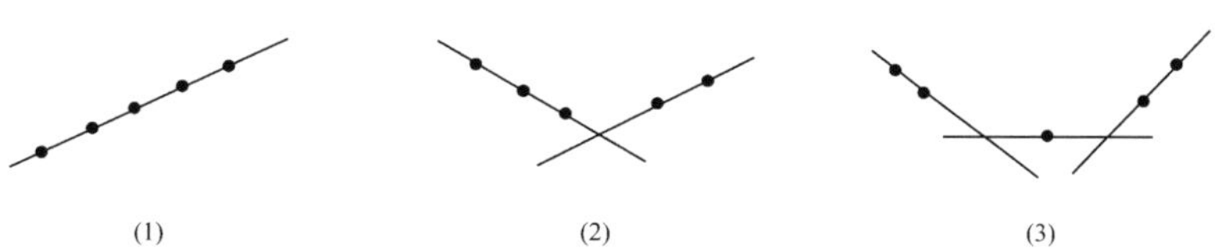

FIGURE 4. The three types of stable curves for (S^2, B) where B has five elements. The two left define one-dimensional strata, and the right defines a point stratum.

For $\mathcal{AM}_{(S^2,A)}$ there are only two isomorphism types (shown in Figure 5). The left picture depicts points belonging to the main component $\mathcal{M}_{(S^2,A)} \subset \mathcal{AM}_{(S^2,A)}$, which is isomorphic to a thrice punctured sphere, while the right picture depicts one of the three single point boundary points.

(I) (II)

FIGURE 5. The two topological homeomorphism types of stable curves for (S^2, A) where A has four elements.

The elements $\alpha \in \mathcal{AM}_f$ that lie in positive dimensional strata must be of the form (1) or (2) in Figure 4. Those of type (1) lie in $\mathcal{M}_f = \mathbb{P}^1 \times \mathbb{P}^1 \setminus \mathcal{L} \cup \mathcal{Z}$ and map under both p_U and p_L to elements of $\mathcal{AM}_{(S^2,A)}$ of type (I). Those of type (2) divide into four subtypes: those that correspond to a partition of the form

(2a) $\{a, b, z\} \cup \{c, \infty\}$;
(2b) $\{a, \infty, z\} \cup \{b, c\}$;
(2c) $\{a, b, c\} \cup \{\infty, z\}$; or
(2d) $\{a, b, \infty\} \cup \{c, z\}$.

For types (2a) and (2b), p_L maps α to an element in $\mathcal{AM}_{(S^2,A)}$ of type (II), while for types (2c) and (2d), p_L maps α to one of type (I). For types (2a) and (2d) p_U maps α to an element of type (II), while for type (2b) and (2c) α could apriori map to an element of type (I) or (II). This is because the critical values ∞ and z lie in the same component. Thus the rational map $F : \mathcal{C}_U \to \mathcal{C}_L$ must have two isomorphic irreducible components in $\mathcal{C}_U$ lying over the unramified component in $\mathcal{C}_L$ upon which the distinguished points will be distributed.

From this we can reduce the types that can be in $\mathcal{AV}$ to (2a), (2d), as well as (2b) and (2c) under the condition that the preimage under F of the distinguished points in the unramified component lie on the same component of $\mathcal{C}_U$. In the allowable case of types (2a) and (2b), we see that for $p_L(\alpha)$ and $p_U(\alpha)$ to be equal the images of the two maps in $\mathcal{AM}_{(S^2,A)}$ must give the partition

$$\{0, 1\} \cup \{x, \infty\}.$$

Thus, the partition given in (2a) can only be

$$\{0, 1, z\} \cup \{y, \infty\}$$

and for (2b) it can only be

$$\{\infty, y, z\} \cup \{0, 1\}.$$

For type (2c), $p_L(\alpha)$ and $p_U(\alpha)$ must be equal, and F defines the isomorphism on stable curves. Taking into account the combinatorics of f, this implies equality of the cross ratios: $(0, \infty; 1, x)$ and $(\infty, 1; x, 0)$, which is false under the assumption that $x \notin \{0, 1, \infty\}$.

Let $\mathcal{A}_1, \mathcal{A}_2 \subset \mathcal{AM}_f$ be the subsets corresponding to the partitions

$$\mathcal{A}_1 : \{\infty, y, z\} \cup \{0, 1\},$$

and

$$\mathcal{A}_2 : \{0, 1, z\} \cup \{y, \infty\}.$$

Then each of these is isomorphic to $\mathrm{Mod}_{0,4} \times \mathrm{Mod}_{(0,3)}$. Furthermore, the closures of $\mathcal{A}_1$ and $\mathcal{A}_2$ in $\mathcal{AM}_f$ intersect at the point corresponding to the partition

$$\{0, 1\} \cup \{z\} \cup \{y, \infty\}.$$

We have shown the following.

PROPOSITION 4.2. *The pure 1-dimensional algebraic set* $\mathcal{V} \cup \mathcal{A}_1 \cup \mathcal{A}_2 \subset \mathcal{AM}_f$ *is contained in* $\mathcal{AV}_{f,\iota}$*, and its complement is a finite set of points (possibly empty).*

REMARK 4.3. We leave the question of whether $\mathcal{AV}_{f,\iota}$ is connected (in this case, and in general) to future study.

4.3. Blowups. In this section, we find a connected pure 1-dimensional algebraic subset $X \subset \mathcal{AV}$ that contains $\mathcal{V}$, and whose complement in $\mathcal{AV}$ is finite.

Let

$$\sigma : \widetilde{\mathbb{P}^1 \times \mathbb{P}^1} \to \mathbb{P}^1 \times \mathbb{P}^1$$

be sequence of blowups defined as follows. First blowup the points $(0,0)$, $(1,1)$ and (∞,∞) to get the exceptional curves E_0, E_1, E_∞. Next blowup the point of intersection $q \in E_\infty \cap \widehat{L}_y$, where $\widehat{L}_y$ is the proper transform of $\{y = \infty\}$. Let E_q be the exceptional divisor. The union of curves is drawn in Figure 6 (compare Figure 1).

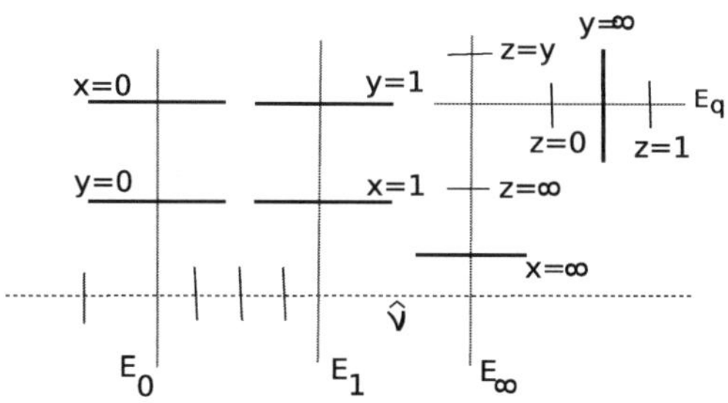

FIGURE 6. The proper transform of $\mathcal{V}$ is drawn as a dotted line.

LEMMA 4.4. *The map* σ *has the following properties:*

(1) σ *restricts to an isomorphism on* $\widetilde{\mathbb{P}^1 \times \mathbb{P}^1} \setminus \sigma^{-1}(Q)$*; and*
(2) *the total transform* $\widetilde{\mathcal{V}}$ *and the proper transform* $\widehat{\mathcal{L} \cup \mathcal{Z}}$ *meet in normal crossing singularities.*
(3) *inclusion induces a surjection on fundamental groups*

$$\pi_1(\mathcal{N}(\widetilde{\mathcal{V}}) \cap \mathcal{M}_f) \to \pi_1(\mathcal{M}_f).$$

PROOF. Properties (1) and (2) follow from the definitions, and property (3) follows from Lemma 3.3. □

Let

$$X = (\widehat{\mathcal{V}} \cup E_\infty \cup E_q) \setminus \widehat{\mathcal{L} \cup \mathcal{Z}}.$$

Then X is connected since the punctures of $\widetilde{\mathcal{V}}$ at intersections with $\widehat{\mathcal{L} \cup \mathcal{Z}}$ occur only in smooth points of $\widetilde{\mathcal{V}}$. Since σ is an isomorphism over $\mathcal{V}$, there is a natural inclusion $\nu : \mathcal{V} \hookrightarrow X$.

LEMMA 4.5. *The inclusion* $\mathcal{V} \hookrightarrow \mathcal{AV}$ *induced by* $\mathcal{D}_{f,\iota} \hookrightarrow \mathcal{AD}_{f,\iota}$ *factors as* $\xi \circ \nu$ *for some embedding*

$$\xi : X \hookrightarrow \mathcal{AV}.$$

PROOF. Recall from Proposition 4.2 the subset $\mathcal{V} \cup \mathcal{A}_1 \cup \mathcal{A}_2 \subset \mathcal{AV}$. In what follows we define embeddings

$$\begin{aligned} \kappa_\infty : E_\infty \setminus \widehat{\mathcal{L} \cup \mathcal{Z}} &\to \mathcal{A}_1 \\ \kappa_q : E_q \setminus \widehat{\mathcal{L} \cup \mathcal{Z}} &\to \mathcal{A}_2 \end{aligned}$$

that together with the projection $\widehat{V} \to \mathcal{V}$ extend to define ξ.

When $\alpha \in \mathcal{M}_f$ approaches a general point of $\mathcal{L} \cup \mathcal{Z}$, it corresponds to two points in $\{0, 1, \infty, y, z\}$ coming together. For the lines $(y = 0,\ y = 1$ and $y = \infty$, the pairs $\{y, 0\}$, $\{y, 1\}$ and $\{y, \infty\}$ approach each other, while simultaneously $\{x, 1\}$, $\{1, \infty\}$ and $\{1, 0\}$ approach each other. Near the lines $x = 0,\ x = 1,\ x = \infty$ the pairs $\{x, 0\})$ (and $\{0, \infty\}$), $\{x, 1\}$ (and $\{0, y\}$) and $\{x, \infty\}$ (and $\{0, 1\}$) approach each other. For z approaching $0, 1, \infty$ or y, we have

$$\begin{array}{rll} \{z, 0\} & \text{near} & 1 - 2x + x^2 - y = 0; x \neq 0, 1; x + y \neq 1 \\ \{z, 1\} & \text{near} & x^2 + y = 1; x \neq 0; x + y \neq 1 \\ \{z, \infty\} & \text{near} & x = 0, x = 1, \text{or } x + y = 1 \\ \{z, y\} & \text{near} & 2xy + x^2 - y - 2x + 1 = 0 \end{array}$$

We choose local coordinates for a neighborhood of E_∞ as follows. Let $\overline{x} = 1/x$ and $\overline{y} = 1/y$ be coordinates for an open neighborhood of $p_\infty = (\infty, \infty)$ in $\mathbb{P}^1 \times \mathbb{P}^1$, so that in the coordinates $(\overline{x}, \overline{y})$ we have $p_\infty = (0, 0)$. Let u (and $\overline{u} = \frac{1}{u}$) be coordinates for E_∞. Then a neighborhood of E_∞ in $\widetilde{\mathbb{P}^1 \times \mathbb{P}^1}$ is isomorphic to the algebraic subset of $\mathbb{C}^2 \times \mathbb{P}^1$ defined by

$$\{(\overline{x}, \overline{y}) \times u \in \mathbb{C}^2 \times \mathbb{P}^1 \ |\overline{x} = u\overline{y}\},$$

and

$$E_\infty = \{(\overline{x}, \overline{y}) \times u \in \mathbb{C}^2 \times \mathbb{P}^1 \mid \overline{x} = \overline{y} = 0\}.$$

With respect to the $\overline{x}, \overline{y}$ coordinates, we have

$$\begin{aligned} z &= -\frac{\overline{x}^4\overline{y}^2(1 - 2x + x^2 - y)^2}{4\overline{x}^4\overline{y}^2 x(x - 1)(x + y - 1)} \\ &= -\frac{(\overline{x}^2\overline{y} - 2\overline{x}\overline{y} + \overline{y} - \overline{x}^2)^2}{4\overline{x}\overline{y}(1 - \overline{x})(\overline{y} + \overline{x} - \overline{x}\overline{y})} \end{aligned}$$

Using the identity $\overline{x} = u\overline{y}$, we have

$$z = -\frac{(u^2\overline{y}^2 - 2u\overline{y} + 1 - u^2\overline{y})^2}{4u\overline{y}(1 - u\overline{y})(1 + u(1 - \overline{y}))}.$$

To each $(u, \overline{y})$, associate the point in $\mathcal{AM}_{(S^2,B)}$ defined by

$$([u, \infty; y, z], *)$$

where the first component is the cross ratio of the points u, ∞, y, z and the second component is the unique triple of points in $\mathbb{P}^1$ up to automorphism. This defines a point in $\mathcal{AM}_{(S^2,B)}$.

Since cross ratio is preserved under automorphisms of $\mathbb{P}^1$ we have

$$[u, \infty : y, z] = [u\overline{y}, \infty, 1, z\overline{y}].$$

As $\overline{y}$ approaches 0, the cross ratio approaches

$$[0,\infty;1,-\frac{1}{4u(1+u)}]$$

and is degenerate only when $u=0$ $(x=\infty)$, $u=-1$ $(z=\infty)$, $u=-\frac{1}{2}$ $(z=y)$, or $u=\infty$ $(y=\infty)$.

We thus have a well-defined embedding:

$$\begin{array}{rcl}\kappa_\infty : E_\infty \setminus \{\widehat{\mathcal{L}\cup\mathcal{Z}}\} & \to & \mathcal{A}_1 \\ u & \mapsto & ([u,\infty;y,z],*).\end{array}$$

Here u corresponds to a line through (∞,∞) in $\mathbb{P}^1\times\mathbb{P}^1$, and we can think of the contracting curve γ as a small loop on this line around (∞,∞).

The intersection of E_∞ with E_q occurs at the point on E_∞ corresponding to $(\overline{y},\overline{u})=(0,0)$, and E_q has a neighborhood parameterized by $((\overline{y},\overline{u}),v)\in\mathbb{C}^2\times\mathbb{P}^1$ where $\overline{y}=v\overline{u}$. Then $\overline{x}=u\overline{y}=v$ and

$$\begin{array}{rcl} z & = & -\dfrac{(1-2x+x^2-y)^2}{4x(x-1)(x+y-1)} \\ & = & -\dfrac{(\overline{u}(1-2\overline{v}+\overline{v}^2)-\overline{v})^2}{4\overline{v}(\overline{v}-1)(\overline{v}+\overline{u}\overline{v}-1)} \end{array}$$

and as $\overline{u}$ goes to 0,

$$z=-\frac{\overline{v}}{4(\overline{v}-1)^2}=-\frac{v}{4(1-v)^2}.$$

Then we have

$$[v,0;1,z]=[1,0;\overline{v},\overline{v}z]=[1,0;\overline{v},-\frac{1}{4(1-v)^2}].$$

The cross ratio is degenerate when $v=\infty$ (where $\widehat{\{y=\infty\}}$ meets E_q), $v=1$ (where E_∞ and E_q intersect), and where $z=0$ and $z=1$ (corresponding to two distinct values of v not equal to 1 or ∞). The points correspond to $v\in E_q\setminus\widehat{\mathcal{L}\cup\mathcal{Z}}$ except at the point where $v=1$.

For this point, consider the 2-component multicurve γ on $S^2\setminus B$ determined by two loops on the planes defined by E_∞ and $\widehat{\{y=\infty\}}$ around $(\overline{y},\overline{u})=(0,0)$. The corresponding point $\alpha_0\in\mathcal{AM}_f$ corresponds to the partition

$$\{y,\infty\}\cup\{z\}\cup\{0,1\}.$$

Define

$$\begin{array}{rcl}\kappa_q : E_q \setminus \{\widehat{\mathcal{L}\cup\mathcal{Z}}\} & \to & \mathcal{A}_2 \\ v & \mapsto & \begin{cases} ([0,1;\infty,\overline{v}],*) & \text{if } v\neq 1 \\ \alpha_0 & \text{if } v=1\end{cases}\end{array}$$

Then κ_∞ and κ_q extend to a morphism $\xi : X\to\mathcal{AV}$. □

PROOF OF THEOREM 1.3. Fix $d_0\in\mathcal{D}_{f,\iota}$, and let $\mathcal{D}_0$ be the connected component of $\mathcal{D}_{f,\iota}$ that contains d_0. Let $\overline{\mathcal{D}}_0$ be the closure of $\mathcal{D}_0$ in $\mathcal{AD}_{f,\iota}$. Then $\overline{\mathcal{D}}_0$ maps to $\widehat{\mathcal{V}}$ under the projection from $\mathcal{AT}_f\to\mathcal{AM}_f$. Let $\widehat{\mathcal{D}}_0\subset\mathcal{AT}_f$ be the connected component of the preimage of $\widehat{\mathcal{V}}$ that intersects $\mathcal{D}_0$. Then, since $\widehat{\mathcal{V}}$ is closed in X and X is connected, we can find a connected component Y_0 of the preimage of X in $\mathcal{AT}_f$ that contains $\widehat{D}_0$, and $\widehat{\mathcal{D}}_0$ is the closure in Y_0 of $\mathcal{D}_0$. That is, $\widehat{\mathcal{D}}_0=\overline{\mathcal{D}}_0$.

We claim that there is an element

$$g \in S_{f,\iota} \cap \mathrm{Image}(\pi_1(N(X) \cap \mathcal{M}_f) \to \pi_1(\mathcal{M}_f)) \setminus \mathrm{Image}(\pi_1(\widehat{V}) \to \pi_1(\mathcal{M}_f)).$$

Let v_0 be the image of d_0 in $\mathcal{V}$, and let γ be a close path starting at v_0 and passing along X to a point near an intersection of $\widehat{\mathcal{Z}}$ with $E_\infty \cup E_q$, forming a small loop around that intersection point, and returning along the original path back to v_0. Then γ is not homotopic in $\mathcal{AM}_f$ to an closed path on $\widehat{V}$ since $\widehat{Z}$ and $\widehat{V}$ do not intersect. Let g be the image of γ in $\pi_1(\mathcal{M}_f)$ after pushing off the boundary of $\mathcal{AM}_f$ into $\mathcal{M}_f$.

Since $g \notin \mathrm{Image}(\pi_1(\widehat{V}) \to \pi_1(\mathcal{M}_f))$, g does not preserve $\widehat{D}_0$, and hence $g(\widehat{\mathcal{D}}_0)$ and $\widehat{\mathcal{D}}_0$ are disjoint. These are the closures of $\mathcal{D}_0$ and $g(\mathcal{D}_0)$ in $\mathcal{AD}_{f,\iota}$ and the claim follows. □

4.4. Connectivity of augmented deformation space. We finish this paper by giving a sufficient condition for $\mathcal{AD}_{f,\iota}$ to be connected in the $\mathrm{Per}_4(0)^*$ case.

By Corollary 3.4, we know that $\widetilde{V}$ has a regular neighborhood U in $\widetilde{\mathbb{P}^1 \times \mathbb{P}^1}$ so that

$$\pi_1(U \cap \mathcal{M}_f) \to \pi_1(\mathcal{M}_f)$$

is surjective.

As a closure of $\mathcal{M}_f$, $\widetilde{\mathbb{P}^1 \times \mathbb{P}^1}$ is birationally equivalent to $\mathcal{AM}_f$ (but not isomorphic). By the birational theory of complex projective surfaces, there is a minimal smooth surface Z with birational morphisms to $\mathcal{AM}_f$ and $\widetilde{\mathbb{P}^1 \times \mathbb{P}^1}$

$$\begin{array}{ccc} & Z & \\ \swarrow & & \searrow \\ \mathcal{AM}_f & & \widetilde{\mathbb{P}^1 \times \mathbb{P}^1}. \end{array}$$

Lifting U to Z and projecting to $\mathcal{AM}_f$ gives a regular neighborhood U' of $\overline{\mathcal{V}} \cup \mathcal{K}$, where $\mathcal{K}$ is a union of boundary curves in $\mathcal{AM}_f$. Since $\mathcal{M}_f$ is a smooth subset of both $\mathcal{AM}_f$ and $\widetilde{\mathbb{P}^1 \times \mathbb{P}^1}$ the two projections of Z are isomorphisms over $\mathcal{M}_f$. Thus

$$U' \cap \mathcal{M}_f = U \cap \mathcal{M}_f$$

and hence

$$\pi_1(U') \to \pi_1(\mathcal{M}_f)$$

is surjective. It follows that U' has a connected preimage in $\mathcal{AT}_f$ and that the preimage of $\overline{V} \cup \mathcal{K}$ in $\mathcal{AT}_f$ is connected.

We have thus shown the following sufficient condition for $\mathcal{AD}_{f,\iota}$ to be connected.

PROPOSITION 4.6. *If $\mathcal{AV}$ is connected and is Zariski dense in $\overline{\mathcal{V}} \cup \mathcal{K}$, then $\mathcal{AD}_{f,\iota}$ is connected.*

References

[1] A. Andreotti and T. Frankel, *The Lefschetz theorem on hyperplane sections*, Ann. of Math. (2) **69** (1959), 713–717, DOI 10.2307/1970034. MR0177422

[2] L. Bers, *Spaces of degenerating Riemann surfaces*, Discontinuous groups and Riemann surfaces (Proc. Conf., Univ. Maryland, College Park, Md., 1973), Princeton Univ. Press, Princeton, N.J., 1974, pp. 43–55. Ann. of Math. Studies, No. 79. MR0361051

[3] X. Buff, A. Epstein, S. Koch, and K. Pilgrim, *On Thurston's pullback map*, Complex dynamics, A K Peters, Wellesley, MA, 2009, pp. 561–583, DOI 10.1201/b10617-20. MR2508269

[4] D. Cheniot, *Une démonstration du théorème de Zariski sur les sections hyperplanes d'une hypersurface projective et du théorème de Van Kampen sur le groupe fondamental du complémentaire d'une courbe projective plane* (French), Compositio Math. **27** (1973), 141–158. MR0366922

[5] P. Deligne and D. Mumford, *The irreducibility of the space of curves of given genus*, Inst. Hautes Études Sci. Publ. Math. **36** (1969), 75–109. MR0262240

[6] A. Douady and J. H. Hubbard, *A proof of Thurston's topological characterization of rational functions*, Acta Math. **171** (1993), no. 2, 263–297, DOI 10.1007/BF02392534. MR1251582

[7] A. Epstein. Transversality in holomorphic dynamics. `http://homepages.warwick.ac.uk/~mases/Transversality.pdf`

[8] W. Fulton, *Algebraic curves. An introduction to algebraic geometry*, W. A. Benjamin, Inc., New York-Amsterdam, 1969. Notes written with the collaboration of Richard Weiss; Mathematics Lecture Notes Series. MR0313252

[9] R. Hartshorne, *Algebraic geometry*, Springer-Verlag, New York-Heidelberg, 1977. Graduate Texts in Mathematics, No. 52. MR0463157

[10] J. Hempel, *Residual finiteness for 3-manifolds*, Combinatorial group theory and topology (Alta, Utah, 1984), Ann. of Math. Stud., vol. 111, Princeton Univ. Press, Princeton, NJ, 1987, pp. 379–396. MR895623

[11] E. Hironaka and S. Koch, *A disconnected deformation space of rational maps*, J. Mod. Dyn. **11** (2017), 409–423, DOI 10.3934/jmd.2017016. MR3668372

[12] J. H. Hubbard and S. Koch, *An analytic construction of the Deligne-Mumford compactification of the moduli space of curves*, J. Differential Geom. **98** (2014), no. 2, 261–313. MR3263519

[13] S. Koch, *Teichmüller theory and critically finite endomorphisms*, Adv. Math. **248** (2013), 573–617, DOI 10.1016/j.aim.2013.08.019. MR3107522

[14] J. Milnor, *Geometry and dynamics of quadratic rational maps*, Experiment. Math. **2** (1993), no. 1, 37–83. With an appendix by the author and Lei Tan. MR1246482

[15] J. Milnor, *On Lattès maps*, Dynamics on the Riemann sphere, Eur. Math. Soc., Zürich, 2006, pp. 9–43, DOI 10.4171/011-1/1. MR2348953

[16] R. Ramadas, *Hurwitz correspondences on compactifications of* $\mathcal{M}_{0,N}$, Adv. Math. **323** (2018), 622–667, DOI 10.1016/j.aim.2017.10.044. MR3725886

[17] N. Selinger, *Thurston's pullback map on the augmented Teichmüller space and applications*, Invent. Math. **189** (2012), no. 1, 111–142, DOI 10.1007/s00222-011-0362-3. MR2929084

[18] E. R. V. Kampen, *On the Fundamental Group of an Algebraic Curve*, Amer. J. Math. **55** (1933), no. 1-4, 255–260, DOI 10.2307/2371128. MR1506962

Email address: `hironaka@math.fsu.edu`

Contemporary Mathematics
Volume **742**, 2020
https://doi.org/10.1090/conm/742/14941

On the geometry of strongly flat semigroups and their generalizations

Tamás László and András Némethi

ABSTRACT. Our goal is to convince the readers that the theory of complex normal surface singularities can be a powerful tool in the study of numerical semigroups, and, in the same time, a very rich source of interesting affine and numerical semigroups. More precisely, we prove that the strongly flat semigroups, which satisfy the maximality property with respect to the Diophantine Frobenius problem, are exactly the numerical semigroups associated with negative definite Seifert homology spheres via the possible 'weights' of the generic S^1–orbit. Furthermore, we consider their generalization to the Seifert rational homology sphere case and prove an explicit (up to a Laufer computation sequence) formula for their Frobenius number. The singularities behind are the weighted homogeneous ones, whose several topological and analytical properties are exploited.

Dedicated to Lê Dũng Tráng

1. Introduction

This note creates a bridge between the theory of numerical semigroups and the theory of complex normal surface singularities. By several examples we suggest how one of the two theories might provide meaningful and enriching questions, ideas and simultaneously powerful tools to the other.

For example, one of the most classical and important problem in the theory of numerical semigroups is the determination of the (minimal set of) generators and also their Frobenius number. These are very difficult algebraic/combinatorial problems. In this note we provide a new tool, based on techniques of singularity theory, which solves these problems for several semigroups (which can be related with singularities).

In theory of surface singularities two semigroups appear very naturally. If (X, o) is a normal surface singularity, and we fix a good resolution of it with n irreducible exceptional curve, then one defines the 'Lipman cone' $\mathcal{S}_{top}$, an affine monoid, submonoid of $\mathbb{N}^n$. It can be determined completely from the combinatorics

2010 *Mathematics Subject Classification.* Primary 32S05, 32S25, 32S50, 20M05, 57M27 Secondary 14Bxx, 14J80.

Key words and phrases. numerical semigroup, strongly flat semigroup, surface singularity, weighted homogeneous singularity, rational homology sphere, Seifert 3-manifold, resolution graph, Lipman cone.

of the resolution graph, in particular, it is a topological invariant associated with a fixed plumbing graph of the oriented (3–manifold) link of the singularity.

The second monoid, $\mathcal{S}_{an}$, still associated with the fixed resolution, is a submonoid of $\mathcal{S}_{top}$, and it is determined by the analytic structure of the singularity. In general is very hard to determine it.

The projection of $\mathcal{S}_{an}$ to any of the coordinates of $\mathbb{N}^n$ (given by the choice of one of the irreducible exceptional curves) provides an interesting numerical semigroup. It is a real challenge to compute its generators, Frobenius number, list its properties (eg. decide if it is symmetric or not).

In this note we take the case of weighted homogeneous surface singularities. Their links are oriented Seifert 3–manifolds. In particular, the minimal good resolution graph (or, the minimal plumbing graph of the link) is star–shaped. We will assume that the link M is rational homology sphere, that is, all the genus–decorations of the graph are zero. We also take the generic case, when the number of 'legs' of the star–shaped graph is ≥ 3. Furthermore, we will choose for the projection the central vertex (which, in Seifert geometry, corresponds to the generic S^1–orbit). We define $\mathcal{S}_M$ as the projection of $\mathcal{S}_{an}$.

Our starting point in the construction of the bridge mentioned above is to show that $\mathcal{S}_M$ appears naturally in the classical theory of numerical semigroups. Indeed, Raczunas and Chrząstowski-Wachtel in [**RChW96**] characterized certain semigroups, which realizes a sharp upper bound estimate for the Diophantine Frobenius problem/number. They called them 'strongly flat semigroups'. In this note we prove that these semigroups can be realized as semigroups of type $\mathcal{S}_M$ associated with Seifert integral homology spheres.

Then, we continue with the study of $\mathcal{S}_M$ in the rational homology sphere case. One of the final goals is the computation the Frobenius number. In order to determine this we define a $\mathcal{S}_M$–module as well (which probably was not considered in semigroup theory). The main tools are the topological properties of Seifert 3–manifolds, Pinkham's theory of weighted homogeneous singularities, Laufer's computations sequences, and results (valid for splice quotient singularities) regarding the generators of $\mathcal{S}_{an}$.

In the case of links of Brieskorn–Hamm complete intersection we determine the minimal set of generators of $\mathcal{S}_M$ as well.

2. The Diophantine Frobenius problem and strongly flat semigroups

2.1. The famous Diophantine Frobenius problem asks to find an explicit formula for the greatest integer not representable as a nonnegative linear form of a given system of d relatively prime integers $1 \leq a_1 \leq \ldots \leq a_d$. The integer defined in this way is called the *Frobenius number* of the system, or of the numerical semigroup $G(a_1, \ldots, a_d)$, generated by the integers from the system itself. It will be denoted by $f_{G(a_1,\ldots,a_d)}$.

The very first result related to this problem is the well-known formula of Sylvester, namely $f_{G(a_1,a_2)} = a_1a_2 - a_1 - a_2$ [**S1882**]. Although several formulas for peculiar systems and general bounds exist in the literature, the problem is still open in full generality. In this note we will recall/discuss almost nothing from the 'classical combinatorial approach' — the interested reader might consult for more details the excellent monograph of Ramírez Alfonsín [**RA05**] —, our goal is to connect the problem with singularity theory, and to show the strength of this new method. From this point of view the following sharp estimate is a good starting point: Raczunas and Chrząstowski-Wachtel [**RChW96**] found the following upper bound using the least common multiple $\mathrm{lcm}(a_1,\dots,a_d)$

$$f_{G(a_1,\dots,a_d)} \leq (d-1)\cdot \mathrm{lcm}(a_1,\dots,a_d) - \sum_{i=1}^{d} a_i. \tag{2.1.1}$$

Moreover, they have characterized the class of semigroups (or system of generators) for which the equality in (2.1.1) holds. These are the so called strongly flat semigroups, which satisfy the property that for every i one can write $a_i = \alpha_1\dots\alpha_{i-1}\alpha_{i+1}\dots\alpha_d$ with each α_i being the greatest common divisor of $a_1,\dots,a_{i-1}, a_{i+1},\dots,a_d$. (From this also follows that $\gcd(\alpha_i,\alpha_j) = \gcd(a_1,\dots,a_d) = 1$ for $i\neq j$.)

In the sequel we will discuss the geometric and topological aspects of strongly flat semigroups and also their natural topological generalizations. During this discussion we try to embed the solution of the Frobenius problem into a more complex package of topological/geometrical invariants — even to relate with more semigroups with richer structure.

3. Normal surface singularities

3.1. Notations and preliminaries. We consider a complex normal surface singularity (X,o) whose link M is a rational homology sphere. Fix a good resolution $\pi:\widetilde{X}\to X$ with dual graph Γ whose vertices are denoted by $\mathcal{V}$. We use notation $E=\pi^{-1}(o)$ for the exceptional divisor and let $\{E_v\}_{v\in\mathcal{V}}$ be its irreducible components. Notice that M is a rational homology sphere if and only if Γ is a tree and all the E_v are rational.

Then $L := H_2(\widetilde{X},\mathbb{Z})$ is a lattice freely generated by the classes of the irreducible exceptional divisors $\{E_v\}_{v\in\mathcal{V}}$, together with the nondegenerate negative definite intersection form $I=(E_v,E_w)_{v,w}$. They are exactly the integral cycles supported on E. The determinant $\det(\Gamma)$ of the graph by convention is $\det(-I)$ (which is always positive).

If L' denotes $H^2(\widetilde{X},\mathbb{Z})$, then the intersection form provides an embedding $L\hookrightarrow L'$ with factor $L'/L\simeq H_1(M,\mathbb{Z})$, denoted by H. In the sequel $[l']$ denotes the class of l'. In fact, $L'\simeq \mathrm{Hom}_{\mathbb{Z}}(L,\mathbb{Z})$, the dual lattice. The form extends to $L\otimes\mathbb{Q}$ hence to L' too (via the natural inclusion $L'\simeq\{l'\in L\otimes\mathbb{Q} : (l',L)\in\mathbb{Z}\}\subset L\otimes\mathbb{Q}$). The module L' over $\mathbb{Z}$ is freely generated by the (anti-)duals $\{E^*_v\}_v$, where we prefer the convention $(E^*_v,E_w)=-1$ for $v=w$, and 0 otherwise.

For $l'_1,l'_2\in L\otimes\mathbb{Q}$ with $l'_i=\sum_v l'_{iv}E_v$ for $i=\{1,2\}$ one writes $l'_1\geq l'_2$ if $l'_{1v}\geq l'_{2v}$ for any $v\in\mathcal{V}$. In particular, l' is effective if $l'\geq 0$. We set also $\min\{l'_1,l'_2\} := \sum_v \min\{l'_{1v},l'_{2v}\}E_v$. Furthermore, if $l'=\sum_v l'_vE_v$ then we write $|l'| := \{v\in\mathcal{V} : l'_v\neq 0\}$ for the *support* of l'.

In several computations we need the E_v–coefficient of E^*_u; this equals $-(E^*_u,E^*_v) = (-I^{-1})_{uv}$, which multiplied by $\det(\Gamma)$ is the determinant of the subgraph obtained from Γ by deleting the shortest path connecting u and v and the adjacent edges.

For more details on normal surface singularities and resolution graphs see eg. [**N99, N05**].

3.1.1. *Characteristic cycles.* The *canonical cycle* $K \in L'$ is defined by the *adjunction formulae*

$$(K + E_v, E_v) + 2 = 0 \quad \text{for all } v \in \mathcal{V}. \tag{3.1.1}$$

Let $Char := \{k \in L' \,:\, (k + l, l) \in 2\mathbb{Z} \text{ for any } l \in L\}$ be the set of characteristic cycles in L'. By the adjunction formulae it can be written as $Char = K + 2L'$. There is a natural action of L on $Char$ given by $l * k := k + 2l$ whose orbits are of type $[k] := k + 2L$. Moreover, H acts freely and transitively on the set of orbits by $[l'] * [k] = [k + 2l']$. Therefore, every orbit can be written in the form $[k] = K + 2(l' + L)$ for some $l' \in L'$ with a fixed group element $[l'] = h$ which indexes the orbit. One has an identification between the orbits of $Char$ and the *spin*^{c}-structures of the link M, see eg. [**N05**].

Associated with any characteristic cycle k we define the Riemann-Roch function

$$\chi_k(l') := -(k + l', l')/2 \quad \text{for any } \; l' \in L'. \tag{3.1.2}$$

For simplicity we will use the notation $\chi := \chi_K$. Notice that $\chi_k(l)$ $(l \in L)$ by index (Riemann-Roch type) formula equals the Euler characteristic of certain line bundles, see e.g. [**N07**, 2.2.8].

A topological type of singularity is called *numerically Gorenstein* if $K \in L$ (this property does not depend on the choice of the resolution graph). An analytic type of singularity is called *Gorenstein* if $K \in L$ and the sheaf of holomorphic 2–forms $\Omega_{\widetilde{X}}$ is isomorphic with $\mathcal{O}_{\widetilde{X}}(K)$.

3.1.2. *Minimal and distinguished cycles.* We define the Lipman cone by $\mathcal{S}'_{top} := \{l' \in L' \mid (l', E_v) \leq 0 \text{ for all } v \in \mathcal{V}\}$, the semigroup (monoid) of anti–nef rational cycles from L'. It is generated over $\mathbb{Z}_{\geq 0}$ by the cycles E^*_v. Using the fact that the intersection form is negative definite (or the determinantal characterizations from 3.1) one shows that all the entries of E^*_v are strict positive. Hence $\mathcal{S}'_{top}$ sits in the first quadrant. Define also $\mathcal{S}_{top} := L \cap \mathcal{S}'_{top}$, the semigroup (monoid) of anti-nef integral cycles, the integral Lipman cone. By its very definition it is an affine semigroup.

For any $h \in H$ there is a unique minimal element of $\{l' \in L' \mid [l'] = h\} \cap \mathcal{S}'_{top}$ (guaranteed eg. by Lemma 5.3.4), which will be denoted by s_h. Furthermore, we can also consider the semi-open cube $\{\sum_v l'_v E_v \in L' \mid 0 \leq l'_v < 1\}$ which contains a unique representative r_h for every h so that $[r_h] = h$. One has $s_h \geq r_h$, however, in general, $s_h \neq r_h$ (this can happen even for star–shaped resolution graphs, see eg. [**N07**, Ex. 4.5.4]).

The minimal cycle s_h defines a distinguished characteristic cycle $k_r := K + 2s_h$ of the corresponding orbit $[K + 2s_h]$ indexed by $h \in H$.

Sometimes it is more convenient to use $Z_K := -K \in L'$ instead of K, eg. if the resolution graph is minimal then $Z_K \in \mathcal{S}'_{top}$ (use the adjunction formulae (3.1.1)).

3.1.3. *Principal and anti-nef cycles.* Any analytic function $f : (X, o) \to (\mathbb{C}, 0)$ determines an effective principal divisor $(\pi \circ f) = (f)_\Gamma + St(f)$, where $(f)_\Gamma$ is a cycle in $L_{>0}$ and $St(f)$ is supported by the strict transform of $\{f = 0\}$. We define the set of principal cycles by

$$\mathcal{S}_{an} := \{(f)_\Gamma \,:\, f \in \mathfrak{m}_{(X,o)}\} \cup \{0\}. \tag{3.1.3}$$

Then, in fact, $(f_\Gamma, E_v) \leq 0$ for any $v \in \mathcal{V}$, hence $\mathcal{S}_{an}$ is a subsemigroup of $\mathcal{S}_{top}$. If (X, o) is rational (and π is arbitrary), or minimally elliptic (and π is minimal and good) then $\mathcal{S}_{an} = \mathcal{S}_{top}$, but in general $\mathcal{S}_{an} \neq \mathcal{S}_{top}$, see eg. [**NNP11**].

In general, for an arbitrary analytic singularity type, it is very difficult problem to decide if an element of $\mathcal{S}_{top}$ is principal or not (this is not a topological/combinatorial problem, the answer definitely might depend on the choice of the analytic structure supported by the fixed resolution graph).

3.2. Seifert 3–manifolds as singularity links.

3.2.1. The negative definite intersection form I together with the collection of genera $\{g(E_v)\}_v$ can be coded in a connected dual resolution graph Γ as well. The link of the singularity can be recovered from the graph by plumbing construction. Furthermore, by a theorem of Grauert [**G62**], any such connected negative definite graph can be realized as dual resolution graph of some analytic singularity. In this note we focus on star–shaped graphs, their associated plumbed 3–manifolds are Seifert 3–manifolds, and they can be analytically realized by weighted homogeneous singularities (however, they might be realized by many other analytic structures as well). First we provide some details and notation regarding the combinatorics of the graph.

3.2.2. *Seifert 3–manifolds.* Assume that a resolution graph Γ is star–shaped with d legs, $d \geq 3$. Each leg is a chain determined by the normalized Seifert invariant (α_i, ω_i), where $0 < \omega_i < \alpha_i$, $\gcd(\alpha_i, \omega_i) = 1$ as follows. If we consider the Hirzebruch–Jung (negative) continued fraction expansion

$$\alpha_i/\omega_i = [b_{i1}, \ldots, b_{i\nu_i}] = b_{i1} - 1/(b_{i2} - 1/(\cdots - 1/b_{i\nu_i})\ldots) \quad (b_{ij} \geq 2),$$

then the ith leg has ν_i vertices, say $v_{i1}, \ldots, v_{i\nu_i}$, with Euler decorations (self–intersection numbers) $-b_{i1}, \ldots, -b_{i\nu_i}$, where v_{i1} is connected by the central vertex denoted by v_0. All these vertices have genus–decorations zero. We also use ω_i' satisfying $\omega_i\omega_i' \equiv 1 \pmod{\alpha_i}$, $0 < \omega_i' < \alpha_i$. (One shows that α_i is the determinant of the ith–leg Γ_i, $\omega_i = \det(\Gamma_i \setminus v_{i1})$, and $\omega_i' = \det(\Gamma_i \setminus v_{i\nu_i})$.)

The central vertex v_0 has an Euler decoration $-b_0$ and genus decoration g.

The plumbed 3–manifold M associated with such a star–shaped graph Γ has a Seifert structure and its associated normalized Seifert invariants is denoted by

$$Sf = (-b_0, g; (\alpha_i, \omega_i)_{i=1}^d).$$

In the sequel we will assume that M is a rational homology sphere, or, equivalently, the central node has $g = 0$.

The orbifold Euler number of M is defined as $e = -b_0 + \sum_i \omega_i/\alpha_i$. The negative definiteness of the intersection form is equivalent with $e < 0$. We also write $\alpha := \operatorname{lcm}(\alpha_1, \ldots, \alpha_d)$.

Let $\mathfrak{h} := |H|$ be the order of $H = H_1(M, \mathbb{Z}) = L'/L$, and let $\mathfrak{o}$ be the order of the class of E_0^* (or of the generic S^1 Seifert–orbit) in H (in the plumbing construction, this orbit is the S^1–fiber over a generic point of E_0). Then one has (see eg. [**Ne81**])

$$\mathfrak{h} = \alpha_1 \cdots \alpha_d |e|, \qquad \mathfrak{o} = \alpha |e|. \tag{3.2.1}$$

If M is an integral homology sphere (that is, Seifert homology sphere) then necessarily all α_i's are pairwise relative prime and (cf. (3.2.1)) $\alpha|e| = 1$. This reads as the Diophantine equation $(b_0 - \sum_i \omega_i/\alpha_i)\alpha = 1$, hence all ω_i's and b_0 are determined uniquely by the α_i's. The corresponding Seifert 3–manifold (link) will be denoted by $\Sigma(\alpha_1, \ldots, \alpha_d)$.

3.2.3. *Weighted homogeneous surface singularities.* Let (X, o) be a normal weighted homogeneous surface singularity. It is the germ at the origin of an affine variety X with good $\mathbb{C}^*$-action, which means that its affine coordinate ring is $\mathbb{Z}_{\geq 0}$–graded: $R_X = \oplus_{\ell \geq 0} R_{X,\ell}$. The dual graph of the minimal good resolution is star–shaped, and the $\mathbb{C}^*$-action of the singularity induces an S^1–Seifert action on the link. In particular, the link M of (X, o) is a negative definite Seifert 3–manifold characterized by its normalized Seifert invariants $Sf = (-b_0, g; (\alpha_i, \omega_i)_{i=1}^d)$. It is known that weighted homogeneous surface singularity with numerically Gorenstein topological type is Gorenstein (in particular, eg., combinatorial consequences of the Gorenstein duality can be read as combinatorial symmetries of the numerically Gorenstein graph).

The formulae below are attributed to Dolgachev, Pinkham and Demazure, hence by some authors are called Dolgachev–Pinkham–Demazure formulae; we will follow here Pinkham's work [**P77**]. By [**P77**] the complex structure is completely recovered by the Seifert invariants and the configuration of points $\{P_i := E_0 \cap E_{i1}\}_{i=1}^d \subset E_0$. (Here E_0 is the irreducible exceptional curve indexed by the central vertex v_0.) In fact, the graded ring of the local algebra of the singularity is given by the formula $R_X = \oplus_{\ell \geq 0} R_{X,\ell} = \oplus_{\ell \geq 0} H^0(E_0, \mathcal{O}_{E_0}(D^{(\ell)}))$ with $D^{(\ell)} := \ell(-E_0|_{E_0}) - \sum_{i=1}^d \lceil \ell\omega_i/\alpha_i \rceil P_i$, where for $r \in \mathbb{R}$ $\lceil r \rceil$ denotes the smallest integer greater than or equal to r.

In particular, when M is a rational homology sphere, ie. $g = 0$, one has $E_0 \simeq \mathbb{P}^1$, hence Pinkham's formula implies that $\dim(R_{X,\ell}) = \max\{0, 1 + N(\ell)\}$ is topological, where $N(\ell)$ is the quasi-linear function

$$N(\ell) := \deg D^{(\ell)} = b_0\ell - \sum_{i=1}^d \left\lceil \frac{\ell\omega_i}{\alpha_i} \right\rceil. \tag{3.2.2}$$

Since $-\lceil x \rceil \leq -x$ one obtains $N(\ell) \leq |e|\ell$, hence $N(\ell) < 0$ for $\ell < 0$. Therefore, if we consider the set $\mathcal{S}_M := \{\ell \in \mathbb{Z}_{\geq 0} | R_{X,\ell} \neq 0\}$, which is a numerical semigroup by the grading property of the algebra, by the above results it can be characterized topologically with the Seifert invariants as follows

$$\mathcal{S}_M = \{\ell \in \mathbb{Z} \mid N(\ell) \geq 0\}. \tag{3.2.3}$$

It is called the *semigroup of* (X, o), *or, of the Seifert rational homology sphere link* M. In particular, the Poincaré series $P_0(t) = \sum \dim_{\mathbb{C}}(R_{X,\ell})t^\ell$ of the graded local algebra is also topological, namely

$$P_0(t) = \sum_{\ell \geq 0} \max\{0, 1 + N(\ell)\} t^\ell. \tag{3.2.4}$$

Its support is exactly $\mathcal{S}_M$. For more (from both analytic and topological points of view), see [**N05, N12**] and [**LN14**]. By a similar pattern, one can define the following series as well:

$$P_0^+(t) = \sum_{\ell \geq 0} \max\{0, -1 - N(\ell)\} t^\ell, \quad \text{and} \quad P_0^{neg}(t) = \sum_{\ell \geq 0} (1 + N(\ell)) t^\ell. \tag{3.2.5}$$

P_0^+ has two interpretations. Firstly, since $E_0 \simeq \mathbb{P}^1$, one has $\max\{0, -1 - N(\ell)\} = \dim H^1(E_0, \mathcal{O}_{E_0}(D^{(\ell)}))$, hence $P_0^+(t) = \sum_{\ell \geq 0} h^1(E_0, \mathcal{O}_{E_0}(D^{(\ell)}))t^\ell$. Since $N(\ell)$ asymptotically behaves like $|e|\ell$ (see also Prop 3.2.11 below), $P_0^+(t)$ is a polynomial. By [**P77**], $P_0^+(1)$ equals the geometric genus p_g of the weighted homogeneous singularity. ¿From topological point of view, $P_0^+(1)$ equals the $(h=0)$–equivariant part of the normalized Seiberg–Witten invariant of the link (cf. [**N05**]).

It is clear that both P_0 and P_0^+ are determined from P_0^{neg}. The point is that both P_0^+ and P_0^{neg} are also determined from the Poincaré series P_0. Indeed, P_0 has a unique 'polynomial + negative degree part' decomposition into a sum of a polynomial and a rational function of negative degree, and in fact, this decomposition is exactly $P_0 = P_0^+ + P_0^{neg}$ [**BN10, NO09**] (for the more general multivariable case see [**LN14, LNN18**]. In particular, P_0 determined all the integers $\{N(\ell)\}_{\ell\geq 0}$ and p_g as well.

The entries $N(\ell)+1$ in the Poincaré series suggest to define for any Seifert rational homology sphere the set

$$\mathcal{M} = \mathcal{M}_M := \{\ell\in\mathbb{Z} \mid 1+N(\ell)\geq 0\}. \tag{3.2.6}$$

Clearly $\mathcal{S}_M\subset\mathcal{M}_M$. Notice that, in general, $\mathcal{M}$ is not a numerical semigroup, consider eg. the Seifert manifold $\Sigma(2,3,7)$ with Seifert invariants $(-1;(2,1),(3,1),(7,1))$, when $-1\in\mathcal{M}$ but $-84\notin\mathcal{M}$. However, it is a module over the semigroup $\mathcal{S}_M$ (if $\ell_1\in\mathcal{S}_M$ and $\ell_2\in\mathcal{M}_M$ then $\ell_1+\ell_2\in\mathcal{M}_M$). Again by the definition of $N(\ell)$ one has $\ell\in\mathcal{M}_M$ for any $\ell\geq (d-1)/|e|$ and $N(\ell)<-1$ for $\ell\ll 0$. Hence, the set $\mathbb{Z}\setminus\mathcal{M}$ is bounded from above and we can define the Frobenius number $f_{\mathcal{M}}$ of $\mathcal{M}$ as the largest integer not in $\mathcal{M}$. Similarly, $\mathcal{M}$ is bounded from below, thus one can define $\min\{\mathcal{M}\}$ to be the smallest integer in the module $\mathcal{M}$.

If $N(\ell)=-1$ then ℓ is neither in the support of $P_0(t)$ nor in the support of $P_0^+(t)$. However, this set (non–coded by these series) is important (eg. from the point of view of sharp cohomology vanishing). It is exactly the difference between the supports of $\mathcal{M}_M$ and $\mathcal{S}_M$.

Furthermore, $\mathbb{Z}_{\geq 0}\setminus\mathcal{M}$ is exactly the support of $P_0^+(t)$, hence

$$(X,o) \text{ is rational } \Leftrightarrow\ p_g=0\ \Leftrightarrow\ P_0^+\equiv 0\ \Leftrightarrow\ \mathbb{Z}_{\geq 0}\subset\mathcal{M}. \tag{3.2.7}$$

3.2.4. *The combinatorial number γ.* Before we recall certain known facts regarding the integers $\{N(\ell)\}_\ell$ (and certain consequences regarding $\mathcal{S}_M$ and $\mathcal{M}_M$) we define a key numerical invariant:

$$\gamma := \frac{1}{|e|}\cdot\Big(d-2-\sum_{i=1}^d\frac{1}{\alpha_i}\Big)\in\mathbb{Q}, \tag{3.2.8}$$

which has a central importance regarding properties of weighted homogeneous surface singularities or Seifert rational homology spheres. It has several interpretations.

First of all, the adjunction formula gives (in any graph) the identity

$$Z_K = E+\sum_v(\delta_v-2)E_v^*, \tag{3.2.9}$$

where $E:=\sum_v E_v$. In a star–shaped graph the E_0 coefficients of all E_v^* associated with end–vertices are computed by $-(E_v^*,E_0^*)=1/(|e|\alpha_v)$ and the E_0 coefficients of E_0^* is $-(E_0^*,E_0^*)=1/|e|$ (cf. [**N05**, (11.1)], or the determinant–formula from 3.1). Hence, the E_0-coefficient of Z_K is exactly $\gamma+1$.

The number $-\gamma$ is called the 'log discrepancy' of E_0, γ the 'exponent' of the weighted homogeneous germ (X,o), and $\mathfrak{o}\gamma$ is the Goto–Watanabe a–invariant (cf. [**GW78**, (3.1.4)]) of the universal abelian covering of (X,o). In [**Ne81**] $e\gamma$ appears as the orbifold Euler characteristic (see also [**NO10**, 3.3.6]).

3.2.5. *More on Z_K.* If (X,o) is ADE, then $Z_K = 0$ and $\gamma = -1$. By [**CR74**], $\gamma|e| = d - 2 - \sum_i 1/\alpha_i$ is negative if and only of $\pi_1(M)$ is finite (this can happen only if $d = 3$ and $\sum_i 1/\alpha_i > 1$). In this case, (X,o) is a quotient singularity, hence rational. Thus, in this case $P_0^+(t) \equiv 0$ too.

Therefore, if (X,o) is not rational then $\gamma \geq 0$, that is, the E_0–coefficient of Z_K is ≥ 1. We claim that in these cases all the coefficients of Z_K are strict positive. (Indeed, Z_K restricted on any leg, say $Z_K|_i$, by adjunction relations satisfies $(Z_K|_i, E_{ik}) \leq 0$, $1 \leq k \leq \nu_i$, with strict inequality for $k = 1$, hence all the coefficients of $Z_K|_i$ are strict positive.) In particular, we also have $Z_K \geq r_{[Z_K]}$.

Let us rename the ith end–vertex by E_i and let us compute the E_i–coefficient of Z_K.

For these one needs for end–vertices i, j the identities $(E_i^*, E_j^*) = (e\alpha_i\alpha_j)^{-1}$ for $i \neq j$ and $(E_i^*, E_i^*) = (e\alpha_i^2)^{-1} - \omega_i'/\alpha_i$ if $i = j$; cf. [**N05**, (11.1)], or the determinant–formula from 3.1. Therefore, by (3.2.9) and a computation,

$$-(Z_K, E_i^*) = 1 + (\gamma - \omega_i')/\alpha_i. \tag{3.2.10}$$

Hence, if Γ is numerically Gorenstein (that is, $Z_K \in L$) then $\gamma \in \mathbb{Z}$ and $\gamma \equiv \omega_i' \pmod{\alpha_i}$ for all i. Hence, in the numerically Gorenstein non–ADE case — when we already know that $\gamma \geq 0$ — by this congruence we get the stronger $\gamma \geq 1$. (If $\gamma = 1$ and Γ is minimally Gorenstein then $\omega_i' = \omega_i = 1$ and $b_0 = d - 2$, hence (X,o) is the 'polygonal' minimally elliptic singularity with $Z_K = E + E_0$.)

3.2.6. *Some inequalities satisfied by $\{N(\ell)\}_\ell$.* The following facts are proved eg. in [**NO10**, Sections 3.1–3.3] (some of them can be verified by direct computations).

PROPOSITION 3.2.11.
(a) $-(\alpha-1)|e| - d \leq N(\ell) - \lceil \ell/\alpha \rceil \alpha|e| \leq -1$. *In particular* $\lim_{\ell\to\infty} N(\ell) = \infty$.
(b) If $\ell > \gamma$ *then* $h^1(E_0, \mathcal{O}_{E_0}(D^{(\ell)})) = 0$, *ie.* $N(\ell) \geq -1$. *In particular,* $\deg P_0^+(t) \leq \max\{0, \gamma\}$.
(c) If (X,o) is numerically Gorenstein (but not ADE) then $N(\gamma) = -2$ *and* $\deg P_0^+(t) = \gamma$.
(d) $N(\alpha) = \alpha(b_0 - \sum_i \omega_i/\alpha_i) = \alpha|e| = \mathfrak{o} > 0$.
(e) $N(\ell + \alpha) = N(\ell) + N(\alpha) = N(\ell) + \mathfrak{o} > N(\ell)$ *for any* $\ell \geq 0$.
(f) $N(\ell) \geq 0$ *for any* $\ell > \alpha + \gamma$.

3.2.7. *The numerically Gorenstein case.* Proposition 3.2.11 implies the following.

COROLLARY 3.2.12. *Assume that Γ is numerically Gorenstein non–ADE.*
(a) Then $f_{\mathcal{M}_M} = \gamma$.
(b) Assume additionally that $\mathfrak{o} = 1$ *(eg. M is an integral homology sphere). Then* $f_{\mathcal{S}_M} = \alpha + \gamma$.

PROOF. *(a)* Use *(b)–(c)* of Prop. 3.2.11. For *(b)* use *(e)–(f)* of Prop. 3.2.11 and $N(\gamma) = -2$. □

The following numerical symmetry is the trace of the Gorenstein symmetry. In the case when M is an integral homology sphere the statement was proved in [**CK14**] too. For a generalization here see Proposition 5.3.8.

PROPOSITION 3.2.13. *If the graph is numerically Gorenstein (that is, $Z_K \in L$), then*

$$N(\ell) + N(\gamma - \ell) = -2 \quad \textit{for any } \ell \in \mathbb{Z}, \tag{3.2.14}$$

PROOF. From (3.2.10) we have $\gamma \equiv \omega_i' \pmod{\alpha_i}$, hence $\omega_i\gamma \equiv \omega_i\omega_i' \equiv 1 \pmod{\alpha_i}$ too. Then

$$N(\gamma-\ell)=b_0(\gamma-\ell)-\sum_i \frac{\omega_i\gamma-1}{\alpha_i}+\sum_i\Big\lceil \frac{1-\omega_i\ell}{\alpha_i}\Big\rceil.$$

Then using $\lceil (1-x)/\alpha_i\rceil + \lceil x/\alpha_i\rceil = 1$ ($x\in\mathbb{Z}$) the statement follows. □

This combined with Corollary 3.2.12*(b)* gives

COROLLARY 3.2.15. *If the graph is numerically Gorenstein non-ADE then*

$$\ell\in\mathcal{M}_M \ \Leftrightarrow\ \gamma-\ell\notin\mathcal{S}_M$$

(3.2.16) $$\{\ell\,:\,N(\ell)=-1\}=\{\ell\,:\,\gamma-\ell\notin\mathcal{S}_M,\ \ell\notin\mathcal{S}_M\}=\mathbb{Z}\setminus\Big((\gamma-\mathcal{S}_M)\cup\mathcal{S}_M\Big).$$

In particular, $\min\{\mathcal{M}_M\}+f_{\mathcal{S}_M}=\gamma$*. This, via Corollary* 3.2.12*(b) for an integral homology sphere M reads as* $\min\{\mathcal{M}_M\}=-\alpha$*.*

(Let us verify 'elegantly', using the first identity above, that $\mathcal{M}$ is an $\mathcal{S}$–module. Since $\mathcal{S}$ is a semigroup one has $(\gamma-\ell\notin\mathcal{S},\ s\in\mathcal{S})\Rightarrow\gamma-\ell-s\notin\mathcal{S}$. This transforms into $(\ell\in\mathcal{M},\ s\in\mathcal{S})\Rightarrow\ell+s\in\mathcal{M}$.)

3.2.8. *Connection with Gorenstein analytic type.* The identity (3.2.14) is the topological 'trace' of Gorenstein duality of algebraic geometry. Indeed, a weighted homogeneous singularity with numerically Gorenstein topological type is automatically Gorenstein. In this case one has $D^{(\ell)}+D^{(\gamma-\ell)}=K_{E_0}$ in $\mathrm{Pic}(E_0)$ (even if g is not zero), which at deg–level (with $g=0$) is (3.2.14). The Gorenstein duality implies that $h^1(\mathcal{O}_{E_0}(D^{(\ell)}))=h^0(\mathcal{O}_{E_0}(D^{(\gamma-\ell)}))$. If $g=0$ then the very same formula follows already from (3.2.14), since $\max\{0,-N(\ell)-1\}=\max\{0,N(\gamma-\ell)+1\}$.

This, for any numerically Gorenstein Seifert rationally homology sphere reads as

$$P_0^+(t)=\sum_{\ell=0}^{\gamma}\dim R_{X,\gamma-\ell}t^{\ell}=\sum_{\ell=0}^{\gamma}\dim R_{X,\ell}t^{\gamma-\ell}.$$

3.2.9. *The connection between $\mathcal{S}_{an}$ and $\mathcal{S}_M$.* Above we defined two semigroups (monoids), $\mathcal{S}_{an}$ is a submonoid of $L_{\geq 0}=\mathbb{N}^{|\mathcal{V}|}$, sitting in the Lipman cone $\mathcal{S}_{top}$. The other is a numerical submonoid of $\mathbf{N}$. The next statement basically follows from [**P77**].

LEMMA 3.2.17. *The elements of $\mathcal{S}_M$ are the E_0–coefficients of the elements of $\mathcal{S}_{an}$.*

PROOF. Associated with the irreducible component E_0 of the exceptional divisor we consider the divisorial filtration $\mathcal{F}:=\{\mathcal{F}(\ell)\}_{\ell\in\mathbb{Z}}$, where $\mathcal{F}(\ell):=\{f\in\mathcal{O}_{(X,0)}\,|\,(f)_\Gamma|_{E_0}\geq\ell\}$ (and $(f)_\Gamma|_{E_0}:=((f)_\Gamma,-E_0^*)$ is the E_0-coefficient of the principal cycle $(f)_\Gamma$ associated with $f\in\mathcal{O}_{(X,0)}$, cf. 3.1.3). On the other hand, by [**P77**], $\mathcal{F}(\ell)=\oplus_{\ell'\geq\ell}R_{X,\ell'}$. Hence $R_{X,\ell}=\mathcal{F}(\ell)/\mathcal{F}(\ell+1)$. Hence $\ell\in\mathcal{S}_M$ if and only if there exists $(f)_\Gamma\in\mathcal{S}_{an}$ with $(f)_\Gamma|_{E_0}=\ell$. □

This fact shows that there is a richer monoid behind $\mathcal{S}_M$, namely $\mathcal{S}_{an}$.

3.2.10. *The combinatorial determination of $\mathcal{S}_{an}$.* Before we state the next result we observe that both monoids $\mathcal{S}_{top}$ and $\mathcal{S}_{an}$ are closed to taking the 'min' in $L\subset\mathbb{Z}^{|\mathcal{V}|}$: namely, if $l_1,l_2\in\mathcal{S}$ then $\min\{l_1,l_2\}\in\mathcal{S}$ too.

The next theorem was proved for 'splice-quotient singularities' [**N12**, Theorem 7.1.2]. Since weighted homogeneous singularities are splice quotient, the theorem applies in our situation. Below we formulate it tailored already to the weighted homogeneous rational homology sphere case.

Let $\mathcal{E}$ be the set of end–vertices. An *integral monomial cycle* has the form $\mathfrak{M} := \sum_{e\in\mathcal{E}} n_e E_e^* \in L$, where each $n_e \in \mathbb{Z}_{\geq 0}$. (Note that the cycles E_e^* usually are *rational* cycles, hence the condition that $\mathfrak{M}$ is integral imposes some 'sub–lattice' condition on the coefficients $\{n_e\}_e$ whenever H is non–trivial.)

THEOREM 3.2.18. [**N12**, Theorem 7.1.2] *Any element of $\mathcal{S}_{an}$ can be obtained as $\min_k\{\mathfrak{M}_k\}$ for certain finitely many integral monomial cycles $\{\mathfrak{M}_k\}_k$.*

4. Strongly flat semigroups and Seifert homology spheres

4.1. In this section let $M = \Sigma(\alpha_1, \ldots, \alpha_d)$ be a Seifert integral homology sphere. Thus, $\alpha_1, \ldots, \alpha_d \geq 2$ ($d \geq 3$) are pairwise relatively prime integers and both b_0 and $(\omega_1, \ldots, \omega_d)$ are uniquely determined by the Diophantine equation $\alpha(b_0 - \sum_{i=1}^d \omega_i/\alpha_i) = 1$.

First note that if we set $a_i := \alpha/\alpha_i$ then the greatest common divisor of $a_1, \ldots, a_{i-1}, a_{i+1}, \ldots, a_d$ is α_i, hence the system $\{a_i\}_{i=1}^d$ generates a strongly flat semigroup $G(a_1, \ldots, a_d)$. Moreover, using the above Diophantine equation one shows that $N(a_i) = 0$. Hence $G(a_1, \ldots, a_d) \subset \mathcal{S}_M$. This fact was already noticed by Can and Karakurt in [**CK14**, Theorem 4.1] too.

Let us provide the analytic interpretation/proof of this statement.

A weighted homogeneous analytic realization is given by a Brieskorn isolated complete intersection. This consists of $d-2$ equations in $(\mathbb{C}^d, 0)$ of type $\sum_{i=1}^d a_{ki} z_i^{\alpha_i} = 0$ ($k = 1, \ldots, d-2$), where the matrix $\{a_{ki}\}_{i,k}$ has full rank.

The strict transforms of the coordinate functions z_i are supported on the end–exceptional curves, their divisors $(z_i)_\Gamma$ are exactly the cycles E_i^* associated with the end–vertices. In particular, these cycles belong to $\mathcal{S}_{an}$. ($N(a_i) = 0$ shows that $\dim(R_{X,a_i}) = 1$, that is, each R_{X,a_i} is a 1–dimensional vector space generated by the coordinate function z_i.) The E_0–coefficient of E_i^* is a_i (cf. [**N05**]). Hence the inclusion $G(a_1, \ldots, a_d) \subset \mathcal{S}_M$ follows from Lemma 3.2.17 as well.

Can and Karakurt in [**CK14**, Theorem 4.1] observed also that the two numerical semigroups on the interval $[0, \gamma]$ agree. Thanks to Theorem 3.2.18 we show that in fact they agree everywhere.

THEOREM 4.1.1. *$\mathcal{S}_M = G(a_1, \ldots, a_d)$, that is, the strongly flat semigroups with at least three generators can be realized as numerical semigroups associated with Seifert integral homology spheres $\Sigma(\alpha_1, \ldots, \alpha_d)$.*

PROOF. We will apply Lemma 3.2.17 and Theorem 3.2.18. By Theorem 3.2.18 the cycles E_i^* generate $\mathcal{S}_{an}$ by the operations + and min. Note that the E_0–coefficient of $\min_k\{\mathfrak{M}_k\}$ is the E_0 coefficient of one of the integral monomial cycles $\mathfrak{M}_k$. On the other hand, since $H = L'/L = 0$, all the dual cycles E_i^* are automatically integral, hence $\mathcal{S}_M$ is generated by the E_0–coefficients of the end dual cycles E_i^*, that is, by the a_i's. □

In this case $\mathrm{lcm}(a_1, \ldots, a_d) = \alpha$, hence the right hand side of (2.1.1) (via (3.2.1) is $\gamma + \alpha$. Hence the two statements $f_{\mathcal{S}_M} = \gamma + \alpha$ and $f_{G(a_1,\ldots,a_n)} = \gamma + \alpha$, from Raczunas and Chrząstowski-Wachtel [**RChW96**] (cf. 2.1) and from Corollary 3.2.12*(b)* respectively, identify.

EXAMPLE 4.1.2. Note that the strongly flat semigroups might be realized another way as well. For example, the strongly flat semigroup $G(6,14,21)$ is the numerical semigroup associated with $\Sigma(2,3,7)$. However, it can be also realized as the semigroup of the Seifert rational homology sphere given by $(-2,0;(2,1),(2,1),(3,1),(3,1),(7,1),(7,1))$.

4.1.1. *The symmetry of* $\mathcal{S}_M$. It is really amazing how naturally appears another symmetry in this integral homology sphere situation, at this time for $\mathcal{S}_M$. First notice that

$$f_{\mathcal{M}_M} = \gamma \text{ and } \min\{\mathcal{M}_M\} = -\alpha. \tag{4.1.3}$$

If $\Gamma \neq -E_8$, then this follows from Corollaries 3.2.12*(a)* and 3.2.15. In the rational case (that is, for $-E_8$) $f_{\mathcal{M}} = \gamma = -1$, and $f_{\mathcal{S}_M} = \gamma + \alpha$ is still valid (use eg. (2.1.1)), and $\min\{\mathcal{M}_M\} = -\alpha$ too (use Proposition 3.2.13).

However, one can do much more. Since $N(\alpha) = \mathfrak{o} = 1$, by Proposition 3.2.11*(e)* $N(\alpha+\gamma-\ell) = N(\gamma-\ell)+1$, hence (3.2.14) reads as

$$N(\ell) + N(\alpha+\gamma-\ell) = -1 \text{ for any } \ell \in \mathbb{Z}. \tag{4.1.4}$$

this implies (compare with (3.2.16))

$$\ell \in \mathcal{S}_M \Leftrightarrow \alpha+\gamma-\ell \notin \mathcal{S}_M. \tag{4.1.5}$$

In particular, $\mathcal{S}_M$ is symmetric. Moreover, one verifies (either by a combination of (3.2.16) and (4.1.5), or by $(N(\ell)\geq 0) \Leftrightarrow (N(-\alpha+\ell)+1\geq 0)$ via Proposition 3.2.11*(e)*) that

(4.1.6)

$\mathcal{M}_M$ is generated as $\mathcal{S}_M$–module by <u>one</u> element, namely by $-\alpha = \min\{\mathcal{M}_M\}$.

In other words, $\mathcal{M}_M = -\alpha + \mathcal{S}_M$.

4.1.2. Lemma 3.2.17 and Theorem 4.1.1 have the following important messages:

(a) The combinatorial approach of the theory of strongly flat semigroups is replaced by topological and algebro-geometrical tools and methods. This connection also gives the possibility of wider applicability of the classical combinatorial theory in topology and singularity theory (and viceversa). It would be interesting to find a similar topological realizability for any $G(a_1,\dots,a_d)$.

(b) As the strongly flat $G(a_1,\dots,a_d)$ is the one–coordinate projection of an affine monoid $\mathcal{S}_{an}$, one should try to find similar embedding of arbitrary numerical semigroups into some canonical affine monoids, which hopefully carry more geometry (and definitely their study is more challenging).

(c) As $G(a_1,\dots,a_d)$ is the E_0–projection of $\mathcal{S}_{an}$, one can ask, what about the other projection? Are they 'interesting' numerical semigroups?

Eg., let us determine the numerical semigroup $\mathcal{S}_{end}$ obtained from $\mathcal{S}_{an}$ associated with $\Sigma(\alpha_1,\dots,\alpha_d)$ projected on the end–vertex associated with index d.

Again, $\mathcal{S}_{an}$ is generated by the duals of the end–vertices $\{E_i^*\}_{i=1}^d$. Their E_d–coefficients were already listed in front of (3.2.10). Namely, the E_d coefficient of E_i^* $(1\leq i<d)$ is $1/(|e|\alpha_i\alpha_d) = \alpha/(\alpha_i\alpha_d)$. Hence the first $(d-1)$ generators generate the usual G–type monoid associated with $\alpha_1,\dots,\alpha_{d-1}$. On the other hand, the E_d–coefficient of E_d^* is $1/(|e|\alpha_d^2)+\omega_d'/\alpha_d$, which after a computation (and using the Diophantine equation) is $\lceil \alpha_1\cdots\alpha_{d-1}/\alpha_d\rceil$. Hence, $\mathcal{S}_{end}$ is generated by

$$\alpha_2\cdots\alpha_{d-1},\ \alpha_1\alpha_3\cdots\alpha_{d-1},\cdots,\ \alpha_1\cdots\alpha_{d-2},\ \lceil \alpha_1\cdots\alpha_{d-1}/\alpha_d\rceil,$$

where all α_i are pairwise relative prime.

Eg., if α_d is very large compared with the others then the last generator is 1, hence merely this generator generates $\mathcal{S}_{end}$, which is $\mathbb{N}$. There are choices when the last generator is already in the monoid generated by the others, hence, as generator can be deleted. And there are also examples when this last generator is the conductor of the monoid generated by the first $d-1$ generators (take eg. $\Sigma(2,3,7)$). So, in general, the computation of the minimal set of generators, and also of the Frobenius number, can be a challenge.

(d) The aim of the forthcoming sections is to generalize the above results to the case of Seifert rational homology spheres: we will determine the Frobenius numbers of $\mathcal{S}_M$ and of the $\mathcal{S}_M$–module $\mathcal{M}_M$. In fact, the module will play a crucial auxiliary role: first we compute its Frobenius number, and then we reduce the computation of $f_{\mathcal{S}}$ to $f_{\mathcal{M}}$ associated with another aide graph.

However, it will turn out that the tools are far to be merely arithmetical, they rely on the mathematical machinery of 'generalized Laufer computation sequences', which originally was used in the computation of analytic invariants of surface singularities, and later in the computation of the lattice cohomology of their links. In order to make the presentation more complete, in the following we review some motivations, definitions and facts from [**N05**, **LN15**] about the lattice cohomology and then we present the special computation sequences needed in the proof.

5. Topology and arithmetics of the quasi-linear function

5.1. $\{N(\ell)\}_\ell$ and Heegaard–Floer homology. As a preamble, let us try first to present only in a few words how the sequence $\{N(\ell)\}_\ell$ from (3.2.2) appears in modern low–dimensional topology.

Heegaard–Floer homology by Ozsváth and Szabó [**OSz04**] is one of the most important and highlighted invariants of 3-manifolds. In the case of negative definite Seifert 3–manifolds the results of [**OSz03**] and [**N05**] have shown that the calculation of Heegaard–Floer homology is purely combinatorial. In fact, it is isomorphic with the lattice cohomology introduced by the second author [**N08**] (see also [**N07**]), which provides a combinatorial recipe for its computation. (In fact, everything generalizes to the case of 'almost rational' graphs, see [**N05**, **N07**].)

To start with, we define the discrete function $\tau : \mathbb{N} \to \mathbb{Z}$ by the recurrence

$$\tau(\ell+1) = \tau(\ell) + 1 + N(\ell) \tag{5.1.1}$$

with initial condition $\tau(0) = 0$. The statement is that this function (in fact, already its 'local minimum and maximum points and values') determines the Heegaard–Floer homology of M.

It is a mystery that this topologically motivated 'difference τ–function' is exactly the quasi-linear function $\ell \mapsto 1 + N(\ell)$ imposed by the (analytic) Poincaré series of the local ring, determined by Pinkham's theory. For more, see sections 5.2–5.3, especially Remark 5.3.2.

This theoretical result initiated several work regarding the concrete properties of the function $N(\ell)$, see e.g. the work of Can and Karakurt in [**CK14**] in the integral homology sphere case regarding the local extrema of τ. Their results have been extended in [**KL15**] and applied to the 'botany problem' for Seifert homology spheres. In this work further deep structural connections with Heegaard–Floer theory are established.

5.2. Lattice cohomology and generalized Laufer computation sequences. The statements of this subsection will not be used later, they have purely informative and motivational role.

5.2.1. *Definition and reduction.* To define the lattice cohomology one first needs a lattice $\mathbb{Z}^n$ with fixed base elements $\{E_i\}_{i=1}^n$, which provide a cubical decomposition of $\mathbb{R}^n = \mathbb{Z}^n \otimes \mathbb{R}$: the 0-cubes are the lattice points $l \in \mathbb{Z}^n$, a q-cube will be denoted by (l, I) for some $I \subset \{1, \ldots, n\}$ with $|I| = q$ and it is determined by its vertices $l + \sum_{j\in J} E_j$ for any $J \subset I$. One also considers a weight function $w : \mathbb{Z}^n \to \mathbb{Z}$ bounded below, and for each cube (l, I) one defines $w(l, I) := \max\{w(v) \ : \ v \text{ vertex of } (l, I)\}$. Then for each integer $N \geq \min w$ one defines the simplicial complex S_N as the union of all cubes with $w(l, I) \leq N$. The lattice cohomology associated with the pair $(\mathbb{Z}^n, w)$ is the $\mathbb{Z}[U]$-modules $\{\mathbb{H}^q(\mathbb{Z}^n, w)\}_{q\geq 0}$ defined by $\mathbb{H}^q(\mathbb{Z}^n, w) := \bigoplus_{N\geq \min w} H^q(S_N, \mathbb{Z})$. The U-action consists of the restriction maps induced by the inclusions $S_N \hookrightarrow S_{N+1}$.

In some cases one takes only part of the lattice, say $(\mathbb{Z}_{\geq 0})^n$, and only those cubes which are supported by this part, a weight function $w : (\mathbb{Z}_{\geq 0})^n \to \mathbb{Z}$, and one defines in the same way $\mathbb{H}^q((\mathbb{Z}_{\geq 0})^n, w)$.

In the case of normal surface singularities, we fix a good resolution Γ and construct the lattice cohomology using the lattice $L \cong \mathbb{Z}^{|\mathcal{V}|}$ and the weight function χ_{k_r} associated with any characteristic orbit $[k]$, usually denoted by $\{\mathbb{H}^*(L, [k])\}_{[k]}$. In fact, it it independent of the choice of the resolution, it depends only on the link and the *spin*^{c}–structure codified by $[k]$. It serves as a cohomological categorification for the Seiberg–Witten invariant, see eg. [**N05, N11**].

The Reduction theorem of [**LN15**] allows to reduce the rank of the lattice and evidences the complexity of the cohomology. In fact, in the case of Seifert rational homology spheres (and in general, in the case of plumbed manifolds associated with 'almost rational' graphs) the higher cohomology $\mathbb{H}^{\geq 1}(L, [k]) = 0$, while $\mathbb{H}^0(L, [k])$ is reduced to the rank–one case $\mathbb{H}^0(\mathbb{Z}_{\geq 0}, w_{[k]})$, where $\ell \mapsto w_{[k]}(\ell)$ $(\ell \geq 0)$ is determined via some special universal cycles, denoted by $\{x_{[k]}(\ell)\}_\ell$ in [**LN15**], [**N05**], and via $w_{[k]}(\ell) = \chi_{k_r}(x_{[k]}(\ell))$. If $[k] = [K]$, then $w_{[k]}(\ell) = \tau(\ell)$ from (5.1.1), see Remark 5.3.2.

We will not provide here any cohomology computation, however by the above comment we wished to inform the reader about the origins of the technical tools from below, and why were they developed. The interested reader might consult the articles mentioned above and the references therein.

5.2.2. *Generalized Laufer computation sequences.*

One of the main technical tools of the computation of the lattice cohomology is the 'computation sequence'. A computation sequence has the form $\{z_i\}_{i=0}^k$, $z_i \in L'$, such that for any i one determines by some algorithmic rule a vertex $v(i) \in \mathcal{V}$ using which one sets $z_{i+1} := z_i + E_{v(i)}$. It is the discrete latticial analogue of a continuous path. It connects z_0 to z_k. The cycle z_k usually has some universal property targeted by the algorithm.

As we already mentioned, we have to compute the simplicial cohomology of the simplicial complexes S_N (a union of latticial cubes). Note also that on S_N we have the restriction of the weight function χ_{k_r}. Recall that in Morse theory, in the presence of a height function f, we wish to contract our space along the flow determined by f (by decreasing f), or to find extremal points as universal points at the 'end' of flow–lines. In our case we proceed similarly, the computation sequences are the 'flow–paths' determined by an initial lattice point and the weight (hight) function χ_{k_r}. In this way we can contract S_N, or we find some 'extremal' lattice points with universal properties.

The first computation sequence in L was introduced by Laufer by a special optimization algorithm (flowing in the direction of the Lipman cone $\mathcal{S}_{top}$), it is χ–decreasing, starts with E and ends at Z_{min}, the Artin's cycle, the minimal non–zero element of the Lipman cone $\mathcal{S}_{top}$ [**L72**]. In this way Laufer succeeded to define a rationality criterion (basically equivalent with the contractibility of all spaces S_N, see also [**N05, N07**]). The tool was successfully used later in the theory of elliptic singularities as well [**L77, N99b, O05**]. For application in lattice cohomology see [**N05, N07, LN15**].

5.2.3. Note that the restriction of χ_{k_r} on L (where $k_r = K + 2s_h$) can be rewritten as the restriction of χ on the shifted lattice $s_h + L = r_h + L$. Indeed, for any $l \in L$ one has $\chi_{k_r}(l) = \chi(l + s_h) - \chi(s_h)$. Below in our computations we will use this second approach. In fact, we will not need all the possible classes, but only those associated with $h = 0$ and $h = [Z_k]$ (if Γ is numerically Gorenstein, then even these two classes agree). For them the associated package of invariants will be compared by duality.

5.3. Universal cycles and their concatenated sequences. The next lemma generalizes similar statements form [**N05**, §7], [**N07**, Prop. 4.3.3] and [**LN15**, §3], see also [**L72**].

LEMMA 5.3.1. *Set $\mathcal{V}^* := \mathcal{V} \setminus \{v_0\}$, and take r_h associated with $h \in H$ as above.*

(1) *For any $\ell \in \mathbb{Z}_{\geq 0}$ there exists a unique minimal (rational) cycle x^ℓ with respect to the following properties:*
 (a) *$m_0(l') = m_0(r_h) + \ell$ and*
 (b) *$l' \in (r_h + L)$ and $(l', E_v) \leq 0$ for any $v \in \mathcal{V}^*$.*
(2) *One has $x^0 \geq r_h$ and $x^{\ell+1} \geq x^\ell + E_0$ for any $\ell \geq 0$.*
(3) *For any $l' \in r_h + L$ satisfying (a) we have $\chi(l') \geq \chi(x^\ell)$.*
(4) *There is a 'generalized Laufer computation sequence' $\{z_i\}_{i\geq 0}$ connecting x^ℓ and $x^{\ell+1}$ as follows: set $z_0 := x^\ell$, $z_1 := x^\ell + E_0$ and assume that z_i ($i \geq 1$) is already constructed. If z_i does not satisfy (b), then we have $(z_i, E_{v(i)}) > 0$ for some $v(i) \in \mathcal{V}^*$ and set $z_{i+1} = z_i + E_{v(i)}$. Otherwise, we necessarily have $z_i = x^{\ell+1}$.*
 Similarly, there is a 'generalized Laufer computation sequence' $\{z_i\}_{i\geq 1}$ connecting r_h and x^0 as follows: set $z_1 := r_h$, and then repeat the definition from the previous case.
(5) *$\chi(x^0) = \chi(r_h)$ and $\chi(x^{\ell+1}) - \chi(x^\ell) = 1 - (x^\ell, E_0)$ for $\ell \geq 0$.*

PROOF. *(1)* The proof is similar to one given by Laufer targeting Z_{min} in [**L72**]. The negative definiteness of the intersection form guarantees the existence of a cycle with properties (a) and (b). Then, it is enough to prove that if x_1, x_2 satisfy (a) and (b) then $x := \min\{x_1, x_2\}$ also satisfies (a) and (b). Indeed, for any $v \in \mathcal{V}^*$ there is at least one index $i \in \{1, 2\}$ such that $E_v \notin |x_i - x|$. Hence, $(x, E_v) = (x_i, E_v) + (x - x_i, E_v) \leq 0$.
(2) Assume that we can write $x^\ell = x_1^\ell - x_2^\ell$ such that $x_i^\ell \geq 0$ for $i \in \{1, 2\}$ and $|x_1^\ell| \cap |x_2^\ell| = \emptyset$. Then for any $E_v \in |x_2^\ell|$ we have $(-x_2^\ell, E_v) \leq (x^\ell, E_v) \leq 0$. In particular, $(x_2^\ell)^2 \geq 0$ which implies $x_2^\ell = 0$ by the negative definiteness of the intersection form. Since $x^\ell \in r_h + L$ we get $x^\ell \geq r_h$.

Finally observe that $x^{\ell+1}-E_0$ satisfy (a)–(b) for ℓ, hence by the minimality of x^ℓ the last inequality follows as well.
(3) Assume first that $l'\leq x^\ell$. Then there is a generalized Laufer computation sequence $\{z_i\}_{i=0}^m$ connecting l' with x^ℓ as follows: set $z_0:=l'$ and assume that z_i is already constructed. If for some $v(i)\in\mathcal{V}^*$ one has $(z_i,E_{v(i)})>0$ continue with $z_{i+1}=z_i+E_{v(i)}$. We claim that along these steps still $z_i\leq x^\ell$. We verify this by induction, we assume that $z_i\leq x^\ell$ and we prove that $z_{i+1}\leq x^\ell$. For this observe that $m_{v(i)}(z_i)=m_{v(i)}(x^\ell)$ cannot happen. Indeed, in such case, $(x^\ell,E_{v(i)})=(x^\ell-z_i,E_{v(i)})+(z_i,E_{v(i)})>0$, a contradiction. Hence necessarily $m_{v(i)}(z_i)<m_{v(i)}(x^\ell)$, ie. $z_{i+1}\leq x^\ell$ too. In particular, the constructed sequence $\{z_i\}_i$ must stop, say at z_m, and $z_m\leq x^\ell$. Then, by the minimality of x^ℓ from part *(1)* we have $z_m=x^\ell$. Moreover, along the computation sequence $\chi(z_{i+1})=\chi(z_i)+1-(z_i,E_{v(i)})\leq\chi(z_i)$ for any $0\leq i<m$. Hence, $\chi(l')\geq\chi(x^\ell)$.

In general, we write $l'=x^\ell-l_1+l_2$ such that $l_1,l_2\in L$ with $l_1\geq 0$, $l_2\geq 0$, both are supported on $\mathcal{V}^*$ and $|l_1|\cap|l_2|=\emptyset$. Then $\chi(l')=\chi(x^\ell-l_1)+\chi(l_2)+(l_1,l_2)-(x^\ell,l_2)$. Since $(l_1,l_2)\geq 0$ by their supports, $-(x^\ell,l_2)\geq 0$ by (b), and $\chi(l_2)\geq 0$ by the fact that it is supported on a rational subgraph (cf. [**A62, A66**]), we get $\chi(l')\geq\chi(x^\ell-l_1)$. On the other hand, we have proved $\chi(x^\ell-l_1)\geq\chi(x^\ell)$.
(4) Use the inequalities from *(2)* and the computation sequence from the proof of *(3)*.
(5) One verifies that the computation sequence from *(4)* is χ–constant starting from $i=1$ (use Laufer's rationality criterion, for details see [**N05**]). Hence $\chi(x^{\ell+1})-\chi(x^\ell)=\chi(x^\ell+E_0)-\chi(x^\ell)$. □

The cycles x^ℓ depend on h (though we did not emphasize it notationally). We will adopt the following notations: if $h=0$ then we write $x(\ell):=x^\ell$ (associated with $h=0$) and for $h=[Z_K]$ we set $x^*(\ell):=x^\ell$ (associated with $h=[Z_K]$).

REMARK 5.3.2. For $h=0$ one verifies (see [**N05**, Prop. 11.11]) that

$$x(\ell)=\ell E_0+\sum_{i=1}^{d}\left\lceil\frac{\ell\omega_i}{\alpha_i}\right\rceil E_{i1}+\text{terms supported on other base elements.}$$

Hence $\chi(x(\ell+1))-\chi(x(\ell))=1-(x(\ell),E_0)=1+N(\ell)$.

REMARK 5.3.3. Regarding the discussion from subsection 5.2 we add the following. For $h=0$ the cycles $\{x(\ell)\}_{\ell\geq 0}$ are exactly the cycles which determine the reduced ('half') lattice $(\mathbb{Z}_{\geq 0},w)$, by $w(\ell)=\chi(x(\ell))=\tau(\ell)$. If $h\neq 0$ then the cycles x^ℓ and $x_{[k]}(\ell)$ (mentioned in section 5.2) and the corresponding weight functions are connected by a shift (given by the identity from 5.2.3).

For the description of cycles x^ℓ for any h see [**N05**, Prop. 11.11] again.

5.3.1. *The computation sequence targeting $\mathcal{S}'$.* There is another/similar computation sequence (valid for any connected negative definite graph), which starts with any element and ends in $\mathcal{S}'$.

LEMMA 5.3.4. *Fix some $h\in H$ and $l'\in r_h+L$.*

1. *There exists a unique minimal element $s(l')$ of $(r_h+L)\cap\mathcal{S}'$.*
2. *$s(l')$ can be found via the following computation sequence $\{z_i\}_i$ connecting l' and $s(l')$: set $z_0:=l'$, and assume that z_i $(i\geq 0)$ is already constructed. If $(z_i,E_{v(i)})>0$ for some $v(i)\in\mathcal{V}$ then set $z_{i+1}=z_i+E_{v(i)}$. Otherwise $z_i\in\mathcal{S}'$ and necessarily $z_i=s(l')$.*

(3) $\chi(z_{i+1}) \leq \chi(z_i)$ *along the (any such) sequence.*

In general the choice of the individual vertex $v(i)$ might not be unique, nevertheless the final output $s(l')$ is unique.

The proof is almost identical with the proof of Lemma 5.3.1 and is left to the reader.

In particular, one can start with the cycle $l' = r_h$ and obtain $s(r_h) = s_h$.

LEMMA 5.3.5. $Z_K \geq s_{[Z_K]}$.

PROOF. If Γ is a minimal resolution (that is, $b_0 > 1$) then by adjunction formulae we get that $Z_K \in \mathcal{S}'$. Hence, by minimality of $s_{[Z_K]}$ one has $Z_K \geq s_{[Z_K]}$.

If $b_0 = 1$ then by Laufer criterion [**L72**] (X, o) is non–rational, ie. $p_g > 0$. Consider the computation sequence $\{z_i\}_{i=0}^m$ from Lemma 5.3.4*(2)* connecting $r_{[Z_K]}$ with $s_{[Z_K]}$. We wish to show inductively that for each term z_i one has $Z_K \geq z_i$ and $h^1(\mathcal{O}_{Z_K - z_i}) = p_g$.

For $i = 0$ is true, since $Z_K \geq r_{[Z_K]}$ (cf. subsection 3.2.5), and also $h^1(\mathcal{O}_{Z_K - r_{[Z_K]}})$ $= h^1(\mathcal{O}_{\lfloor Z_K \rfloor}) = p_g$ (by Kodaira or Grauert–Riemenschneider type vanishing $h^1(\widetilde{X}, \mathcal{O}_{\widetilde{X}}(-\lfloor Z_K \rfloor)) = 0$).

Assume that for some i one has $Z_K \geq z_i$, $h^1(\mathcal{O}_{Z_K - z_i}) = p_g$, and $z_i < s_{[Z_K]}$. Then the next term in the sequence is $z_{i+1} = z_i + E_{v(i)}$ with $(E_{v(i)}, z_i) > 0$.

Assume first that the $E_{v(i)}$–coefficients of $Z_K - z_i$ is positive. Then $Z_K \geq z_{i+1}$ holds. Furthermore, in the cohomological long exact sequence of

$$0 \to \mathcal{O}_{E_{v(i)}}(-Z_K + z_{i+1}) \to \mathcal{O}_{Z_K - z_i} \to \mathcal{O}_{Z_K - z_{i+1}} \to 0$$

the Chern number $(E_{v(i)}, -Z_K + z_{i+1}) > -2$, hence $h^1(\mathcal{O}_{E_{v(i)}}(-Z_K + z_{i+1})) = 0$, which implies $h^1(\mathcal{O}_{Z_K - z_{i+1}}) = h^1(\mathcal{O}_{Z_K - z_i}) = p_g$.

Assume next the opposite: $m_{v(i)}(Z_K - z_i) = 0$. Then $0 < (E_{v(i)}, z_i) = (E_{v(i)}, z_i - Z_K) + (E_{v(i)}, Z_K) \leq E_{v(i)}^2 + 2$, hence necessarily $v(i) = v_0$ (since for other vertices $E_{v(i)}^2 + 2 \leq 0$). This means that $Z_K - z_i \in L_{\geq 0}$ is supported on the legs, but since these subgraphs are strings supporting rational singularities, this fact contradicts with $h^1(\mathcal{O}_{Z_K - z_i}) = p_g > 0$. □

5.3.2. *Duality properties.* Consider the cycle $\{x(\ell)\}_{\ell \geq 0}$ and $\{x^*(\ell)\}_{\ell \geq 0}$ defined after Lemma 5.3.1. Let us define $\delta := \min\{\ell : x^*(\ell) \in \mathcal{S}'\}$.

Then consider the concatenated sequences from Lemma 5.3.1 connecting $r_h \mapsto x^0 \mapsto x^1 \mapsto \cdots \mapsto x^\delta$ (if $h = 0$ then this is the one–element sequence 0). This (with special choices of the vertices $v(i)$) satisfies the requirements of the computation sequence from Lemma 5.3.4. This shows that

$$x^*(\delta) = s_{[Z_K]} \quad \text{and} \quad \delta = m_0(s_{[Z_K]} - r_{[Z_K]}). \tag{5.3.6}$$

Furthermore, since this concatenated sequence has the property $\chi(z_{i+1}) \leq \chi(z_i)$ by Lemma 5.3.4*(3)*,

$$(x^*(\ell), E_0) > 0 \text{ for all } \ell = 0, \ldots, \delta - 1, \text{ however } (x^*(\delta), E_0) \leq 0. \tag{5.3.7}$$

Set $\Delta := m_0(Z_K - r_{[Z_K]})$. By Lemma 5.3.5 one has $\Delta \geq \delta$.

PROPOSITION 5.3.8. *(a) The following 'duality property' holds:*

$$\chi(x^*(\ell)) = \chi(x(\Delta - \ell)) \quad \textit{for } 0 \leq \ell \leq \Delta.$$

(b) Let $N^*(\ell)$ be defined similarly as $N(\ell)$, but for the x^*–sequence, namely, $N^*(\ell) := -(x^*(\ell), E_0)$, cf. Lemma 5.3.1*(5). Then*

$$N(\ell) + N^*(\Delta - 1 - \ell) = -2 \text{ for any } 0 \leq \ell \leq \Delta - 1.$$

(If $Z_K \in L$ then $x(\ell) = x^*(\ell)$, hence $N(\ell) = N^*(\ell)$ too, and we recover Prop. 3.2.13.*)*

PROOF. (a) Note that $m_0(Z_K - x^*(\ell)) = \Delta - \ell = m_0(x(\Delta - \ell))$. Hence, by Lemma 5.3.1*(3)* applied for $x(\Delta - \ell)$ gives $\chi(Z_K - x^*(\ell)) \geq \chi(x(\Delta - \ell))$. Symmetrically, $m_0(x^*(\ell)) = m_0(Z_K - x(\Delta - \ell))$, hence by the same argument applied for $x^*(\ell)$ gives $\chi(x^*(\ell)) \leq \chi(Z_K - x(\Delta - \ell))$. Finally observe that $\chi(Z_K - y) = \chi(y)$.

(b) By Remark 5.3.2, Lemma 5.3.1*(5)* and part (a) of this Proposition one has

$$\begin{aligned} 1 + N(\ell) &= \chi(x(\ell+1)) - \chi(x(\ell)) = \chi(x^*(\Delta - \ell - 1)) - \chi(x^*(\Delta - \ell)) \\ &= -1 + (x^*(\Delta - \ell - 1), E_0). \end{aligned} \tag{5.3.9}$$

Then, by definition, the last term is $-1 - N^*(\Delta - \ell - 1)$. □

5.4. The Frobenius number of $\mathcal{M}$. Let M be a negative definite Seifert rational homology sphere and we consider the $\mathcal{S}$-module $\mathcal{M}$ associated with it. Recall that $p_g = P_0^+(1)$. Then by (3.2.5) and (3.2.6) we obtain that $\mathbb{Z}_{\geq 0} \subset \mathcal{M}$ if and only if $p_g = 0$, cf. (3.2.7).

THEOREM 5.4.1. *Assume that Γ is not rational and set $s := m_0(s_{[Z_K]})$. Then*

$$f_{\mathcal{M}} = \Delta - \delta - 1 = \gamma - s = m_0(Z_K - s_{[Z_K]}) - 1 \geq 1.$$

In particular, $\gamma = s + f_{\mathcal{M}} \geq s + 1 \geq 1$.

PROOF. Consider again the identities from (5.3.9). The last term, by (5.3.7), is ≥ 0 for $\Delta - \ell - 1 \in \{0, \ldots, \delta - 1\}$, while it is < 0 for $\Delta - \ell - 1 = \delta$. This means that $\{\gamma - s + 1, \ldots, \gamma - m_0(r_h)\} \subset \mathcal{M}$, but $\gamma - s \notin \mathcal{M}$. Furthermore, $\ell \in \mathcal{M}$ for any $\ell > \gamma$ by Proposition 3.2.11*(b)*. □

COROLLARY 5.4.2. *Assume that Γ is non–rational.*

(a) If Γ is numerically Gorenstein, then $s_{[Z_K]} = 0$, hence $f_{\mathcal{M}} = \gamma$. This is compatible with Corollary 3.2.12*(a) and with the integral homology sphere case from* [**CK14**].

(b) Assume that $\gamma \in \mathbb{Z}$, but $Z_K \notin L$ (see eg. $Sf = (-1, 0; (6,1)_{i=1}^4)$). Then $s_{[Z_K]} \geq r_{[Z_K]} \neq 0$. But $s_{[Z_K]}$ being a non-zero element of $\mathcal{S}'$, all its E_v–coefficients are strict positive. Hence $s > 0$ and $f_{\mathcal{M}} < \gamma$.

(c) If (X, o) is weighted homogeneous normal surface singularity then $\deg P_0^+ = \gamma - s$.

REMARK 5.4.3. The authors know no closed formula for the number s in terms of the Seifert invariants. By the definition it is the E_0–coefficient of $s_{[Z_K]}$ (hence, it can be determined from the E_v^*-coefficients of $s_{[Z_K]}$ too). However, the only fact known about $s_{[Z_K]}$ is that the E_v^*-coefficients are determined as a solution of a system of Diophantine quasipolynomial inequalities (see [**N05**, Proposition 11.5]). Therefore, the formula for the Forbenius number given by the Theorem 5.4.1 is explicit up to the number s given by a Laufer computation sequence associated with the class $[Z_K]$ (or the solution of that system of inequalities).

EXAMPLE 5.4.4. Let us denote by $M_{(70)}$ the Seifert rational homology sphere associated with the following plumbing graph:

$\Gamma_{(70)}$:

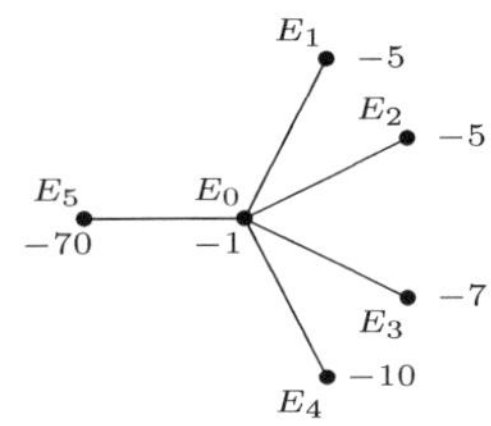

with $Sf = (-1, 0; (5,1), (5,1), (7,1), (10,1), (70,1))$

We write all the cycles $l' = \sum_{i=0}^{5} l'_i E_i$ in the form $l' = (l'_0, l'_1, \dots, l'_5)$. By solving the adjunction equations (3.1.1) we get $Z_K = (47/6, 13/6, 13/6, 11/6, 19/12, 13/12)$, hence $\gamma = 41/6$ and $\Gamma_{(70)}$ is not numerically Gorenstein. Clearly

$$r_{[Z_K]} = (5/6, 1/6, 1/6, 5/6, 7/12, 1/12).$$

$s_{[Z_K]}$ is calculated by the Laufer computation sequence from section 5.3 and we get

$$s_{[Z_K]} = r_{[Z_K]} + 3E_0 + E_1 + E_2 = (23/6, 7/6, 7/6, 5/6, 7/12, 1/12).$$

By Theorem 5.4.1 the Frobenius number of the module $\mathcal{M}_{(70)}$ is $\gamma - s = 41/6 - 23/6 = 3$.

6. The semigroup $\mathcal{S}_M$

6.1. Semigroups generated by periodic subadditive functions. We consider the function $f_M : \mathbb{Z}_{\geq 0} \to \mathbb{Z}_{\geq 0}$, $f_M(\ell) = \frac{1}{|e|} \sum_{i=1}^{d} \{-\omega_i \ell/\alpha_i\}$, where we define the rational part as $x := x - [x]$ for any $x \in \mathbb{Q}$. Then it is a subadditive function with period $\alpha = \mathrm{lcm}(\alpha_1, \dots, \alpha_d)$ in the following sense:

$$f(\ell_1 + \ell_2) \leq f(\ell_1) + f(\ell_2), \ \ f(0) = 0, \ \ \text{and} \ \ f(\ell + \alpha) = f(\ell) \ \ \text{for every } \ell_1, \ell_2, \ell \in \mathbb{Z}_{\geq 0}.$$

Usually, such a function defines a semigroup $\mathcal{S}_f = \{\ell \in \mathbb{Z}_{\geq 0} \ : \ f(\ell) \leq \ell\}$, see [**R07**]. In this case, for the above function f_M, the semigroup associated with f_M is exactly the numerical semigroup $\mathcal{S}_M$.

Again, by [**R07**], the Apéry set of $\mathcal{S}_f$ with respect to an element $\alpha \in \mathcal{S}$ consists of $\lceil (f(\ell) - \ell)/\alpha \rceil \alpha + \ell$ for $\ell = 0, \dots, \alpha - 1$. Thus, in our case, (where $\alpha = \gcd\{\alpha_i\}$ as in the previous sections)

$$Ap(\mathcal{S}_M, \alpha) = \{\lceil -N(\ell)/\mathfrak{o} \rceil \alpha + \ell \, : \, \ell = 0, \dots, \alpha - 1\}.$$

Then the Frobenius number can be expressed by Selmer's formula as

$$f_{\mathcal{S}_M} = \max\{Ap(\mathcal{S}_M, \alpha)\} - \alpha = \max_{\ell=0}^{\alpha-1} \{\lceil -N(\ell)/\mathfrak{o} \rceil \alpha + \ell\} - \alpha.$$

Furthermore, the number of gaps $\#\{\mathbb{N} \setminus \mathcal{S}_M\}$ is

$$\frac{1}{\alpha} \sum_{w \in Ap(\mathcal{S}_M, \alpha)} w - \frac{\alpha - 1}{2} = \sum_{\ell=0}^{\alpha-1} \left\lceil - \frac{N(\ell)}{\mathfrak{o}} \right\rceil.$$

Though the above formula gives a method to compute $f_{\mathcal{S}_M}$, still is difficult to use it to get an explicit closed–form expression in terms of Seifert invariants.

The goal of this section is to find a different expression for $f_{\mathcal{S}_M}$ in terms of the lattice L.

6.2. The Frobenius number of $\mathcal{S}$. The idea is to construct a new manifold out of the original one, and to prove that the Frobenius number of the module associated with the new manifold coincides with the Frobenius number of the semigroup associated with the original Seifert rational homology sphere. Since the Frobenius number of modules is determined in Theorem 5.4.1, we get the desired $f_{\mathcal{S}_M}$ as well.

6.2.1. We consider a negative definite Seifert rational homology sphere M with Seifert invariants $(-b_0;(\alpha_i,\omega_i)_{i=1}^d)$ and its associated semigroup $\mathcal{S}_M$ defined by the quasi-linear function $N(\ell)$.

Let $M_{(n)}$ be a newly defined Seifert rational homology sphere characterized by the Seifert invariants $(-b_0;(\alpha_i,\omega_i)_{i=1}^d),(n,1))$ for some integer $n\geq 1/|e|$. By this choice of n the new graph is also negative definite since its orbifold Euler number $e_{(n)}=e+1/n<0$. We can associate with $M_{(n)}$ the module $\mathcal{M}_{(n)}$ defined by equation (3.2.6). Then we have the following comparison result.

PROPOSITION 6.2.1. *For big enough $n\gg 0$, $\mathcal{M}_{(n)}$ becomes a semigroup independent of n, and it coincides with $\mathcal{S}_M$.*

PROOF. Let $N_{(n)}(\ell)$ be the quasi-linear function associated with $\mathcal{M}_{(n)}$. Then $N(\ell)=N_{(n)}(\ell)+\lceil \ell/n\rceil$ (†). Therefore, for $\ell>0$, $\ell\in\mathcal{M}_{(n)}$ implies $\ell\in\mathcal{S}_M$.

For the opposite inclusion we need some preparation. First we observe that

$$N(\ell)\geq t\cdot\mathfrak{o}-1 \quad \text{for any} \quad \ell>\gamma-s+t\cdot\alpha \quad\text{and}\quad t\in\mathbb{Z}_{\geq 0}, \tag{6.2.2}$$

Indeed, one has $N(\alpha)=\mathfrak{o}$, hence the $t=0$ case is true since $\ell>\gamma-s$ implies $N(\ell)\geq -1$ by Theorem 5.4.1. Then we can proceed by induction using $N(\ell+\alpha)=N(\ell)+\mathfrak{o}$, cf. Proposition 3.2.11*(e)*.

This system implies that for $\gamma-s+t\cdot\alpha<\ell\leq\gamma-s+(t+1)\cdot\alpha$ (‡) one has $1+N_{(n)}(\ell)=1+N(\ell)-\lceil\ell/n\rceil\geq t\cdot\mathfrak{o}-\lceil\ell/n\rceil$. Then we choose n sufficiently large such that $t\cdot\mathfrak{o}-\lceil\ell/n\rceil$ is non–negative for any $t\geq 1$ and for any ℓ satisfying (‡), and additionally $n\geq\gamma-s+\alpha$.

Now we can conclude. Take $\ell\in\mathcal{S}_M$. If $0<\ell\leq\gamma-s+\alpha$ then $\ell\leq n$ too, hence by the expression (†) one has $\ell\in\mathcal{M}_{(n)}$. Otherwise, $1+N_{(n)}(\ell)\geq t\cdot\mathfrak{o}-\lceil\ell/n\rceil\geq 0$. Hence $\ell\in\mathcal{M}_{(n)}$ again. □

6.2.2. *The $b_0\geq d$ case.* Since $N(1)=b_0-\sum_i\lceil\omega_i/\alpha_i\rceil=b_0-d$ we obtain that $\mathcal{S}_M=\mathbb{N}\Leftrightarrow N(1)\geq 0\Leftrightarrow b_0\geq d$.

Maybe is worth to mention that if $b_0\geq d$ then the fundamental cycle of Γ (the minimal element of $\mathcal{S}\setminus\{0\}$) is reduced , hence (X,o) is rational by [**L72**] (singularities with reduced fundamental cycle are called 'minimal rational').

In fact, in such cases, the graph of $M_{(n)}$ is also rational (for a proof see [**GN18**, Th. 4.1.3]). Hence we can also argue as follows: $\mathcal{M}_{(n)}=\mathbb{N}$ by (3.2.7) and $\mathcal{S}_M=\mathcal{M}_{(n)}$ by Proposition 6.2.1.

6.2.3. *The $b_0<d$ case.* By the above discussion if $b_0<d$ then $f_{\mathcal{S}_M}$ is well–defined and is ≥ 1.

We denote by $\check{s}$ the E_0-coefficient of $s_{[Z_K+E_0^*]}$, where (via the standard notation) $s_{[Z_K+E_0^*]}$ is the unique minimal element in $\mathcal{S}'$ associated with $[Z_K+E_0^*]\in H$.

THEOREM 6.2.3. *If $b_0<d$ then*

$$f_{\mathcal{S}_M}=\gamma+\frac{1}{|e|}-\check{s}.$$

EXAMPLE 6.2.4. (1) If $Z_K \in L$ and $\mathfrak{o} = 1$ (eg. if M is an integral homology sphere) then $1/|e| = \alpha$, $\check{s} = 0$, hence $f_{\mathcal{S}_M} = \gamma + \alpha$, compare with Section 4.

(2) If $\mathfrak{o} = 1$ then $[E_0^*] = 0$, hence $\check{s} = s$. Therefore $f_{\mathcal{S}_M} = \gamma + \alpha - s$.

(3) If $Z_K \in L$ then $1/|e| = \alpha/\mathfrak{o}$. (Note also that always is true that $m_0(E_0^*) = -(E_0^*, E_0^*) = \prod_i \alpha_i/|H| = 1/|e|$.) Since $\check{s} = m_0(s_{[E_0^*]})$ and E_0^* is already in $\mathcal{S}'$, we get $E_0^* - s_{[E_0^*]} \in L_{\geq 0}$, hence $1/|e| \geq \check{s}$. In particular, $f_{\mathcal{S}_M} = \gamma + m_0(E_0^* - s_{[E_0^*]}) \geq \gamma$. This combined with Corollary 3.2.15 gives $\min\{\mathcal{M}_M\} = \gamma - f_{\mathcal{S}_M} = \check{s} - 1/|e| = -m_0(E_0^* - s_{[E_0^*]})$.

(4) In general, $s_{[Z_K+E_0^*]} = s_{[s_{[Z_K]}+E_0^*]}$, but $s_{[Z_K]} + E_0^* \in \mathcal{S}'$, hence $s_{[Z_K]} + E_0^* - s_{[Z_K+E_0^*]} \in L_{\geq 0}$, which implies $s + 1/|e| \geq \check{s}$. That is, $f_{\mathcal{S}_M} \geq \gamma - s$. Note that $\gamma - s = f_{\mathcal{M}_M}$. This shows that $f_{\mathcal{S}_M} - f_{\mathcal{M}_M} = m_0(s_{[Z_K]} + E_0^* - s_{[Z_K+E_0^*]})$.

EXAMPLE 6.2.5. Before we proceed to the proofs, let us illustrate the previous statements by further discussion on Example 5.4.4.

Let M be the Seifert rational homology sphere with

$$Sf = (-1, 0; (5,1), (5,1), (7,1), (10,1)).$$

The claim is that $n = 70$ is big enough in order to have Proposition 6.2.1 to be valid. Thus, the module $\mathcal{M}_{(70)}$ of the manifold $M_{(70)}$ is a semigroup, same as $\mathcal{S}_M$, and the calculations from Example 5.4.4 gives $f_{\mathcal{S}_M} = 3$. We are going to present the calculation using the formula of Theorem 6.2.3 too.

Associated with M, the adjunction formulae (3.1.1) give

$$Z_K = (24/5, 39/25, 39/25, 7/5, 32/25),$$

where $l' = (l'_0, l'_1, \ldots, l'_4)$ represents the cycle $l' = \sum_{i=0}^4 l'_i E_i$ (see the graph from Example 5.4.4). Hence, $\gamma = 19/5$ and one can also calculate $e = -5/14$. On the other hand, by the calculation of $E_0^* = (14/5, 14/25, 14/25, 2/5, 7/25)$ we deduce that $r_{[Z_K+E_0^*]} = (3/5, 3/25, 3/25, 4/5, 14/25)$. Then by running a general Laufer computation sequence (cf. Lemma 5.3.4) we find that $s_{[Z_K+E_0^*]} = r_{[Z_K+E_0^*]} + 3E_0 + E_1 + E_2$, hence $\check{s} = 18/5$. Finally, we get $f_{\mathcal{S}_M} = 19/5 + 14/5 - 18/5 = 3$.

6.3. The proof of Theorem 6.2.3. We denote by $\Gamma_{(n)}$ the plumbing graph associated with $M_{(n)}$ and let $L_{(n)}$ and $L'_{(n)}$ be the associated lattices as in 3.1. One can consider the (homological) inclusion operator $j_{(n)} : L \to L_{(n)}$ identifying naturally the corresponding E-base elements indexed by $\mathcal{V}(\Gamma)$ in the two lattices. This preserves the intersection forms. Moreover, it extends naturally to the rational cycles as well, in particular to $j_{(n)} : L' \to L'_{(n)}$. Additionally, denote by E_+ the new base element of $L_{(n)}$ associated with the newly created vertex v_+ of $\Gamma_{(n)}$.

Set the notation $E^*_{v,(n)}$ for the dual base elements in $L'_{(n)}$ associated with $j_{(n)}(E_v)$. Let $j^*_{(n)} : L'_{(n)} \to L'$ be the (cohomological) dual operator of $j_{(n)}$, defined by $j^*_{(n)}(E^*_{v,(n)}) = E^*_v$ for $v \in \mathcal{V}(\Gamma)$ and $j^*_{(n)}(E^*_+) = 0$. Then for every $l' \in L'_{(n)}$ and $l \in L$ one has the projection formula

$$(j^*_{(n)}(l'), l)_\Gamma = (l', j_{(n)}(l))_{\Gamma_{(n)}}. \tag{6.3.1}$$

This imply that $j^*_{(n)}(E_+) = -E_0^* \in L'$ (usually $\notin L$). Hence $j^*_{(n)}(E_+ + j_{(n)}E_0^*) = 0$. One also sees that $j^*_{(n)}(j_{(n)}E_v) = E_v$.

Set $H_{(n)} := L'_{(n)}/L_{(n)} = H_1(M_{(n)}, \mathbb{Z})$. Denote by $Z_{K(n)}$ the anti-canonical cycle in $L'_{(n)}$ and consider also the cycle $s_{[Z_{K(n)}]}$ associated with the class $[Z_{K(n)}] \in H_{(n)}$.

LEMMA 6.3.2 ([**LSz16**], [**LNN17**]). *For any* $h \in H_{(n)}$ *one has* $j^*_{(n)}(s_h) = s_{[j^*_{(n)}(s_h)]} \in L'$. *In particular,* $j^*_{(n)}(s_{Z_{K(n)}}) = s_{[j^*_{(n)}(s_{Z_{K(n)}})]}$.

By a computation (based on adjunction formulae) one verifies that

$$Z_{K(n)} = j_{(n)}Z_K + c_{(n)} \cdot (E_+ + j_{(n)}E_0^*), \quad \text{where} \quad c_{(n)} = \frac{n+\gamma-1}{n-1/|e|}. \tag{6.3.3}$$

The inequality $c_{(n)} < 1$ is equivalent with $|e|\gamma < |e| - 1$, or $d + \sum_i(\omega_i - 1)/\alpha_i < b_0 + 1$. This last inequality implies $b_0 \geq d$. Hence, by our assumption $b_0 < d$ we automatically have $c_{(n)} \geq 1$. Since $c_{(n)} < 2$ for $n \gg 0$, we get that

$$r_{Z_{K(n)}} - (c_{(n)} - 1)E_+ \in j_{(n)}(L \otimes \mathbb{Q}). \tag{6.3.4}$$

Analyzing the algorithm from Lemma 5.3.4 with $n \gg 0$ we obtain that in the computation sequence connecting $r_{Z_{K(n)}}$ with $s_{Z_{K(n)}}$ we never add E_+, hence (6.3.4) is valid for $s_{Z_{K(n)}}$ too. In particular, there exists $l_{(n)} \in L \otimes \mathbb{Q}$ such that

$$s_{Z_{K(n)}} = (c_{(n)} - 1)E_+ + j_{(n)}(l_{(n)}). \tag{6.3.5}$$

Since $Z_{K_{(n)}} - E_+ - r_{Z_{K(n)}} \in j_{(n)}(L)$, by the above statement $Z_{K_{(n)}} - E_+ - s_{Z_{K(n)}} \in j_{(n)}(L)$ too, hence $[j^*_{(n)}s_{Z_{K(n)}}] = [j^*_{(n)}(Z_{K(n)} - E^+)]$. This last term via (6.3.3) (and the properties of $j^*_{(n)}$ listed above) is $[Z_K + E_0^*]$. Hence Lemma 6.3.2 reads as $j^*_{(n)}s_{Z_{K(n)}} = s_{[Z_K+E_0^*]}$. This applied to (6.3.5) gives

$$s_{[Z_K+E_0^*]} = j^*_{(n)}((c_{(n)} - 1)E_+ + j_{(n)}(l_{(n)})) = -(c_{(n)} - 1)E_0^* + l_{(n)}. \tag{6.3.6}$$

We wish to compute $s_{(n)}$, the $j_{(n)}E_0$–coefficient of $s_{Z_{K(n)}}$. By (6.3.5) it is the E_0–coefficient of $l_{(n)}$, which by (6.3.6) is $\check{s} + (c_{(n)} - 1)/|e|$.

On the other hand, by a computation, the γ–invariant of $M_{(n)}$ satisfies $\gamma_{(n)} = \gamma + c_{(n)}/|e|$.

Then, by Theorem 5.4.1 one has $f_{\mathcal{M}_{(n)}} = \gamma_{(n)} - s_{(n)} = \gamma + c_{(n)}/|e| - (\check{s} + (c_{(n)} - 1)/|e|) = \gamma + 1/|e| - \check{s}$. Then apply Proposition 6.2.1.

7. Is $\mathcal{S}_M$ symmetric?

7.1. In the integral homology sphere case we have seen (cf. Section 4) that $\mathcal{S}_M$ is symmetric: $\ell \in \mathcal{S}_M \Leftrightarrow f_{\mathcal{S}_M} - \ell \notin \mathcal{S}_M$. It is natural to ask whether this fact extends to the case of rational homology spheres or not. The next example shows that the answer in general is no.

EXAMPLE 7.1.1. Take the Seifert 3-manifold given by

$$(-1, 0; (4, 1), (4, 1), (4, 1), (10, 1), (40, 1)).$$

Then $Z_K = (18, 5, 5, 5, 13/5, 7/5)$ and $E_0^* = (8, 2, 2, 2, 4/5, 1/5)$. Hence $r_{[Z_K+E_0^*]} = (0, 0, 0, 0, 2/5, 3/5)$. By the algorithm from Lemma 5.3.4

$$s_{[Z_K+E_0^*]} = (4, 1, 1, 1, 2/5, 3/5),$$

thus $\check{s} = 4$. Furthermore, $1/|e| = 8$ and $\gamma = 17$. Therefore, by Theorem 6.2.3, $f_{\mathcal{S}_M} = 17 + 8 - 4 = 21$.

On the other hand, by a computation, $N(4) = N(7) = N(10) = N(11) = N(14) = N(17) = -1$, hence the elements 4,7,10,11, 14, 17 do not belong to $\mathcal{S}_M$ but (4,17), (7,14), (10,11) are symmetric with respect to $f_{\mathcal{S}_M}$.

Usually symmetries in algebraic geometry (and algebra) are induced by some Gorenstein property. Hence, we can try to restrict ourselves to the numerically Gorenstein topological type, and ask the same question: is $\mathcal{S}_M$ symmetric? In this case $f_{\mathcal{S}_M}$ is computed in Example 6.2.4*(c)*, $\min\{\mathcal{M}_M\} + f_{\mathcal{S}_M} = \gamma$ by Corollary 3.2.15, and $N(\min\{\mathcal{M}_M\} + \ell) + N(f_{\mathcal{S}_M} - \ell) = -2$ for any $\ell \in \mathbb{Z}$ by (3.2.14).

LEMMA 7.1.2. *$\mathcal{M}_M = \min\{\mathcal{M}_M\} + \mathcal{S}_M$ if and only if $\mathcal{S}_M$ is symmetric.*

PROOF. Consider the following equivalences:

$$\ell \in \mathcal{S}_M \;\Leftrightarrow\; \min\{\mathcal{M}_M\} + \ell \in \min\{\mathcal{M}_M\} + \mathcal{S}_M$$

$$f_{\mathcal{S}_M} - \ell \notin \mathcal{S}_M \;\Leftrightarrow\; N(f_{\mathcal{S}_M} - \ell) + 1 \leq 0 \;\Leftrightarrow\; N(\min\{\mathcal{M}_M\} + \ell) + 1 \geq 0$$
$$\Leftrightarrow\; \min\{\mathcal{M}_M\} + \ell \in \mathcal{M}_M.$$

Then compare the first and last terms. □

For numerically Gorenstein graphs the symmetry $N(\ell) + N(\gamma - \ell) = -2$ can be interpreted as a consequence of the Gorenstein property of weighted homogeneous singularities, however, it is not clear at all if this fact implies any of the properties from the above Lemma.

The point is that $\mathcal{S}_M$ in general is not symmetric even if we restrict ourselves to the numerically Gorenstein case.

EXAMPLE 7.1.3. Consider the Seifert 3-manifold given by

$$(-2, 0; (2,1), (2,1), (3,1), (3,1), (7,1), (7,1), (84,1)).$$

Then $Z_K = (86, 43, 43, 29, 29, 13, 13, 2)$, hence Γ is numerically Gorenstein. By a computation $1/|e| = 28$, and $\gamma = 85$. Furthermore,

$$E_0^* = s_{[E_0^*]} = (28, 14, 14, 28/3, 28/3, 4, 4, 1/3),$$

hence $\check{s} = 1/|e| = 28$. In particular, $f_{\mathcal{S}_M} = \gamma = 85$.

On the other hand, $N(6) = N(85 - 6) = -1$, hence 6 and 85-6 are symmetric but none of them belong to $\mathcal{S}_M$. (The same is true for the pairs (12, 85-12), (14, 85-14) and (18, 85-18).

(In fact, in this case $\min\{\mathcal{M}_M\} = 0$, but $\mathcal{M}_M \neq \mathcal{S}_M$. Moreover, the graph can also be interpreted as $\Gamma_{(84)}$ associated with its subgraph, whose consequences can be tested by the interested reader.)

8. Problem: classification of $\mathcal{S}_M$ for Seifert rational homology sphere

8.1. We start this section by the following question regarding the semigroups $\mathcal{S}_M$ associated with Seifert rational homology sphere.

QUESTION 1. How can the numerical semigroups $\mathcal{S}_M$ be classified based on the peculiar geometry/topology of Seifert rational homology spheres?

As we have already shown is section 4 the above question can be completely answered in the case of Seifert homology spheres. Indeed, in that case, $\mathcal{S}_M$ is completely characterized by its minimal set of generators, which are exactly the Seifert invariants of the manifold. However, in general, for nontrivial H, the characterization is more difficult: for the description of $\mathcal{S}_{an}$ we need to understand the structure (generators) of the integral monomial cycles.

This leads to the following more precise task.

QUESTION 2. Determine the minimal set of generators of $\mathcal{S}_M$ in terms of the Seifert invariants!

In the sequel we clarify this problem completely in the case of Brieskorn–Hamm rational homology spheres.

8.2. Brieskorn–Hamm rational homology spheres. We consider positive integers $a_i \geq 2$ $(i = 1, \ldots, n)$ for some fixed $n \geq 3$. For a generic set of complex numbers $c = \{c_{ji}\}_{i=1,\ldots,n}^{j=1,\ldots,n-2}$ we consider the Brieskorn–Hamm isolated complete intersection singularity $(X_c(a_1, \ldots, a_n), 0)$ where

$$X_c(a_1, \ldots, a_n) := \{z \in \mathbb{C}^n \mid c_{j1}z_1^{a_1} + \cdots + c_{jn}z_n^{a_n} = 0 \ \text{ for every } \ 1 \leq j \leq n-2\}.$$

Then the link $M = \Sigma(a_1, \ldots, a_n)$ of $(X_c(a_1, \ldots, a_n), 0)$ is independent of the choice of c [**H72**]. It is an oriented Seifert 3-manifold with Seifert invariants

$$Sf = (-b_0, g; \underbrace{(\alpha_1, \omega_1), \ldots, (\alpha_1, \omega_1)}_{s_1}, \ldots, \underbrace{(\alpha_n, \omega_n), \ldots, (\alpha_n, \omega_n)}_{s_n}),$$

where all the entries are determined explicitly from the integers $\{a_i\}_i$, for details see [**JN83, NR77**].

Furthermore, by [**NN02**, Proposition 6.3] one has a complete characterization of the cases when M is a rational homology sphere. Namely, $g = 0$ if and only if the set $(a_1, \ldots, a_n)$ (after a possible permutation) has exactly one of the following forms:

(i) $(a_1, \ldots, a_n) = (m \cdot p_1, m \cdot p_2, p_3, \ldots, p_n)$, where the integers $\{p_i\}_{i=1}^n$ are pairwise relative prime, and $\gcd(m, p_i) = 1$ for any $i \geq 3$. In this case $\alpha_i = p_i$ for all i. Moreover, $s_1 = s_2 = 1$ and $s_i = m$ for any $i \geq 3$. If we set $q_i := \operatorname{lcm}(a_1, \ldots, a_n)/a_i$ then ω_i is determined from the congruence $\omega_i q_i \equiv -1 \pmod{\alpha_i}$, while b_0 is expressed from

$$\mathfrak{o} = |e| \cdot \alpha_1 \ldots \alpha_n = 1 \tag{8.2.1}$$

using the fact that $[E_0^*]$ is trivial in H.

(ii) $(a_1, \ldots, a_n) = (2^c \cdot p_1, 2p_2, 2p_3, p_4, \ldots, p_n)$ where p_i are odd and pairwise relative prime, and $c \geq 1$. In this case $\alpha_1 = 2^{c-1}p_1$ and $\alpha_i = p_i$ for $i \geq 2$. Furthermore, $s_i = 2$ for $i \leq 3$ and $s_i = 4$ for $i \geq 4$. One determines ω_i similarly as in (i), using the fact that in this case $\mathfrak{o} = 2$.

THEOREM 8.2.2. *The minimal set of generators for the semigroup $\mathcal{S}_M$ of Brieskorn–Hamm rational homology spheres corresponding to the above cases are*

(i) $\alpha_2 \ldots \alpha_n$, $\alpha_1\alpha_3 \ldots \alpha_n$ *and* $m \cdot \alpha_1 \ldots \alpha_{i-1}\alpha_{i+1} \ldots \alpha_n$ *for every* $3 \leq i \leq n$;

(ii) $\alpha_1 \ldots \alpha_{i-1}\alpha_{i+1} \ldots \alpha_n$ *for* $i \in \{1, 2, 3\}$ *and* $2 \cdot \alpha_1 \ldots \alpha_{i-1}\alpha_{i+1} \ldots \alpha_n$ *for* $4 \leq i \leq n$.

PROOF. We extend the proof of Theorem 4.1.1. The fact that the elements of the semigroup $\mathcal{S}_M$ are the E_0–coefficients of the principal cycles from $\mathcal{S}_{an}$ is valid also for any weighted homogeneous singularity rational homology sphere link. Thus, we have $\mathcal{S}_M = \{(l, -E_0^*) : l \in \mathcal{S}_{an}\}$. On the other hand, by the combinatorial characterization [**N12**, Theorem 7.1.2] (see Theorem 3.2.18 here) $l \in \mathcal{S}_{an}$ if and only if $l = \min_k \sum_{e \in \mathcal{E}} n_e^{(k)} E_e^*$ for a finite set $\{\sum_{e \in \mathcal{E}} n_e^{(k)} E_e^*\}_k \subset L$, where $n_e^{(k)} \in \mathbb{Z}_{\geq 0}$ and $\mathcal{E}$ is the set of end-vertices of the graph Γ. Thus, first we have to characterize the set of integers $\{n_e\}_{e \in \mathcal{E}}$ which provide an integral cycle $\sum_{e \in \mathcal{E}} n_e E_e^* \in L$.

We discuss case (i). Let us denote by E_1^*, E_2^* and E_{ij}^* ($3 \leq i \leq n$ and $1 \leq j \leq m$) the duals of the end–vertices. In fact, the classes $[E_0^*]$, $[E_1^*]$ and $[E_2^*]$ are trivial in H, moreover one has (cf. [**NN02**, pg. 301])

$$H \simeq \oplus_{i\geq 3}\langle [E_{ij}^*], 1 \leq j \leq m \ : \ \alpha_i[E_{ij}^*] = 0 \ \text{ for all } j, \ \text{and} \ \sum_j [E_{ij}^*] = 0\rangle \simeq \oplus_{i\geq 3}(\mathbb{Z}_{\alpha_i})^{m-1}.$$

This implies that the generating monomial cycle are E_1^*, E_2^*, and for $i \geq 3$ cycle of type $\sum_{j=1}^m n_{ij}E_{ij}^* \in L$ ($n_{ij} \in \mathbb{Z}_{\geq 0}$). This last condition says that $\sum_{j=1}^m n_{ij}[E_{ij}^*] = 0$ in H, hence by the above description of H, $\sum_{j=1}^m n_{ij}E_{ij}^* \in L$ (with all $n_{ij} \in \mathbb{Z}_{\geq 0}$) if and only if

$$(8.2.3) \quad (n_{ij})_j \in \mathbb{Z}\langle (\alpha_i, 0, \ldots, 0), \ldots, (0, \ldots, 0, \alpha_i), (1, 1, \ldots, 1)\rangle, \qquad (n_{ij} \in \mathbb{Z}_{\geq 0}).$$

Hence, there exists integers $k_0, k_1, \ldots, k_m$ such that $n_{ij} = k_j\alpha_i + k_0$. This implies that all the α_i–remainders are the same, say k_0', and, in fact, $n_{ij} = k_j'\alpha_i + k_0'$ for non–negative integers $k_0', k_1', \ldots, k_m'$.

This implies that $\sum_j n_{ij} = A\alpha_i + Bm$ for non–negative integers A and B, and the pair $A = 0$ and $B = 1$ can be realized.

Write $\hat{\alpha}_i := \alpha_1 \ldots \alpha_{i-1}\alpha_{i+1} \ldots \alpha_n$. Then using the identity (8.2.1) and formulae and the standard formulae for $-(E_v^*, E_u^*)$ we get

$$(E_i^*, -E_0^*) = \hat{\alpha}_i \ (i = 1, 2), \quad (E_{ij}^*, -E_0^*) = \hat{\alpha}_i \ (i \geq 3).$$

Hence, the generators of $\mathcal{S}_M$ are $(E_i^*, -E_0^*) = \hat{\alpha}_i$ for $i = 1, 2$, and $(\sum_j n_{ij}E_{ij}^*, -E_0^*)$ whenever $i \geq 3$ and $(n_{ij})_j$ satisfies (8.2.3). This last term has the form $(A\alpha_i + Bm)\hat{\alpha}_i = A\prod_i \alpha_i + Bm\hat{\alpha}_i$. Since $\prod_i \alpha_i$ is already generated by $\hat{\alpha_1}$, the generators for $i \geq 3$ can be replace by $m\hat{\alpha}_i$ (which can be realized).

Though in case (ii) the order of $[E_0^*]$ is 2, hence one has more congruences, the proof goes similarly. Therefore we omit it and invite the interested reader to check the details. □

Acknowledgments. TL was supported by ERCEA Consolidator Grant 615655 – NMST and by the Basque Government through the BERC 2014-2017 program and by Spanish Ministry of Economy and Competitiveness MINECO: BCAM Severo Ochoa excellence accreditation SEV-2013-0323. He is very grateful to the members of BCAM for the warm hospitality and for providing with him an excellent research environment during his postdoctoral stay in Bilbao.

AN was partially supported by NKFIH Grant K112735.

TL and AN were supported by NKFIH Grant "Élvonal" (Frontier) KKP 126683.

References

[A62] M. Artin, *Some numerical criteria for contractability of curves on algebraic surfaces*, Amer. J. Math. **84** (1962), 485–496, DOI 10.2307/2372985. MR146182

[A66] M. Artin, *On isolated rational singularities of surfaces*, Amer. J. Math. **88** (1966), 129–136, DOI 10.2307/2373050. MR199191

[BN10] G. Braun and A. Némethi, *Surgery formula for Seiberg-Witten invariants of negative definite plumbed 3-manifolds*, J. Reine Angew. Math. **638** (2010), 189–208, DOI 10.1515/CRELLE.2010.007. MR2595340

[CK14] M. B. Can and Ç. Karakurt, *Calculating Heegaard-Floer homology by counting lattice points in tetrahedra*, Acta Math. Hungar. **144** (2014), no. 1, 43–75, DOI 10.1007/s10474-014-0432-2. MR3267169

[CR74] P. E. Conner and F. Raymond, *Holomorphic Seifert fiberings*, Proceedings of the Second Conference on Compact Transformation Groups (Univ. Massachusetts, Amherst, Mass., 1971), Springer, Berlin, 1972, pp. 124–204. Lecture Notes in Math., Vol. 299. MR0590802

[EN85] D. Eisenbud and W. Neumann, *Three-dimensional link theory and invariants of plane curve singularities*, Annals of Mathematics Studies, vol. 110, Princeton University Press, Princeton, NJ, 1985. MR817982

[GW78] S. Goto and K. Watanabe, *On graded rings. I*, J. Math. Soc. Japan **30** (1978), no. 2, 179–213, DOI 10.2969/jmsj/03020179. MR494707

[GN18] E. Gorsky and A. Némethi, *On the set of L-space surgeries for links*, Adv. Math. **333** (2018), 386–422, DOI 10.1016/j.aim.2018.05.009. MR3818081

[G62] H. Grauert, *Über Modifikationen und exzeptionelle analytische Mengen* (German), Math. Ann. **146** (1962), 331–368, DOI 10.1007/BF01441136. MR0137127

[H72] H. A. Hamm, *Exotische Sphären als Umgebungsränder in speziellen komplexen Räumen* (German), Math. Ann. **197** (1972), 44–56, DOI 10.1007/BF01427951. MR0314073

[JN83] M. Jankins and W. D. Neumann, *Lectures on Seifert manifolds*, Brandeis Lecture Notes, vol. 2, Brandeis University, Waltham, MA, 1983. MR741334

[KL15] Ç. Karakurt and T. Lidman, *Rank inequalities for the Heegaard Floer homology of Seifert homology spheres*, Trans. Amer. Math. Soc. **367** (2015), no. 10, 7291–7322, DOI 10.1090/S0002-9947-2014-06451-9. MR3378830

[L72] H. B. Laufer, *On rational singularities*, Amer. J. Math. **94** (1972), 597–608, DOI 10.2307/2374639. MR330500

[L77] H. B. Laufer, *On minimally elliptic singularities*, Amer. J. Math. **99** (1977), no. 6, 1257–1295, DOI 10.2307/2374025. MR568898

[LN14] T. László and A. Némethi, *Ehrhart theory of polytopes and Seiberg-Witten invariants of plumbed 3-manifolds*, Geom. Topol. **18** (2014), no. 2, 717–778, DOI 10.2140/gt.2014.18.717. MR3180484

[LN15] T. László and A. Némethi, *Reduction theorem for lattice cohomology*, Int. Math. Res. Not. IMRN **11** (2015), 2938–2985, DOI 10.1093/imrn/rnu015. MR3373041

[LNN17] László, T., Nagy, J. and Némethi, A.: Surgery formulae for the Seiberg–Witten invariant of plumbed 3-manifold, *Rev. Math. Complut.* (2019). https://doi.org/10.1007/s13163-019-00297-z.

[LNN18] T. László, J. Nagy, and A. Némethi, *Combinatorial duality for Poincaré series, polytopes and invariants of plumbed 3-manifolds*, Selecta Math. (N.S.) **25** (2019), no. 2, Art. 21, 31, DOI 10.1007/s00029-019-0468-9. MR3917945

[LSz16] László, T. and Szilágyi, Zs.: On Poincaré series associated with links of normal surface singularities, *Trans. of AMS* (2019). https://doi.org/10.1090/tran/7802.

[LSz17] T. László and Zs. Szilágyi, *Non-normal affine monoids, modules and Poincaré series of plumbed 3-manifolds*, Acta Math. Hungar. **152** (2017), no. 2, 421–452, DOI 10.1007/s10474-017-0726-2. MR3682892

[N99] A. Némethi, *Five lectures on normal surface singularities*, Low dimensional topology (Eger, 1996/Budapest, 1998), Bolyai Soc. Math. Stud., vol. 8, János Bolyai Math. Soc., Budapest, 1999, pp. 269–351. With the assistance of Ágnes Szilárd and Sándor Kovács. MR1747271

[N99b] A. Némethi, *"Weakly" elliptic Gorenstein singularities of surfaces*, Invent. Math. **137** (1999), no. 1, 145–167, DOI 10.1007/s002220050327. MR1703331

[N05] A. Némethi, *On the Ozsváth-Szabó invariant of negative definite plumbed 3-manifolds*, Geom. Topol. **9** (2005), 991–1042, DOI 10.2140/gt.2005.9.991. MR2140997

[N07] A. Némethi, *Graded roots and singularities*, Singularities in geometry and topology, World Sci. Publ., Hackensack, NJ, 2007, pp. 394–463, DOI 10.1142/9789812706812_0013. MR2311495

[N08] A. Némethi, *Lattice cohomology of normal surface singularities*, Publ. Res. Inst. Math. Sci. **44** (2008), no. 2, 507–543, DOI 10.2977/prims/1210167336. MR2426357

[N11] A. Némethi, *The Seiberg-Witten invariants of negative definite plumbed 3-manifolds*, J. Eur. Math. Soc. (JEMS) **13** (2011), no. 4, 959–974, DOI 10.4171/JEMS/272. MR2800481

[N12] A. Némethi, *The cohomology of line bundles of splice-quotient singularities*, Adv. Math. **229** (2012), no. 4, 2503–2524, DOI 10.1016/j.aim.2012.01.003. MR2880230

[NNP11] A. Némethi, W. D. Neumann, and A. Pichon, *Principal analytic link theory in homology sphere links*, Topology of algebraic varieties and singularities, Contemp. Math., vol. 538, Amer. Math. Soc., Providence, RI, 2011, pp. 377–387, DOI 10.1090/conm/538/10614. MR2777831

[NN02] A. Némethi and L. I. Nicolaescu, *Seiberg-Witten invariants and surface singularities*, Geom. Topol. **6** (2002), 269–328, DOI 10.2140/gt.2002.6.269. MR1914570

[NO09] A. Némethi and T. Okuma, *On the Casson invariant conjecture of Neumann-Wahl*, J. Algebraic Geom. **18** (2009), no. 1, 135–149, DOI 10.1090/S1056-3911-08-00493-1. MR2448281

[NO10] A. Némethi and T. Okuma, *The embedding dimension of weighted homogeneous surface singularities*, J. Topol. **3** (2010), no. 3, 643–667, DOI 10.1112/jtopol/jtq019. MR2684515

[Ne81] W. D. Neumann, *Abelian covers of quasihomogeneous surface singularities*, Singularities, Part 2 (Arcata, Calif., 1981), Proc. Sympos. Pure Math., vol. 40, Amer. Math. Soc., Providence, RI, 1983, pp. 233–243. MR713252

[NR77] W. D. Neumann and F. Raymond, *Seifert manifolds, plumbing, μ-invariant and orientation reversing maps*, Algebraic and geometric topology (Proc. Sympos., Univ. California, Santa Barbara, Calif., 1977), Lecture Notes in Math., vol. 664, Springer, Berlin, 1978, pp. 163–196. MR518415

[O05] T. Okuma, *Numerical Gorenstein elliptic singularities*, Math. Z. **249** (2005), no. 1, 31–62, DOI 10.1007/s00209-004-0688-2. MR2106969

[OSz04] P. Ozsváth and Z. Szabó, *Holomorphic disks and topological invariants for closed three-manifolds*, Ann. of Math. (2) **159** (2004), no. 3, 1027–1158, DOI 10.4007/annals.2004.159.1027. MR2113019

[OSz03] P. Ozsváth and Z. Szabó, *On the Floer homology of plumbed three-manifolds*, Geom. Topol. **7** (2003), 185–224, DOI 10.2140/gt.2003.7.185. MR1988284

[P77] H. Pinkham, *Normal surface singularities with C^* action*, Math. Ann. **227** (1977), no. 2, 183–193, DOI 10.1007/BF01350195. MR0432636

[RChW96] M. Raczunas and P. Chrząstowski-Wachtel, *A Diophantine problem of Frobenius in terms of the least common multiple*, Discrete Math. **150** (1996), no. 1-3, 347–357, DOI 10.1016/0012-365X(95)00199-7. Selected papers in honour of Paul Erdős on the occasion of his 80th birthday (Keszthely, 1993). MR1392743

[RA05] J. L. Ramírez Alfonsín, *The Diophantine Frobenius problem*, Oxford Lecture Series in Mathematics and its Applications, vol. 30, Oxford University Press, Oxford, 2005. MR2260521

[R07] J. C. Rosales, *Subadditive periodic functions and numerical semigroups*, J. Algebra Appl. **6** (2007), no. 2, 305–313, DOI 10.1142/S0219498807002223. MR2316425

[RGS09] J. C. Rosales and P. A. García-Sánchez, *Numerical semigroups*, Developments in Mathematics, vol. 20, Springer, New York, 2009. MR2549780

[S1882] J. J. Sylvester, *On Subvariants, i.e. Semi-Invariants to Binary Quantics of an Unlimited Order*, Amer. J. Math. **5** (1882), no. 1-4, 79–136, DOI 10.2307/2369536. MR1505319

[O08] T. Okuma, *The geometric genus of splice-quotient singularities*, Trans. Amer. Math. Soc. **360** (2008), no. 12, 6643–6659, DOI 10.1090/S0002-9947-08-04559-5. MR2434304

Present Address: Alfréd Rényi Institute of Mathematics, Hungarian Academy of Sciences, Reáltanoda utca 13-15, H-1053, Budapest, Hungary
Old: BCAM - Basque Center for Applied Math., Mazarredo, 14 E48009 Bilbao, Basque Country – Spain
Email address: laszlo.tamas@renyi.mta.hu

Alfréd Rényi Institute of Mathematics, Hungarian Academy of Sciences, Reáltanoda utca 13-15, H-1053, Budapest, Hungary
ELTE - University of Budapest, Dept. of Geometry, Budapest, Hungary
BCAM - Basque Center for Applied Math., Mazarredo, 14 E48009 Bilbao, Basque Country – Spain
Email address: nemethi.andras@renyi.mta.hu

Contemporary Mathematics
Volume **742**, 2020
https://doi.org/10.1090/conm/742/14942

Characteristic cycles and the relative local Euler obstruction

David B. Massey

ABSTRACT. In this paper, we investigate the local Euler obstruction and the relative local Euler obstruction in terms of constructible complexes of sheaves, characteristic cycles, and vanishing cycles. The fundamental tool that we use is the notion of a characteristic complex for an analytic space embedded in affine space.

1. Introduction

The local Euler obstruction, defined by MacPherson in [**17**] in 1974 has been studied by many researchers (see, for instance, [**12**], [**5**], [**16**], [**4**], and [**9**]) and is, at this point, a standard piece of data associated to a singular point of a complex analytic space. The local Euler obstruction is an obstruction to extending a stratified radial vector field to a non-zero lift in the Nash modification.

The relative local Euler obstruction, defined in [**3**], is an analog of the local Euler obstruction for a complex analytic function $f : X \to \mathbb{C}$ at a point $\mathbf{p}$ which is a stratified isolated critical point of f. This is again defined in an obstruction-theoretic way; it is an obstruction to extending the conjugate of the stratified gradient vector field of f to a non-zero lift in the Nash modification. This relative concept is beginning to be studied by a number of other researchers; see, for instance, [**8**], [**7**], [**30**], and [**1**].

Our own contribution to [**3**] appeared in the last section of that paper, where we used derived category techniques to extend the definition of the relative local Euler obstruction to functions with arbitrary critical loci, but - in [**3**] - we referred to this generlaized concept as the *defect* of f. We also gave an algorithm for calculating the defect of f via a process similar to how we defined the Lê cycles, Lê numbers, and their generalizations in [**18**], [**19**], [**20**], and especially in Remark 1.6 of Part IV of [**25**].

In this paper, we recall our earlier characterization of the local Euler obstruction in terms of *characteristic complexes* and recall and re-derive some standard properties of the local Euler obstruction. We then recall our general definition/characterization of the defect of f - which we now take as the general definition

2010 *Mathematics Subject Classification.* 32B15, 32C35, 32C18, 32B10.
Key words and phrases. characteristic cycle, constructible complexes, local Euler obstruction, relative local Euler obstruction.

of the relative local Euler obstruction - in terms of vanishing cycles and characteristic complexes. Finally, in Theorem 5.9, we prove a number of basic properties for the relative local Euler obstruction and give some examples.

We must begin with a section on the basics of characteristic cycles. Throughout this section and much of this paper, we must assume that the reader is familiar with fundamental aspects of the derived category of bounded constructible complexes of sheaves, perverse sheaves, and the nearby and vanishing cycles. Good references for the theory are [**14**], [**6**], and [**29**].

We thank Jörg Schürmann for valuable comments on the first version of this paper.

2. Characteristic Cycles

Basic references for much of what we write below are [**13**] and [**25**]. A general reference for details of this section is [**26**].

Throughout this paper, we fix a base ring R that is a regular, Noetherian ring with finite Krull dimension (e.g., $\mathbb{Z}$, $\mathbb{Q}$, or $\mathbb{C}$). This implies that every finitely-generated R-module has finite projective dimension (in fact, it implies that the projective dimension of the module is at most $\dim R$). In fact, in later sections, we fix our base as $\mathbb{Z}$.

We let $\mathcal{U}$ be an open neighborhood of the origin of $\mathbb{C}^{n+1}$, and let X be a closed, analytic subset of $\mathcal{U}$. We let $\mathbf{z} := (z_0, \dots, z_n)$ be coordinates on $\mathcal{U}$.

Recall that the *complex link*, $\mathbb{L}_{X,\mathbf{p}}$, of X at $\mathbf{p}$ is the Milnor fiber of a generic affine linear form, restricted to X, at $\mathbf{p}$. That is, the complex link is

$$\mathbb{L}_{X,\mathbf{p}} := B^\circ_\epsilon(\mathbf{p}) \cap X \cap V(L-b),$$

where $B^\circ_\epsilon(\mathbf{p})$ is an open ball in $\mathcal{U}$ of radius ϵ, where $0<\epsilon \ll 1$, centered at $\mathbf{p}$, L is a generic affine linear form which is zero at $\mathbf{p}$, and b is a complex number such that $0<|b|\ll\epsilon$. The homotopy-type of the complex link is an analytic invariant of the germ of X at $\mathbf{p}$.

Let $\mathfrak{S}$ be a complex analytic Whitney stratification of X, with connected strata. Let $\mathbf{F}^\bullet$ be a bounded complex of sheaves of R-modules on X, which is constructible with respect to $\mathfrak{S}$. For each $S\in\mathfrak{S}$, we let $d_S := \dim S$, and let $(\mathbb{N}_{X,S}, \mathbb{L}_{X,S})$ denote *complex Morse data for S in X*, consisting of a normal slice and complex link of S in X. Recall that, if $\mathbf{p}\in S$, then $\mathbb{L}_{X,S}$ is the complex link of the normal slice to S at $\mathbf{p}$, i.e., $\mathbb{L}_{X,S} = \mathbb{L}_{\mathbb{N}_{X,S},\mathbf{p}}$. The homeomorphism-type of the pair $(\mathbb{N}_{X,S}, \mathbb{L}_{X,S})$ is independent of the choices.

DEFINITION 2.1. *For each $S\in\mathfrak{S}$ and each integer k, the isomorphism-type of the R-module $m^k_S(\mathbf{F}^\bullet) := \mathbb{H}^{k-d_S}(\mathbb{N}_{X,S}, \mathbb{L}_{X,S}; \mathbf{F}^\bullet)$ is independent of the choice of $(\mathbb{N}_{X,S}, \mathbb{L}_{X,S})$; we refer to $m^k_S(\mathbf{F}^\bullet)$ as the degree k Morse module of S with respect to $\mathbf{F}^\bullet$.*

REMARK 2.2. The shift by d_S above is present so that perverse sheaves can have non-zero Morse modules in only degree 0.

We also remark that, up to isomorphism, $m^k_S(\mathbf{F}^\bullet)$ can be obtained in terms of vanishing cycles. To accomplish this, select any point $\mathbf{p}\in S$. Consider an analytic function $\tilde{g} : (\mathcal{U}', \mathbf{p}) \to (\mathbb{C}, 0)$ on some open neighborhood of $\mathbf{p}$ in $\mathcal{U}$ such that $d_{\mathbf{p}}\tilde{g}$

is a nondegenerate covector (in the sense of [**13**]), and such that $\mathbf{p}$ is a (complex) nondegenerate critical point of $\tilde{g}_{|_{\mathcal{U}'\cap S}}$. Let $g := \tilde{g}_{|_{\mathcal{U}'\cap X}}$. Then, $m_S^k(\mathbf{F}^\bullet)$ is isomorphic to the stalk cohomology $H^k(\phi_g[-1]\mathbf{F}^\bullet)_{\mathbf{p}}$.

Note that, if $\mathbf{0}$ is a point-stratum, then $m_{\mathbf{0}}^k(\mathbf{F}^\bullet) \cong H^k(\phi_{\mathcal{L}}[-1]\mathbf{F}^\bullet)_{\mathbf{0}}$, where $\mathcal{L}$ is the restriction to X of a generic linear form $\tilde{\mathcal{L}}$. In particular, if $\mathbf{0}$ is a point-stratum which is not an isolated point of X,

$$m_{\mathbf{0}}^k(\mathbb{Z}_X^\bullet) \cong H^k(\phi_{\mathcal{L}}[-1]\mathbb{Z}_X^\bullet)_{\mathbf{0}} \cong \mathbb{H}^k(\mathbb{N}_{X,\mathbf{0}}, \mathbb{L}_{X,\mathbf{0}}; \mathbb{Z}_X^\bullet) \cong \widetilde{H}^{k-1}(\mathbb{L}_{X,\mathbf{0}}; \mathbb{Z}),$$

where, in the final term on the right, $\widetilde{H}$ denotes the usual reduced singular cohomology of the complex link of X at $\mathbf{0}$.

Finally, if $\mathbf{p}$ is in a stratum S_0, and $\mathbb{N}_{X,S_0}$ is a normal slice to S_0 in X at $\mathbf{p}$, then $\mathbb{N}_{X,S_0}$ is stratified by

$$\{\mathbb{N}_{X,S_0} \cap S \mid S \in \mathfrak{S}\},$$

though the strata need not be connected. Nonetheless, if $Y := \mathbb{N}_{X,S_0}$ and $\hat{S} := Y \cap S$, it is trivial that, up to homeomorphism, $(\mathbb{N}_{Y,\hat{S}}, \mathbb{L}_{Y,\hat{S}})$ (at any point of $\hat{S}$) is given by $(\mathbb{N}_{X,S}, \mathbb{L}_{X,S})$, simply because transverse intersections of transverse intersections are transverse intersections.

While [**13**] is a basic reference for the set-up above, Goresky and MacPherson use Teissier's result on page 455 of [**31**] to know that there is such a non-degenerate covector in a normal slice; we use this result again in Proposition 2.7.

Now, for any analytic submanifold $M \subseteq \mathcal{U}$, we denote the conormal space

$$\{(\mathbf{p}, \omega) \in T^*\mathcal{U} \mid \omega(T_{\mathbf{p}}M) \equiv 0\}$$

by $T_M^*\mathcal{U}$, and will typically be interested in its closure $\overline{T_M^*\mathcal{U}}$ in $T^*\mathcal{U}$.

DEFINITION 2.3. *Suppose that R is an integral domain.*

Define $c_S(\mathbf{F}^\bullet) := \sum_{k\in\mathbb{Z}}(-1)^k \operatorname{rank}(m_S^k(\mathbf{F}^\bullet))$, and define the **characteristic cycle of $\mathbf{F}^\bullet$ (in $T^*\mathcal{U}$)** *to be the analytic cycle*

$$\mathrm{CC}(\mathbf{F}^\bullet) := \sum_{S\in\mathfrak{S}} c_S(\mathbf{F}^\bullet)\left[\overline{T_S^*\mathcal{U}}\right].$$

We write $c_{\mathbf{0}}(\mathbf{F}^\bullet)$ in place of $c_{\{\mathbf{0}\}}(\mathbf{F}^\bullet)$, and let $c_{\mathbf{0}}(\mathbf{F}^\bullet) = 0$ if $\{\mathbf{0}\} \notin \mathfrak{S}$.

The underlying set $|\mathrm{CC}(\mathbf{F}^\bullet)| = \bigcup_{c_S(\mathbf{F}^\bullet)\neq 0} \overline{T_S^*\mathcal{U}}$ is the **characteristic variety of $\mathbf{F}^\bullet$ (in $T^*\mathcal{U}$)**.

For $0 \leq k \leq \dim X$, let

$$\mathrm{CC}_{\geq k}(\mathbf{F}^\bullet) \ := \sum_{S\in\mathfrak{S}, \dim S\geq k} c_S(\mathbf{F}^\bullet)\left[\overline{T_S^*\mathcal{U}}\right].$$

Throughout this paper, whenever we refer to $c_S(\mathbf{F}^\bullet)$ or $\mathrm{CC}(\mathbf{F}^\bullet)$, we assume that the base ring is an integral domain, even if we do not explicitly state this.

REMARK 2.4. There are various conventions for the signs involved in the characteristic cycle. In fact, our definition above uses a different convention than we used in our earlier works. Our current definition is the most desirable when working with perverse sheaves. In hopes of avoiding confusion with our earlier work, we have changed our notation for the characteristic cycle. Note that, using our current convention, the characteristic cycle is not changed by extending $\mathbf{F}^\bullet$ by zero to all of $\mathcal{U}$.

We give some basic, easy properties of the characteristic cycle concern how they work with shifting, constant sheaves, distinguished triangles, and the Verdier dual $\mathcal{D}\mathbf{F}^\bullet$. The proofs of (1), (2), and (3) below are trivial, and we leave them to the reader. The proof of (4) is not so trivial, and holds only in the complex analytic context and uses that the non-degenerate covectors are $\mathbb{C}^*$-invariant in $T^*_S X$; see (5.11), page 281, of [**29**].

PROPOSITION 2.5.

(1) $\mathrm{CC}(\mathbf{F}^\bullet[j]) = (-1)^j \mathrm{CC}(\mathbf{F}^\bullet)$.

(2) *If X is a pure-dimensional (e.g., connected) complex manifold, then*

$$\mathrm{CC}(\mathbf{R}^\bullet_X) = (-1)^{\dim X}[T^*_X\mathcal{U}],$$

i.e., $\mathrm{CC}(\mathbf{R}^\bullet_X[\dim X]) = [T^*_X\mathcal{U}]$.

(3) *If* $\mathbf{A}^\bullet \to \mathbf{B}^\bullet \to \mathbf{C}^\bullet \xrightarrow{[1]} \mathbf{A}^\bullet$ *is a distinguished triangle in* $D^b_c(X)$, *then* $\mathrm{CC}(\mathbf{B}^\bullet) = \mathrm{CC}(\mathbf{A}^\bullet) + \mathrm{CC}(\mathbf{C}^\bullet)$.

(4) $\mathrm{CC}(\mathbf{F}^\bullet) = \mathrm{CC}(\mathcal{D}\mathbf{F}^\bullet)$.

For calculating the characteristic cycle of the constant sheaf, the following is very useful:

COROLLARY 2.6. *Suppose that Y and Z are closed analytic subsets of X such that $X = Y \cup Z$. Then,*

$$\mathrm{CC}(\mathbf{R}^\bullet_X) = \mathrm{CC}(\mathbf{R}^\bullet_Y) + \mathrm{CC}(\mathbf{R}^\bullet_Z) - \mathrm{CC}(\mathbf{R}^\bullet_{Y\cap Z}).$$

PROOF. Let $j : Y \hookrightarrow X$, $k : Z \hookrightarrow X$, and $l : Y \cap Z \hookrightarrow X$ denote the respective inclusions. Then, there is a canonical distinguished triangle

$$\mathbf{R}^\bullet_X \to j_*j^*\mathbf{R}^\bullet_X \oplus k_*k^*\mathbf{R}^\bullet_X \to l_*l^*\mathbf{R}^\bullet_X \xrightarrow{[1]} \mathbf{R}^\bullet_X.$$

As the pull-back of the constant sheaf is the constant sheaf, and as the characteristic cycle is unaffected by extensions by zero, the desired conclusion follows immediately from Item 3 of Proposition 2.5. □

It is also easy to describe how the characteristic cycle of normal slices to strata depend on the original characteristic cycle. We use the set-up and remark at the end of Remark 2.2.

PROPOSITION 2.7. *Suppose that $\mathbf{p}$ is in a stratum S_0 of $\mathfrak{S}$, and $\mathbb{N}_{X,S_0}$ is a normal slice to S in X at $\mathbf{p}$, then $\mathbb{N}_{X,S_0}$ is stratified by*

$$\{\mathbb{N}_{X,S_0} \cap S \mid S \in \mathfrak{S}\},$$

though the strata need not be connected. For $S \in \mathfrak{S}$, let $\hat{S} := S \cap \mathbb{N}_{X,S_0}$.

Then, we have an equality of Morse modules $m^k_{\hat{S}}(\mathbf{F}^\bullet_{|\mathbb{N}_{X,S_0}}[-\dim S_0]) = m^k_S(\mathbf{F}^\bullet)$, which implies that

$$\mathrm{CC}(\mathbf{F}^\bullet_{|\mathbb{N}_{X,S_0}}[-\dim S_0]) = \sum_{S\in\mathfrak{S}} c_S(\mathbf{F}^\bullet)\left[\overline{T^*_{S\cap\mathbb{N}_{X,S_0}}\mathcal{U}}\right],$$

(Note that a summand would be 0 unless $S_0 \subseteq \overline{S}$.)

PROOF. Let $Y := \mathbb{N}_{X,S_0}$. Using the end of Remark 2.2, we find

$$m^k_{\hat{S}}(\mathbf{F}^\bullet_{|_Y}) = \mathbb{H}^{k-\dim \hat{S}}(\mathbb{N}_{Y,\hat{S}}, \mathbb{L}_{Y,\hat{S}}; \mathbf{F}^\bullet_{|_Y}) =$$

$$\mathbb{H}^{k-\dim S+\dim S_0}(\mathbb{N}_{X,S}, \mathbb{L}_{X,S}; \mathbf{F}^\bullet) = m_S^{k+\dim S_0}(\mathbf{F}^\bullet),$$

i.e., $m^k_{\hat{S}}(\mathbf{F}^\bullet_{|_Y}[-\dim S_0]) = m^k_S(\mathbf{F}^\bullet)$. The conclusions follow at once. □

Note that

$$\left[\overline{T^*_{S\cap \mathbb{N}_{X,S_0}}\mathcal{U}}\right] = \sum_{C\subset S\cap \mathbb{N}_{X,S_0}} \left[\overline{T^*_C\mathcal{U}}\right],$$

where C runs over the connected components of $S\cap \mathbb{N}_{X,S_0}$.

The following proposition is immediate from formula (5.6) on page 277 of [**29**].

PROPOSITION 2.8. *Suppose that R is a principal ideal domain. Let X and Y be analytic spaces, let $\pi_1 : X\times Y \to X$ and $\pi_2 : X \to Y$ denote the projections. Let $\mathfrak{S}$ and $\mathfrak{S}'$ be Whitney stratifications of X and Y, respectively. Let $\mathbf{A}^\bullet$ and $\mathbf{B}^\bullet$ be bounded, complexes of sheaves on X and Y, respectively, which are constructible with respect to $\mathfrak{S}$ and $\mathfrak{S}'$, respectively. Let $\mathbf{A}^\bullet \overset{L}{\boxtimes} \mathbf{B}^\bullet := \pi_1^*\mathbf{A}^\bullet \overset{L}{\otimes} \pi_2^*\mathbf{B}^\bullet$.*

Then, $\mathbf{A}^\bullet \overset{L}{\boxtimes} \mathbf{B}^\bullet$ is constructible with respect to the product stratification

$$\{S\times S' \mid S\in\mathfrak{S},\ S'\in\mathfrak{S}'\}$$

and, for all $S\in\mathfrak{S}$ and $S'\in\mathfrak{S}'$,

$$m^k_{S\times S'}\big(\mathbf{A}^\bullet \overset{L}{\boxtimes} \mathbf{B}^\bullet\big) =$$

$$\bigoplus_{i+j=k} m^i_S(\mathbf{A}^\bullet)\otimes m^j_{S'}(\mathbf{B}^\bullet) \ \oplus \bigoplus_{i+j=k+1} \mathrm{Tor}\,\big(m^i_S(\mathbf{A}^\bullet), m^j_{S'}(\mathbf{B}^\bullet)\big).$$

Consequently,

$$c_{S\times S'}\big(\mathbf{A}^\bullet \overset{L}{\boxtimes} \mathbf{B}^\bullet\big) = c_S(\mathbf{A}^\bullet)\cdot c_{S'}(\mathbf{B}^\bullet).$$

We also have the following simple result, well-known to experts, whose proof we include for completeness. Recall the definition of $\mathrm{CC}_{\geq k}(\mathbf{F}^\bullet)$ from Definition 2.3.

PROPOSITION 2.9. *Suppose that $\mathbf{A}^\bullet$ and $\mathbf{B}^\bullet$ are bounded, constructible complexes of sheaves on the d-dimensional analytic space X. Suppose that $\mathfrak{S}$ is a Whitney stratification with respect to which both $\mathbf{A}^\bullet$ and $\mathbf{B}^\bullet$ are constructible (which always exists).*

Then, $\mathrm{CC}_{\geq k}(\mathbf{A}^\bullet) = \mathrm{CC}_{\geq k}(\mathbf{B}^\bullet)$ if and only if, for all $S\in\mathfrak{S}$ such that $\dim S\geq k$, for all $\mathbf{p}\in S$, there is an equality of Euler characteristics of the stalk cohomology $\chi(\mathbf{A}^\bullet)_\mathbf{p} = \chi(\mathbf{B}^\bullet)_\mathbf{p}$.

In particular, $\mathrm{CC}(\mathbf{A}^\bullet) = \mathrm{CC}(\mathbf{B}^\bullet)$ if and only if, for all $\mathbf{p}\in X$, $\chi(\mathbf{A}^\bullet)_\mathbf{p} = \chi(\mathbf{B}^\bullet)_\mathbf{p}$.

PROOF. The proof is by downward induction on k. Certainly the result is trivial for $k = d$. Now suppose that $k_0 \geq 0$ and that the statement is true for all k such that $k_0+1\leq k\leq d$; we wish to show that the statement is true for $k = k_0$.

Let $S_0 \in \mathfrak{S}$ be a stratum of dimension k_0, and let $\mathbf{p}_0 \in S_0$. For each stratum S of dimension greater than or equal to k_0+1, let $\mathbf{p}_S$ denote a point of S. If we let $\mathbf{F}^\bullet$ be $\mathbf{A}^\bullet$ or $\mathbf{B}^\bullet$, then

$$c_{S_0}(\mathbf{F}^\bullet) := \chi(\mathbb{N}_{S_0}, \mathbb{L}_{S_0}; \mathbf{F}^\bullet[-k_0]) = \chi(\mathbb{N}_{S_0}; \mathbf{F}^\bullet[-k_0]) - \chi(\mathbb{L}_{S_0}; \mathbf{F}^\bullet[-k_0]) =$$
$$(-1)^{k_0}\Big\{\chi(\mathbf{F}^\bullet)_{\mathbf{p}_0} - \sum_{S, \dim S \geq k_0+1} \chi\big(\mathbb{L}_{S_0} \cap S\big) \cdot \chi(\mathbf{F}^\bullet)_{\mathbf{p}_S}\Big\}.$$

Note that our inductive hypothesis implies that the summation on the right above is the same whether $\mathbf{F}^\bullet$ equals $\mathbf{A}^\bullet$ or $\mathbf{B}^\bullet$.

Therefore, $c_{S_0}(\mathbf{A}^\bullet) = c_{S_0}(\mathbf{B}^\bullet)$ if and only if $\chi(\mathbf{A}^\bullet)_{\mathbf{p}_0} = \chi(\mathbf{B}^\bullet)_{\mathbf{p}_0}$, and we are finished. □

Below, we use ψ_f and ϕ_f to denote the nearby and vanishing cycles along f, respectively; we also frequently subtract from f the possibly non-zero constant $f(\mathbf{p})$ when working at a point $\mathbf{p}$. Combined with Proposition 2.9, what we prove below is the well-known fact that the constructible functions given by taking the Euler characteristics of the nearby and vanishing cycles of a complex along a function f depend only on f and the constructible function given by taking the Euler characteristics of the stalks of the complex.

COROLLARY 2.10. *Suppose that* $\mathrm{CC}(\mathbf{A}^\bullet) = \mathrm{CC}(\mathbf{B}^\bullet)$ *and that we have a complex analytic* $f : X \to \mathbb{C}$. *Then, for all,* $\mathbf{p} \in X$,

$$\chi(\psi_{f-f(\mathbf{p})}\mathbf{A}^\bullet)_{\mathbf{p}} = \chi(\psi_{f-f(\mathbf{p})}\mathbf{B}^\bullet)_{\mathbf{p}} \quad \text{and} \quad \chi(\phi_{f-f(\mathbf{p})}\mathbf{A}^\bullet)_{\mathbf{p}} = \chi(\phi_{f-f(\mathbf{p})}\mathbf{B}^\bullet)_{\mathbf{p}}.$$

PROOF. For convenience, we shall assume that $f(\mathbf{p}) = 0$. Let $F_{f,\mathbf{p}}$ denote the Milnor fiber of f at $\mathbf{p}$. Once again, choose a Whitney stratification $\mathfrak{S}$ with respect to which both $\mathbf{A}^\bullet$ and $\mathbf{B}^\bullet$ are constructible and, for each $S \in \mathfrak{S}$, select a $\mathbf{p}_S \in S$.

Then,

$$\chi(\psi_f \mathbf{A}^\bullet)_{\mathbf{p}} = \chi(F_{f,\mathbf{p}}; \mathbf{A}^\bullet) = \sum_{S \in \mathfrak{S}} \chi(F_{f,\mathbf{p}} \cap S) \cdot \chi(\mathbf{A}^\bullet)_{\mathbf{p}_S}.$$

By the proposition, this also equals $\chi(\psi_f \mathbf{B}^\bullet)_{\mathbf{p}}$.

The result about the vanishing cycles follows immediately since

$$\chi(\phi_f \mathbf{A}^\bullet)_{\mathbf{p}} = \chi(\psi_f \mathbf{A}^\bullet)_{\mathbf{p}} - \chi(\mathbf{A}^\bullet)_{\mathbf{p}} = \chi(\psi_f \mathbf{B}^\bullet)_{\mathbf{p}} - \chi(\mathbf{B}^\bullet)_{\mathbf{p}} = \chi(\phi_f \mathbf{B}^\bullet)_{\mathbf{p}}.$$

□

3. Characteristic Complexes

For the remainder of this paper, we fix our base ring to be $\mathbb{Z}$.

Some classical constructions in the study of singular spaces, such as calculating the polar varieties and polar multiplicities of Lê and Teissier and the Nash modification, deal with contributions from only the smooth strata of X. From our point of view, these are results where the underlying complex of sheaves is a *characteristic complex.*

Note that, while our definition of the characteristic cycle in this paper is not what we used in [**3**], our definition below of a characteristic complex has been adjusted in such a way that the same complexes here and in [**3**] are characteristic complexes.

DEFINITION 3.1. *Let $X = \bigcup_i X_i$ be the decomposition of X into its irreducible components.*

We say that a complex of sheaves $\mathbf{K}^\bullet$ on X is a **characteristic complex for** *X provided that*

$$\mathrm{CC}(\mathbf{K}^\bullet) = \sum_i (-1)^{\dim X_i} \left[\overline{T^*_{(X_i)_{\mathrm{reg}}}\mathcal{U}}\right].$$

Thus, $\mathbf{K}^\bullet$ is a characteristic complex if and only if there exists a complex analytic Whitney stratification $\mathfrak{S}$ of X, with connected strata, with respect to which $\mathbf{K}^\bullet$ is constructible and such that, for all $S \in \mathfrak{S}$, the Euler characteristic $c_S(\mathbf{K}^\bullet)$ of the Morse modules of S with respect to $\mathbf{K}^\bullet$ is zero **unless** *S is an open dense subset of one of the $(X_i)_{\mathrm{reg}}$, in which case*

$$c_S(\mathbf{K}^\bullet) = (-1)^{\dim S} = (-1)^{\dim X_i}.$$

REMARK 3.2. Note that Remark 2.2 implies that, if $\mathbf{p}$ is not an isolated point in X and $\mathbf{K}^\bullet$ is a characteristic complex for X, then for a generic choice of (restricted) linear forms $\mathcal{L}$, $\chi(\phi_{\mathcal{L}-\mathcal{L}(\mathbf{p})}\mathbf{K}^\bullet)_\mathbf{p} = 0$, i.e., $\chi(\mathbf{K}^\bullet)_\mathbf{p} = \chi(\psi_{\mathcal{L}-\mathcal{L}(\mathbf{p})}\mathbf{K}^\bullet)_\mathbf{p}$.

EXAMPLE 3.3. Suppose that X is a complex manifold with connected components $\{X_i\}_i$. Then it is immediate that $\mathbb{Z}^\bullet_X$ is characteristic complex for X.

PROPOSITION 3.4. *Let $X = \bigcup_i X_i$ be the decomposition of X into its irreducible components. Suppose that, for each i, $\mathbf{K}^\bullet_i$ is a characteristic complex for X_i and let $\widehat{\mathbf{K}}^\bullet_i$ denote the extension by zero of $\mathbf{K}^\bullet_i$ to all of X. Then, $\bigoplus_i \widehat{\mathbf{K}}^\bullet_i$ is a characteristic complex for X.*

PROOF. This is immediate from Item 3 of Proposition 2.5. □

We wish to give a simple, but important, example of characteristic complexes.

EXAMPLE 3.5. Suppose that X is pure-dimensional, of dimension > 0, and has a single singular point at the origin in $\mathcal{U}$. Then,

$$\mathrm{CC}\left(\mathbb{Z}^\bullet_X\right) = (-1)^{\dim X}\left[\overline{T^*_{X_{\mathrm{reg}}}\mathcal{U}}\right] + b\left[T^*_\mathbf{0}\mathcal{U}\right],$$

where the final part of Remark 2.2 tells us that $b = 1 - \chi(\mathbb{L}_{X,\mathbf{0}})$.

To produce a characteristic complex for X, we must modify $\mathbb{Z}^\bullet_X$ at the origin. This is easy.

If $b = 0$, there is nothing to do; $\mathbb{Z}^\bullet_X$ would be a characteristic complex.

If $b < 0$, let $\mathbf{A}^\bullet$ be the extension by zero to all of X of the constant sheaf $\mathbb{Z}^{-b}$ on the point-set $\{\mathbf{0}\}$. Then, $\mathbb{Z}^\bullet_X \oplus \mathbf{A}^\bullet$ is a characteristic complex for X.

If $b > 0$, let $\mathbf{A}^\bullet$ be the extension by zero to all of X of the shifted constant sheaf $\mathbb{Z}^b[1]$ on the point-set $\{\mathbf{0}\}$. Then, $\mathbb{Z}^\bullet_X \oplus \mathbf{A}^\bullet$ is a characteristic complex for X.

Note that, in each of these cases, the stalk cohomology at the origin of the resulting characteristic complex is $\chi(\mathbb{L}_{X,\mathbf{0}})$.

The following proposition is well-known to experts in the form: CC yields a surjection from the Grothendieck group of constructible complexes to the group of Lagrangian cycles. For completeness, we give the proof, which is basically induction on the construction given in Example 3.5.

PROPOSITION 3.6. *Let $\mathfrak{S}$ be any Whitney stratification of X, with connected strata. Then, there exists a characteristic complex on X which is constructible with respect to $\mathfrak{S}$. In particular, characteristic complexes exist for all X.*

PROOF. This proof is contained in Lemma 3.1 of [**21**]. However, we wish to sketch it here.

Note that Proposition 3.4 implies that we need deal only with the case where X is irreducible. Hence, we assume that X is irreducible of dimension d.

Let $\mathfrak{S}$ be a Whitney stratification of X, with connected strata.

For every stratum $S \in \mathfrak{S}$ and every non-negative integer v, let $\mathbf{U}^\bullet_{S,v}$ denote the extension by zero to all of X of $\left(\mathbb{Z}^\bullet_S\right)^v$ so that $c_S(\mathbf{U}^\bullet_{S,v}) = (-1)^{\dim S} v$ (where c_S is the coefficient of $\left[T^*_S\mathcal{U}\right]$ in the characteristic cycle). If v is a negative integer, define $\mathbf{U}^\bullet_{S,v} := \mathbf{U}^\bullet_{S,-v}[1]$ so that, again, $c_S(\mathbf{U}^\bullet_{S,v}) = (-1)^{\dim S} v$.

Now we construct a characteristic complex as a direct sum, canceling out conormal cycles over lower-dimensional strata. Let

$$\mathbf{K}^\bullet_d = \mathbf{K}^\bullet_{\geq d} := \mathbf{U}^\bullet_{X_{\mathrm{reg}},1}.$$

Note that $\mathbf{K}^\bullet_d$ is also constructible with respect to $\mathfrak{S}$ and, if $S \in \mathfrak{S}$ has dimension d, then $c_S(\mathbf{K}^\bullet_d) = (-1)^d$.

Now we need cancel out the contributions to the characteristic cycle from lower-dimensional strata.

Let

$$\mathbf{K}^\bullet_{d-1} := \bigoplus_{S \in \mathfrak{S}, \dim S = d-1} \mathbf{U}^\bullet_{S, -c_S(\mathbf{K}^\bullet_d)},$$

so that $\mathbf{K}^\bullet_{\geq d-1} := \mathbf{K}^\bullet_d \oplus \mathbf{K}^\bullet_{d-1}$ has the property that, for $S \in \mathfrak{S}$ of dimension at least $d-1$,

$$c_S\left(\mathbf{K}^\bullet_{\geq d-1}\right) = \begin{cases} (-1)^d, & \text{if } \dim S = d; \\ 0, & \text{if } \dim S = d-1. \end{cases}$$

Continuing in this manner, we produce $\mathbf{K}^\bullet := \mathbf{K}^\bullet_{\geq 0}$ which is a characteristic complex for X. □

PROPOSITION 3.7. *Let X and Y be analytic spaces, let $\pi_1 : X \times Y \to X$ and $\pi_2 : X \to Y$ denote the projections. Let $\mathbf{K}^\bullet_X$ and $\mathbf{K}^\bullet_Y$ be characteristic complexes for X and Y, respectively. Let $\mathbf{K}^\bullet_X \overset{L}{\boxtimes} \mathbf{K}^\bullet_Y := \pi_1^*\mathbf{K}^\bullet_X \overset{L}{\otimes} \pi_2^*\mathbf{K}^\bullet_Y$.*

Then, $\mathbf{K}^\bullet_X \overset{L}{\boxtimes} \mathbf{K}^\bullet_Y$ is a characteristic complex for $X \times Y$.

PROOF. This is immediate from Item 8 of Proposition 2.8. □

4. The Local Euler Obstruction

Our primary interest in characteristic complexes lies in their relationship with the local Euler obstruction, defined by MacPherson in [**17**]. We let $\mathrm{Eu}_{\mathbf{p}} X$ denote the local Euler obstruction of X at $\mathbf{p}$, and first recall some well-known results, which appear either explicitly or implicitly in [**17**]. Note that we slightly extend the usual definition of the local Euler obstruction to possibly non-pure-dimensional spaces by adding over the irreducible components.

PROPOSITION 4.1.

(1) *The local Euler obstruction is, in fact, local, i.e., if* $\mathcal{W}$ *is an open neighborhood of* $\mathbf{p}$ *in* X*, then* $\operatorname{Eu}_{\mathbf{p}} X = \operatorname{Eu}_{\mathbf{p}} \mathcal{W}$.

(2) *If* $\mathbf{p}$ *is a smooth point of* X*, then* $\operatorname{Eu}_{\mathbf{p}} X = 1$.

(3) *If* $(\mathbf{x}, \mathbf{y}) \in X \times Y$*, then* $\operatorname{Eu}_{(\mathbf{x},\mathbf{y})}(X \times Y) = (\operatorname{Eu}_{\mathbf{x}} X)(\operatorname{Eu}_{\mathbf{y}} Y)$.

(4) *If* $\mathbf{p} \in X$ *and* X_i *denotes the local irreducible components of* X *at* $\mathbf{p}$*, then* $\operatorname{Eu}_{\mathbf{p}} X = \sum_i \operatorname{Eu}_{\mathbf{p}} X_i$.

(5) $\operatorname{Eu}_{\mathbf{x}} X$ *is a constant function of* $\mathbf{x}$ *along the strata of any Whitney stratification of* X *(which has connected strata).*

There is also the important result:

THEOREM 4.2. *(Brylinski, Dubson, Kashiwara, [**5**]) Suppose that* $\mathbf{A}^\bullet$ *on* X *is constructible with respect a Whitney stratification* $\mathfrak{S}$*, and that*

$$\operatorname{CC}(\mathbf{A}^\bullet) = \sum_{S \in \mathfrak{S}} c_S(\mathbf{A}^\bullet) \left[\overline{T^*_S \mathcal{U}}\right].$$

Then, for all $\mathbf{p} \in X$*,*

$$\chi(\mathbf{A}^\bullet)_{\mathbf{p}} = \sum_{S \in \mathfrak{S}} (-1)^{\dim S} c_S(\mathbf{A}^\bullet) \operatorname{Eu}_{\mathbf{p}} \overline{S},$$

where we set $\operatorname{Eu}_{\mathbf{p}}(\overline{S}) = 0$ *if* $\mathbf{p} \notin \overline{S}$.

From this, we immediately conclude the fundamental relationship between characteristic complexes and the local Euler obstruction:

COROLLARY 4.3. *Let* $\mathbf{K}^\bullet$ *be a characteristic complex for* X*. Let* $\mathbf{p} \in X$*. Then,*

$$\operatorname{Eu}_{\mathbf{p}} X = \chi(\mathbf{K}^\bullet)_{\mathbf{p}}.$$

REMARK 4.4. We note, as in [**28**], Remark 0.1, that much of what we have written can be described just using the language of constructible functions. Suppose that, for every constructible complex of sheaves $\mathbf{F}^\bullet$, we let

$$\alpha_{\mathbf{F}^\bullet} : X \to \mathbb{Z}$$

denote the constructible function $\alpha_{\mathbf{F}^\bullet}(\mathbf{x}) := \chi(\mathbf{F}^\bullet)_{\mathbf{x}}$.

Then, it is well-known that $\alpha_{()}$ yields a surjection from the Grothendieck group of constructible complexes on X to the group of constructible functions on X.

As $\mathbf{p} \mapsto \operatorname{Eu}_{\mathbf{p}} X$ is a constructible function, our definition of a characteristic complex $\mathbf{K}^\bullet$ was precisely designed so that $\alpha_{\mathbf{K}^\bullet}$ equals the local Euler obstruction function.

As a corollary to Corollary 4.3, and using the additivity of the Euler characteristic of hypercohomology over complex stratifications, we recover the formula of [**2**]:

COROLLARY 4.5. *Let $\mathfrak{S}$ be a complex analytic Whitney stratification, with connected strata, of X. For each $S \in \mathfrak{S}$, let $\mathbf{p}_S$ denote a point in S. Let $\mathbf{p}$ be an arbitrary non-isolated point of X and let $\mathbb{L}_{X,\mathbf{p}}$ denote the complex link of X at $\mathbf{p}$.*

Then,

$$\operatorname{Eu}_{\mathbf{p}} X = \sum_{S \in \mathfrak{S}} \chi(\mathbb{L}_{X,\mathbf{p}} \cap S) \operatorname{Eu}_{\mathbf{p}_S} X.$$

PROOF. For convenience, we assume that $\mathbf{p} = \mathbf{0}$ and that $\mathcal{L}$ is a generic linear form. Let $\mathbf{K}^\bullet$ be a characteristic complex for X. As $\mathbf{p}$ is not an isolated point of X, Remark 3.2 implies that $\chi(\mathbf{K}^\bullet)_\mathbf{0} = \chi(\psi_\mathcal{L}\mathbf{K}^\bullet)_\mathbf{0}$.

Thus, using the additivity of the Euler characteristic of hypercohomology over complex stratifications, we find

$$\operatorname{Eu}_\mathbf{0} X = \chi(\mathbf{K}^\bullet)_\mathbf{0} = \chi(\psi_\mathcal{L}\mathbf{K}^\bullet)_\mathbf{0} = \chi\left(\mathbb{H}^*(\mathbb{L}_{X,\mathbf{0}}; \mathbf{K}^\bullet)\right) =$$
$$\sum_S \chi(\mathbb{L}_{X,\mathbf{0}} \cap S)\chi(\mathbf{K}^\bullet)_{\mathbf{p}_S} = \sum_{S\in\mathfrak{S}} \chi(\mathbb{L}_{X,\mathbf{0}} \cap S) \operatorname{Eu}_{\mathbf{p}_S} X.$$

□

Before we leave this section, we wish to give a known example of how Corollary 4.5 enables one to calculate local Euler obstructions.

EXAMPLE 4.6. Suppose that $\mathbf{p}$ is an isolated singular point of X. Then, every point in the complex link, $\mathbb{L}_{X,\mathbf{p}}$, is a smooth point of X and, hence, the local Euler obstruction of X at each point of $\mathbb{L}_{X,\mathbf{p}}$ is 1. Thus, we conclude from Corollary 4.5 that

$$\operatorname{Eu}_\mathbf{p} X = \chi(\mathbb{L}_{X,\mathbf{p}}).$$

In particular, suppose that X is a curve, and $\mathbf{p} \in X$. Then the complex link $\mathbb{L}_{X,\mathbf{p}}$, is a finite collection of points; the number of points is simply the multiplicity, $\operatorname{mult}_\mathbf{p} X$, of X at $\mathbf{p}$. Thus, for a curve, we conclude that

$$\operatorname{Eu}_\mathbf{p} X = \chi(\mathbb{L}_{X,\mathbf{p}}) = \operatorname{mult}_\mathbf{p} X.$$

Now suppose that X is 2-dimensional in a neighborhood of $\mathbf{p} \in X$. Let S's denote 1-dimensional Whitney strata which contain $\mathbf{p}$ in their closures. Let $\mathbf{p}_S$ denote an arbitrary point of S near $\mathbf{p}$. Then, we leave it as an exercise for the reader to use Corollary 4.5 to conclude that

$$\operatorname{Eu}_\mathbf{p} X = \chi(\mathbb{L}_{X,\mathbf{p}}) + \sum_S (\operatorname{mult}_\mathbf{p} \overline{S})(\operatorname{mult}_{\mathbf{p}_S} X - 1).$$

5. The Relative Local Euler Obstruction

We now wish to discuss the *relative local Euler obstruction*, as was introduced in [**3**].

Recall that $\mathcal{U}$ is an open neighborhood of the origin of $\mathbb{C}^{n+1}$, X is a closed, analytic subset of $\mathcal{U}$. We let $\mathbf{z} := (z_0, \dots, z_n)$ be coordinates on $\mathcal{U}$. We identify the cotangent space $T^*\mathcal{U}$ with $\mathcal{U} \times \mathbb{C}^{n+1}$ by mapping $(\mathbf{p}, w_0 d_\mathbf{p} z_0 + \cdots + w_n d_\mathbf{p} z_n)$ to $(\mathbf{p}, (w_0, \dots, w_n))$. Let $\pi : T^*\mathcal{U} \to \mathcal{U}$ denote the projection.

Suppose that we have $\mathbf{p} \in X$ and a complex analytic $f : X \to \mathbb{C}$. We let $\tilde{f}$ be a local extension of f at $\mathbf{p}$ to an open neighborhood of $\mathbf{p}$ in $\mathcal{U}$; we assume now that

$\mathcal{U}$ is re-chosen to be this (possibly) smaller open neighborhood of $\mathbf{p}$. We also let $d\tilde{f}$ denote the section of the cotangent bundle to $\mathcal{U}$ given by $d\tilde{f}(\mathbf{x}) = (\mathbf{x}, d_{\mathbf{x}}\tilde{f})$; we let $\operatorname{im}(d\tilde{f})$ denote the image of this section in $T^*\mathcal{U}$.

Thus, in coordinates,

$$\operatorname{im}(d\tilde{f}) = V\left(w_0 - \frac{\partial \tilde{f}}{\partial z_0}, \dots, w_n - \frac{\partial \tilde{f}}{\partial z_n}\right).$$

Note that π, restricted to $\operatorname{im}(d\tilde{f})$, is an isomorphism onto $\mathcal{U}$, with inverse given by $\mathbf{x} \mapsto (\mathbf{x}, d_{\mathbf{x}}\tilde{f})$. In particular, we have an isomorphism

$$\overline{T^*_{X_{\mathrm{reg}}}\mathcal{U}} \cap \operatorname{im}(d\tilde{f}) \cong \pi\left(\overline{T^*_{X_{\mathrm{reg}}}\mathcal{U}} \cap \operatorname{im}(d\tilde{f})\right).$$

In [**22**], we gave a name to this last analytic set:

DEFINITION 5.1. *The* **conormal-regular critical locus**, $\Sigma_{\mathrm{cnr}}f$, *of* f *is defined to be*

$$\Sigma_{\mathrm{cnr}}f := \pi\left(\overline{T^*_{X_{\mathrm{reg}}}\mathcal{U}} \cap \operatorname{im}(d\tilde{f})\right) = \left\{\mathbf{p} \in \mathcal{U} \mid (\mathbf{p}, d_{\mathbf{p}}\tilde{f}) \in \overline{T^*_{X_{\mathrm{reg}}}\mathcal{U}}\right\}.$$

Below, when we take intersection cycles and numbers, we will always be in the case of proper intersections inside the complex analytic manifold $T^*\mathcal{U}$ or inside $\mathcal{U}$ itself. In this case, there are well-defined intersection cycles (**not** cycles classes modulo rational equivalence); see [**10**].

Assuming that X is pure-dimensional, the relative local Euler obstruction, $\operatorname{Eu}_{\mathbf{p}} f$, is defined as an obstruction to extending the conjugate of the stratified gradient vector field of f to a non-zero lift in the Nash modification, provided that $\mathbf{p}$ is a stratified isolated critical point of f; see [**3**].

In Corollary 5.4 of [**3**], we show:

PROPOSITION 5.2. *Suppose that* X *is pure-dimensional and that* $f : X \to \mathbb{C}$ *has a stratified isolated critical point at* $\mathbf{p}$. *Let* $\mathbf{K}^\bullet$ *be a characteristic complex for* X. *Then,* $(\mathbf{p}, d_{\mathbf{p}}\tilde{f})$ *is an isolated point in the intersection* $\overline{T^*_{X_{\mathrm{reg}}}\mathcal{U}} \cap \operatorname{im}(d\tilde{f})$ *(equivalently,* $\mathbf{p}$ *is an isolated point in* $\Sigma_{\mathrm{cnr}}f$*) and*

$$\operatorname{Eu}_{\mathbf{p}} f = \chi\big(\phi_{f-f(\mathbf{p})}[-1]\mathbf{K}^\bullet\big)_{\mathbf{p}} = (-1)^{\dim X}\left(\overline{T^*_{X_{\mathrm{reg}}}\mathcal{U}} \cdot \operatorname{im}(d\tilde{f})\right)_{(\mathbf{p}, d_{\mathbf{p}}\tilde{f})}.$$

Note that, in the case where X *is affine space, this intersection number is the Milnor number of* $f - f(\mathbf{p})$ *at* $\mathbf{p}$.

We can use Proposition 5.2 as the basis for generalizing the definition of the relative local Euler obstruction to (possibly) non-isolated critical points of functions on spaces which need not be pure-dimensional. In [**3**], we referred to this as the *defect* $D_{f,X}$.

DEFINITION 5.3. *Let* $\mathbf{K}^\bullet$ *be a characteristic complex for* X. *Then, we define the* **relative local Euler obstruction of** f *at* $\mathbf{p} \in X$ *to be*

$$\operatorname{Eu}_{\mathbf{p}} f := \chi\big(\phi_{f-f(\mathbf{p})}[-1]\mathbf{K}^\bullet\big)_{\mathbf{p}}.$$

Note that $\operatorname{Eu}_{\mathbf{p}} f$ *is well-defined by Corollary* 2.10.

REMARK 5.4. Note that the definition immediately implies that, if c is a constant, then $\operatorname{Eu}_{\mathbf{p}} f = \operatorname{Eu}_{\mathbf{p}}(f+c)$. Also, note that, in [**3**], we referred to this generalized relative local Euler obstruction as the *defect* $D_{f,X}$.

In terms of constructible functions, what we have done is to define the relative local Euler obstruction to be the constructible shifted vanishing cycle function along f of the local Euler obstruction.

The relative local Euler obstruction is related to the local Euler obstructions of strata and the Euler characteristics of the intersections of the various strata with the Milnor fiber, $F_{f-f(\mathbf{p}),\mathbf{p}}$, of $f - f(\mathbf{p})$ at $\mathbf{p}$ via the following theorem. We proved this theorem, in slightly different terms, in [**3**], but the proof is very short, and we include it for completeness.

THEOREM 5.5. *Let $\mathfrak{S}$ be a complex analytic stratification of X such that the local Euler obstruction of X is constant along the strata (e.g., a Whitney stratification with connected strata). For each $S \in \mathfrak{S}$, let $\mathbf{p}_S$ be a point in S. Then, for $\mathbf{p} \in X$,*

$$\operatorname{Eu}_{\mathbf{p}} f = \operatorname{Eu}_{\mathbf{p}} X - \sum_{S \in \mathfrak{S}} \chi(F_{f-f(\mathbf{p}),\mathbf{p}} \cap S) \operatorname{Eu}_{\mathbf{p}_S} X.$$

PROOF. Let $\mathbf{K}^\bullet$ be a characteristic complex for X. Then,

$$\operatorname{Eu}_{\mathbf{p}} f = \chi\big(\phi_{f-f(\mathbf{p})}[-1]\mathbf{K}^\bullet\big)_{\mathbf{p}}$$

which, by the canonical distinguished triangle relating the nearby and vanishing cycles, gives us

$$\operatorname{Eu}_{\mathbf{p}} f = -\chi\big(\mathbf{K}^\bullet[-1]\big)_{\mathbf{p}} + \chi\big(\psi_{f-f(\mathbf{p})}[-1]\mathbf{K}^\bullet\big)_{\mathbf{p}} =$$

$$\operatorname{Eu}_{\mathbf{p}} X - \chi\big(\mathbb{H}^*(F_{f-f(\mathbf{p})}; \mathbf{K}^\bullet)\big)_{\mathbf{p}} = \operatorname{Eu}_{\mathbf{p}} X - \sum_{S \in \mathfrak{S}} \chi(F_{f-f(\mathbf{p}),\mathbf{p}} \cap S) \operatorname{Eu}_{\mathbf{p}_S} X.$$

□

EXAMPLE 5.6. We should note that, given Theorem 5.5, our definition of the relative local Euler obstruction in terms of vanishing cycles implicitly appears in 4.6 of [**27**].

The next two examples contain known results; see, especially, [**30**].

EXAMPLE 5.7. Consider the case where $f : (\mathcal{U}, \mathbf{0}) \to (\mathbb{C}, 0)$ is a non-locally constant function on an open subset $\mathcal{U} \subseteq \mathbb{C}^{n+1}$. Then, Theorem 5.5 tells us that

$$\operatorname{Eu}_{\mathbf{0}} f = 1 - \chi(F_{f,\mathbf{0}}) = -\widetilde{\chi}(F_{f,\mathbf{0}}),$$

where $\widetilde{\chi}$ denotes the Euler characteristic of the reduced cohomology.

In particular, if $\mathbf{0}$ is an **isolated** critical point of $f : (\mathcal{U}, \mathbf{0}) \to (\mathbb{C}, 0)$, then

$$\operatorname{Eu}_{\mathbf{0}} f = (-1)^{n+1} \mu_{\mathbf{0}}(f),$$

where $\mu_{\mathbf{0}}(f)$ is the Milnor number of f at $\mathbf{0}$.

EXAMPLE 5.8. Now suppose that $\mathbf{0}$ is an isolated singular point of X, and we have $f : (X, \mathbf{0}) \to (\mathbb{C}, 0)$. Then, Theorem 5.5 tells us that

$$\mathrm{Eu}_{\mathbf{0}} f = \mathrm{Eu}_{\mathbf{0}} X - \chi(F_{f,\mathbf{0}}) = \chi(\mathbb{L}_{X,\mathbf{0}}) - \chi(F_{f,\mathbf{0}}).$$

In particular, if f itself is a generic linear form, then $\mathrm{Eu}_{\mathbf{0}} f = 0$. We shall see in Item 2 below that is true for arbitrarily singular spaces.

As another particular case, suppose that X is a curve and that $\tilde{f}$ is a local extension of f to the ambient affine space such that $\dim_{\mathbf{0}} X \cap V(\tilde{f}) = 0$. Then, we conclude that

$$\mathrm{Eu}_{\mathbf{0}} f = \mathrm{mult}_{\mathbf{0}} X - (X \cdot V(\tilde{f}))_{\mathbf{0}}.$$

We now give a number of basic properties of the relative local Euler obstruction.

THEOREM 5.9. *Let $X = \bigcup_i X_i$ be the decomposition of X into its irreducible components, and let f_i denote the restriction of f to X_i.*

The relative local Euler obstruction has the following properties:

(1) *If f is constant in a neighborhood of $\mathbf{p}$, then $\mathrm{Eu}_{\mathbf{p}} f = \mathrm{Eu}_{\mathbf{p}} X$.*

(2) *If $\mathbf{p} \notin \Sigma_{\mathrm{cnr}} f$, then $\mathrm{Eu}_{\mathbf{p}} f = 0$. In particular, if $\mathbf{p}$ is not an isolated point of X, and L is the restriction of a generic linear form to X, then $\mathrm{Eu}_{\mathbf{p}} L = 0$.*

(3) *$\mathrm{Eu}_{\mathbf{p}} f = \sum_i \mathrm{Eu}_{\mathbf{p}} f_i$, where we set $\mathrm{Eu}_{\mathbf{p}} f_i = 0$ if $\mathbf{p} \notin X_i$.*

(4) *If $\mathbf{p}$ is an isolated point in $\Sigma_{\mathrm{cnr}} f$, then*

$$\mathrm{Eu}_{\mathbf{p}} f = \sum_i (-1)^{\dim X_i} \left(\overline{T^*_{(X_i)_{\mathrm{reg}}} \mathcal{U}} \cdot \mathrm{im}(d\tilde{f}) \right)_{(\mathbf{p}, d_{\mathbf{p}}\tilde{f})}.$$

(5) *Let $\mathbf{q} \in Y$ and suppose that we have a complex analytic function $g : Y \to \mathbb{C}$. Let $f \boxplus g$ denote the function from $X \times Y$ to $\mathbb{C}$ given by*

$$(f \boxplus g)(x, y) = f(x) + g(y).$$

Then,

$$\mathrm{Eu}_{(\mathbf{p},\mathbf{q})}(f \boxplus g) = \mathrm{Eu}_{\mathbf{p}} f \cdot \mathrm{Eu}_{\mathbf{q}} g.$$

PROOF. Item 1 follows at once from Theorem 5.5 or, alternatively from Corollary 4.3 and Definition 5.3.

Both Items 2 and 4 essentially follow from the vanishing cycle index theorem of Ginsburg [**11**], and Lê [**15**], but those papers require stronger hypotheses. However, the full results appear in 4.5 and 4.6 of Sabbah [**27**], and also in Corollary 0.3 of Schürmann [**28**].

We can also conclude the results from looking ahead to Theorem 6.1 in Section 6.

Item 3:

Suppose that, for each i, $\mathbf{K}_i^\bullet$ is a characteristic complex for X_i and let $\widehat{\mathbf{K}}_i^\bullet$ denote the extension by zero of $\mathbf{K}_i^\bullet$ to all of X. Then, by Proposition 3.4, $\bigoplus_i \widehat{\mathbf{K}}_i^\bullet$ is a characteristic complex for X. Thus,

$$\mathrm{Eu}_{\mathbf{p}} f = \chi\Big(\phi_{f-f(\mathbf{p})}[-1] \bigoplus_i \widehat{\mathbf{K}}_i^\bullet\Big)_{\mathbf{p}} = \sum_i \chi(\phi_{f-f(\mathbf{p})}[-1]\widehat{\mathbf{K}}_i^\bullet)_{\mathbf{p}} =$$

$$\sum_i \chi\big(\phi_{f_i - f_i(\mathbf{p})}[-1]\mathbf{K}_i^\bullet\big)_{\mathbf{p}} = \sum_i \mathrm{Eu}_{\mathbf{p}} f_i.$$

Item 5:

Let us assume, without loss of generality, that $f(\mathbf{p}) = 0$ and $g(\mathbf{q}) = 0$. Let $\mathbf{K}_X^\bullet$ and $\mathbf{K}_Y^\bullet$ be characteristic complexes for X and Y, Then, we know from Proposition 3.7 that $\mathbf{K}_X^\bullet \overset{L}{\boxtimes} \mathbf{K}_Y^\bullet$ is a characteristic complex for $X \times Y$.

Now, the derived category version of the Sebastiani-Thom Theorem which we proved in [**24**] (but, here, using the more-usual definition/shift on the vanishing cycles) tells us that

$$H^k\big(\phi_{f \boxplus g}[-1](\mathbf{K}_X^\bullet \overset{L}{\boxtimes} \mathbf{K}_Y^\bullet)\big)_{(\mathbf{p},\mathbf{q})} \cong H^k\big(\phi_f[-1]\mathbf{K}_X^\bullet \overset{L}{\boxtimes} \phi_g[-1]\mathbf{K}_Y^\bullet\big)_{(\mathbf{p},\mathbf{q})} \cong$$

$$\bigoplus_{i+j=k} H^i(\phi_f[-1]\mathbf{K}_X^\bullet)_{\mathbf{p}} \otimes H^j(\phi_g[-1]\mathbf{K}_Y^\bullet)_{\mathbf{q}} \oplus$$
$$\bigoplus_{i+j=k+1} \mathrm{Tor}\,\big(H^i(\phi_f[-1]\mathbf{K}_X^\bullet)_{\mathbf{p}}, H^j(\phi_g[-1]\mathbf{K}_Y^\bullet)_{\mathbf{q}}\big).$$

Item 5 follows.

□

REMARK 5.10. We naturally refer to Item 5 above as the *Sebastiani-Thom property* of the relative local Euler obstruction.

We should also remark that the quantity

$$\Big(\overline{T^*_{(X_i)_{\mathrm{reg}}}\mathcal{U}} \cdot \mathrm{im}(d\tilde{f})\Big)_{(\mathbf{p}, d_{\mathbf{p}}\tilde{f})}$$

in Item 5 above can be characterized more geometrically as the number of nondegenerate critical points of a small perturbation of f by a generic linear form L which occur near the origin on $(X_i)_{\mathrm{reg}}$; see Theorem 3.2 of [**21**] and Theorem 1.1 of [**23**]. Thus, Item 5 is a significant generalization of Proposition 2.3 of [**30**], as we do not require that f has a stratified isolated critical point.

EXAMPLE 5.11. Let (x, y, z) be coordinates on $\mathcal{U} := \mathbb{C}^3$, and let (w_0, w_1, w_2) be the corresponding cotangent coordinates. Let

$$X := V(xy) = V(x) \cup V(y) \subseteq \mathbb{C}^3.$$

Let $\tilde{f} : \mathcal{U} \to \mathbb{C}$ be given by $\tilde{f}(x, y, z) := x + y^2 + yz$, and let f be the restriction of $\tilde{f}$ to X.

Note that f does **not** possess a stratified isolated singularity at $\mathbf{0}$, since $\Sigma X = V(x, y) = \{\mathbf{0}\} \times \mathbb{C}$ and $f_{|_{V(x,y)}} \equiv 0$. However, we claim that **$(\mathbf{0}, d_{\mathbf{0}}\tilde{f})$ is** an isolated point in $\overline{T^*_{X_{\mathrm{reg}}}\mathcal{U}} \cap \mathrm{im}\, d\tilde{f}$.

We find

$$\overline{T^*_{X_{\mathrm{reg}}}\mathcal{U}} = V(x, w_1, w_2) \cup V(y, w_0, w_2)$$

and

$$\mathrm{im}\, d\tilde{f} = V(w_0 - 1, w_1 - 2y - z, w_2 - y).$$

Therefore, $\mathrm{im}\, d\tilde{f}$ does not intersect $V(y, w_0, w_2)$, and intersects $V(x, w_1, w_2)$ in the single point $V(x, y, z, w_0 - 1, w_1, w_2)$.

Hence, Item 4 of Theorem 5.9 tells us that

$$\mathrm{Eu}_{\mathbf{0}} f = (-1)^2 \Big(\overline{T^*_{(X)_{\mathrm{reg}}}\mathcal{U}} \cdot \mathrm{im}(d\tilde{f})\Big)_{(\mathbf{0}, d_{\mathbf{0}}\tilde{f})} =$$

$$\Big(\big(V(x,w_1,w_2)+V(y,w_0,w_2)\big)\cdot V(w_0-1,w_1-2y-z,w_2-y)\Big)_{(\mathbf{0},1,0,0)}=1.$$

EXAMPLE 5.12. Suppose that Y is a curve through the origin in $\mathbb{C}^2$. Let $g:\mathbb{C}^2\to C$ be such that $\dim_{\mathbf{0}} Y\cap V(g)=0$. Let $X:=Y\times\mathbb{C}$, where we use z as the coordinate on this new copy of $\mathbb{C}$.

Consider the function $f:(X,\mathbf{0})\to(\mathbb{C},0)$ given by $f(x,y,z)=g(x,y)+z^b$, for some positive integer b.

Then, by Item 5 of Theorem 5.9 - the Sebastiani-Thom property - combined with Example 5.7 and the last part of Example 5.8 - we find that

$$\mathrm{Eu}_{\mathbf{0}} f=\big(\mathrm{mult}_{\mathbf{0}} Y-(Y\cdot V(g))_{\mathbf{0}}\big)(1-b).$$

6. Calculating the relative local Euler obstruction

Here, we recall the algorithm and result which is described in Section 6 of [**3**].

Once again, let $X=\bigcup_i X_i$ be the decomposition of X into its irreducible components, and let f_i denote the restriction of f to X_i. Recall from Item 4 of Theorem 5.9 that $\mathrm{Eu}_{\mathbf{p}} f=\sum_i \mathrm{Eu}_{\mathbf{p}} f_i$. Therefore, calculating $\mathrm{Eu}_{\mathbf{p}} f$ boils down to needing to calculate in the case where X is irreducible.

Thus, suppose throughout this section that X is irreducible, embedded in $\mathcal{U}$, an open subset of $\mathbb{C}^{n+1}$.

Recall our previous set-up:

We let $\mathbf{z}:=(z_0,\dots,z_n)$ be coordinates on $\mathcal{U}$, we identify the cotangent space $T^*\mathcal{U}$ with $\mathcal{U}\times\mathbb{C}^{n+1}$ by mapping $(\mathbf{p},w_0d_{\mathbf{p}}z_0+\cdots+w_nd_{\mathbf{p}}z_n)$ to $(\mathbf{p},(w_0,\dots,w_n))$, and let $\pi:T^*\mathcal{U}\to\mathcal{U}$ denote the projection.

Assume that $\mathbf{p}\in X$, that $\tilde f:\mathcal{U}\to\mathbb{C}$ is a complex analytic function, and that f is the restriction of $\tilde f$ to X. We let $d\tilde f$ denote the section of the cotangent bundle to $\mathcal{U}$ given by $d\tilde f(\mathbf{x})=(\mathbf{x},d_{\mathbf{x}}\tilde f)$; we let $\mathrm{im}(d\tilde f)$ denote the image of this section in $T^*\mathcal{U}$.

Thus, in coordinates,

$$\mathrm{im}(d\tilde f)=V\left(w_0-\frac{\partial\tilde f}{\partial z_0},\dots,w_n-\frac{\partial\tilde f}{\partial z_n}\right).$$

Our method of calculation requires a "generic" choice of coordinates for the ambient, affine space $\mathcal{U}$. This choice of coordinates is made as follows. Refine a Whitney stratification of X to a stratification $\mathcal{W}:=\{W_j\}$ such that $\mathcal{W}$ satisfies Thom's a_f condition and such that $f^{-1}(0)$ is a union of strata of $\mathcal{W}$ (we are not assuming that $\mathcal{W}$ is still a Whitney stratification).

Choose the first coordinate z_0 so that the hyperplane $z_0^{-1}(0)$ transversely intersects, in some neighborhood of the origin, all positive-dimensional strata of $\{W_j\}$. Then, there is an induced stratification (on the germ at the origin) of $X\cap z_0^{-1}(0)$ given by $\{W_j\cap z_0^{-1}(0)\}$. We choose z_1 so that $z_1^{-1}(0)$ transversely intersects, in some neighborhood of the origin, all positive-dimensional strata of $\{W_j\cap z_0^{-1}(0)\}$. We continue in this inductive manner to produce $z_0,\dots,z_{n+1}$. We call such a coordinate choice *prepolar* (at the origin).

Prepolar coordinates are not as generic as possible, but they are generic enough for our purposes. Being prepolar at the origin implies that the coordinates are also prepolar at each point in a neighborhood of the origin, and we assume that we are in such a neighborhood throughout the remainder of this section.

Assuming that the coordinates are prepolar for f at the origin, all of the intersections that we write below are proper in $T^*\mathcal{U}$ (resp., in $\mathcal{U}$) in a neighborhood of $(\mathbf{p}, d_{\mathbf{p}}\tilde{f})$ (resp., $\mathbf{p}$).

The algorithm is as follows:

The cycle $\left[\overline{T^*_{X_{\mathrm{reg}}}\mathcal{U}}\right]$ can be written as a sum of purely $(n+1)$-dimensional cycles

$$\left[\overline{T^*_{X_{\mathrm{reg}}}\mathcal{U}}\right] = \widehat{\Gamma}^{n+1}_{f,\mathbf{z}} + \widehat{\Lambda}^{n+1}_{f,\mathbf{z}},$$

where no component of $\widehat{\Gamma}^{n+1}_{f,\mathbf{z}}$ is contained in $\operatorname{im}(d\tilde{f})$ (i.e. in $\overline{T^*_{X_{\mathrm{reg}}}\mathcal{U}} \cap \operatorname{im}(d\tilde{f})$) and every component of $\widehat{\Lambda}^{n+1}_{f,\mathbf{z}}$ is contained in $\operatorname{im}(d\tilde{f})$.

Now, we define $\widehat{\Gamma}^{k}_{f,\mathbf{z}}$ and $\widehat{\Lambda}^{k}_{f,\mathbf{z}}$ by downward induction. If we have defined the purely $(k+1)$-dimensional cycle $\widehat{\Gamma}^{k+1}_{f,\mathbf{z}}$, then the hypersurface

$$\left[V\left(w_{k+1} - \frac{\partial \tilde{f}}{\partial z_{k+1}}\right)\right]$$

properly intersects $\widehat{\Gamma}^{k+1}_{f,\mathbf{z}}$ inside U, and therefore there is a well-defined, purely k-dimensional intersection cycle

$$\widehat{\Gamma}^{k+1}_{f,\mathbf{z}} \cdot \left[V\left(w_{k+1} - \frac{\partial \tilde{f}}{\partial z_{k+1}}\right)\right],$$

which we can decompose as

$$\widehat{\Gamma}^{k+1}_{f,\mathbf{z}} \cdot \left[V\left(w_{k+1} - \frac{\partial \tilde{f}}{\partial z_{k+1}}\right)\right] =: \widehat{\Gamma}^{k}_{f,\mathbf{z}} + \widehat{\Lambda}^{k}_{f,\mathbf{z}},$$

where no component of $\widehat{\Gamma}^{k}_{f,\mathbf{z}}$ is contained in $\operatorname{im}(d\tilde{f})$ and every component of $\widehat{\Lambda}^{k}_{f,\mathbf{z}}$ is contained in $\operatorname{im}(d\tilde{f})$.

As the $\widehat{\Lambda}^{k}_{f,\mathbf{z}}$ are contained in $\operatorname{im}(d\tilde{f})$, the projection, π, maps each $\widehat{\Lambda}^{k}_{f,\mathbf{z}}$ isomorphically onto a cycle in $\mathcal{U}$; we let $\Lambda^{k}_{f,\mathbf{z}} := \pi_*\big(\widehat{\Lambda}^{k}_{f,\mathbf{z}}\big)$ (this is the proper projection of a cycle). We refer to $\Lambda^{k}_{f,\mathbf{z}}$ as the *k-dimensional Lê-Vogel cycle.*

Note that in the case where $X = \mathcal{U}$, $\left[\overline{T^*_{X_{\mathrm{reg}}}\mathcal{U}}\right] = \mathcal{U} \times \{0\}$, and the Lê-Vogel cycles coincide with the Lê cycles of [**18**], [**19**], and [**20**].

Now, $\Lambda^{k}_{f,\mathbf{z}}$ properly intersects the linear subspace $V(z_0 - p_0, \dots, z_{k-1} - p_{k-1})$ at $\mathbf{p}$, and we define the *k-dimensional Lê-Vogel number of f at* $\mathbf{p}$ to be the intersection number

$$\lambda^{k}_{f,\mathbf{z}}(\mathbf{p}) := \big(\Lambda^{k}_{f,\mathbf{z}} \cdot V(z_0 - p_0, \dots, z_{k-1} - p_{k-1})\big)_{\mathbf{p}}.$$

THEOREM 6.1. (Theorem 6.2 of [**3**]) *Suppose that X is irreducible, and let $\mathbf{K}^\bullet$ be a characteristic complex for X.*

Let $s := \dim_{\mathbf{p}} \Sigma_{\mathrm{cnr}} f$ (where we set $s = -\infty$ if $\mathbf{p} \notin \Sigma_{\mathrm{cnr}} f$), and assume that the coordinates $\mathbf{z}$ are prepolar for f at $\mathbf{0}$.

Then, the Lê-Vogel numbers $\lambda^k_{f,\mathbf{z}}(\mathbf{p})$ are zero if $k > s$, and

$$\mathrm{Eu}_{\mathbf{p}} f = \chi(\phi_f[-1]\mathbf{K}^\bullet)_{\mathbf{p}} = (-1)^{\dim X} \sum_{k=0}^{s} (-1)^k \lambda^k_{f,\mathbf{z}}(\mathbf{p}).$$

In particular, when $s = 0$, the only Lê-Vogel number which is possibly non-zero is $\lambda^0_{f,\mathbf{z}}(\mathbf{p})$, and $\lambda^0_{f,\mathbf{z}}(\mathbf{p}) = \left(\overline{T^*_{X_{\mathrm{reg}}}\mathcal{U}} \cdot \mathrm{im}(d\tilde{f})\right)_{(\mathbf{p}, d_{\mathbf{p}}\tilde{f})}$; thus, when $s = 0$,

$$\mathrm{Eu}_{\mathbf{p}} f = (-1)^{\dim X}\left(\overline{T^*_{X_{\mathrm{reg}}}\mathcal{U}} \cdot \mathrm{im}(d\tilde{f})\right)_{(\mathbf{p}, d_{\mathbf{p}}\tilde{f})}.$$

The serious weakness of Theorem 6.1 is that there is no effective way to obtain the a_f stratification of X which we need in order to know if our coordinates are prepolar or not. The case where $\mathbf{p}$ is an isolated point of $\Sigma_{\mathrm{cnr}} f$ is nice because the resulting formula is independent of the coordinates. However, we can also handle special 1-dimensional cases fairly easily.

EXAMPLE 6.2. Suppose X is irreducible at $\mathbf{p}$, of dimension d, and suppose that $\dim_{\mathbf{p}} \Sigma X \leq 1$. Then, near $\mathbf{p}$, there is a Whitney stratification of X consisting of the connected components of X_{reg}, the connected components of $\Sigma X - \{\mathbf{p}\}$, and $\{\mathbf{p}\}$.

Now, suppose also that $\dim_{\mathbf{p}} \overline{\Sigma(f_{|X_{\mathrm{reg}}})} \leq 1$. Let $\Sigma := \Sigma X \cup \Sigma(f_{|X_{\mathrm{reg}}})$. Then there is an a_f stratification consisting of the connected components of $X_{\mathrm{reg}} - V(f)$, the connected components of $V(f) - \Sigma$, the connected components of $\Sigma - \{\mathbf{p}\}$, and $\{\mathbf{p}\}$.

Then, one easily sees that the requirement that $(z_0, z_1, \ldots, z_n)$ be prepolar is equivalent to requiring that, near $\mathbf{p}$,

- $V(z_0 - p_0)$ contains no irreducible component of Σ;
- for $0 \leq i \leq d-1$, $V(z_0 - p_0, \ldots, z_i - p_i)$ transversely intersects X_{reg} near $\mathbf{p}$; and
- for $0 \leq i \leq d-1$, $\Sigma\left(f_{|X_{\mathrm{reg}} \cap V(z_0 - p_0, \ldots, z_i - p_i)}\right) = \emptyset$.

Let us consider a specific example.

Let $\mathcal{U} := \mathbb{C}^3$ and let $X := V(y^2 - x^3) \subseteq \mathcal{U}$, where we use coordinates (t, x, y) on $\mathcal{U}$. Thus, X is a cross-product of a cusp and $\mathbb{C}$, and has a Whitney stratification consisting of $\{X - V(x,y), V(x,y)\}$.

We wish to determine $\overline{T^*_{X_{\mathrm{reg}}}\mathcal{U}}$. Using (t, x, y, w_0, w_1, w_2) for coordinates on $T^*\mathcal{U} \cong \mathcal{U} \times \mathbb{C}^3$, one looks at the vanishing of $y^2 - x^3$ and the 2×2 minors of the matrix

$$\begin{pmatrix} 0 & -3x^2 & 2y \\ w_0 & w_1 & w_2 \end{pmatrix},$$

and disposes of those irreducible components which lie over $\Sigma X = V(x,y)$.

We find that, as analytic sets,

$$\overline{T^*_{X_{\mathrm{reg}}}\mathcal{U}} = \overline{V(y^2 - x^3, w_0x^2, w_0y, 2w_1y + 3w_2x^2) - V(x,y)} = \overline{V(y^2 - x^3, w_0, 2w_1y + 3w_2x^2) - V(x,y)}.$$

Thus, as cycles,

$$(\dagger) \qquad V(y^2 - x^3, w_0, 2w_1y + 3w_2x^2) = \overline{T^*_{X_{\mathrm{reg}}}\mathcal{U}} + mV(x, y, w_0),$$

for some positive integer m.

In fact, it is easy to show that

$$\overline{T^*_{X_{\rm reg}}\mathcal{U}} = V(y^2 - x^3, w_0, 2w_1y + 3w_2x^2, 4w_1^2 - 9w_2^2x),$$

but, as this is not defined by a regular sequence, it is somewhat problematic to deal with this in the intersections below, so (†) is more useful.

Let $\tilde{f} : \mathcal{U} \to \mathbb{C}$ be given by $\tilde{f}(t, x, y) = 2y - 3tx + t^3$, and let $f := \tilde{f}_{|X}$. Then one easily checks that

$$\overline{\Sigma(f_{|X_{\rm reg}})} = V(x - t^2, y - t^3),$$

and (t, x, y) are prepolar coordinates at $\mathbf{0}$. Thus, we may use the method of Theorem 6.1 to calculate the relative local Euler obstruction.

We will suppress the reference to the coordinate system (t, x, y) in the subscripts below.

We have

$$\mathrm{im}(d\tilde{f}) = V\Big(w_0 + 3(x - t^2),\ w_1 + 3t,\ w_2 - 2\Big),$$

and one easily finds that

$$\overline{T^*_{X_{\rm reg}}\mathcal{U}} \cap \mathrm{im}(d\tilde{f}) = V\left(y - t^3, x - t^2, w_0, w_1 + 3t, w_2 - 2\right),$$

which is 1-dimensional. Note for later that this implies that $\widehat{\Lambda}^2_f = 0$.

We wish to proceed with the algorithm described just before Theorem 6.1, but we get slightly "tricky" in the first intersection, to avoid the problem that $\overline{T^*_{X_{\rm reg}}\mathcal{U}}$ is not defined by a regular sequence.

We begin:

$$\widehat{\Gamma}^3_f = \overline{T^*_{X_{\rm reg}}\mathcal{U}}.$$

Now, we use (†):

$$V(y^2 - x^3, w_0, 2w_1y + 3w_2x^2) \cdot V(w_2 - 2) =$$

$$\overline{T^*_{X_{\rm reg}}\mathcal{U}} \cdot V(w_2 - 2)\ +\ mV(x, y, w_0) \cdot V(w_2 - 2).$$

Thus, we conclude that

$$(\ddagger) \qquad V(y^2 - x^3, w_0, w_1y + 3x^2, w_2 - 2) = \widehat{\Gamma}^2_f + mV(x, y, w_0, w_2 - 2),$$

where we have used our earlier observation that $\widehat{\Lambda}^2_f$ must equal 0.

We wish to investigate the purely 2-dimensional cycle

$$Y := V(y^2 - x^3, w_0, w_1y + 3x^2, w_2 - 2).$$

As sets, $V(w_1) \cap Y = V(x, y, w_0, w_1, w_2 - 2)$, which is 1-dimensional. Therefore, we may calculate the cycle structure of Y by looking at the structure where $w_1 \neq 0$.

Via this approach, we find

$$V(y^2 - x^3, w_0, w_1y + 3x^2, w_2 - 2) =$$

$$V(27y + w_1^3, 9x - w_1^2, w_0, w_2 - 2) + 3V(x, y, w_0, w_2 - 2).$$

Now (‡) implies that

$$\widehat{\Gamma}^2_f = V(27y + w_1^3, 9x - w_1^2, w_0, w_2 - 2).$$

We proceed:

$$\widehat{\Gamma}^2_f \cdot V(w_1+3t) = V(27y+w_1^3, 9x-w_1^2, w_0, w_2-2)\cdot V(w_1+3t) =$$

$$V\left(y-t^3, x-t^2, w_0, w_1+3t, w_2-2\right),$$

which as a set equals $\overline{T^*_{X_{\mathrm{reg}}}\mathcal{U}} \cap \operatorname{im}(d\tilde{f})$.

Therefore,

$$\widehat{\Gamma}^1_f = 0 \text{ and } \widehat{\Lambda}^1_f = V\left(y-t^3, x-t^2, w_0, w_1+3t, w_2-2\right).$$

Since $\widehat{\Gamma}^1_f = 0$, we have that $\widehat{\Lambda}^0_f = 0$. Thus there is only one non-zero Lê-Vogel cycle:

$$\Lambda^1_f = \pi_*\big(\widehat{\Lambda}^1_f\big) = \pi_*\big(V(y-t^3, x-t^2, w_0, w_1+3t, w_2-2)\big) = V(y-t^3, x-t^2),$$

with corresponding Lê-Vogel number at the origin:

$$\lambda^1_f(\mathbf{0}) = \big(V(y-t^3, x-t^2)\cdot V(t)\big)_{\mathbf{0}} = 1.$$

Finally, Theorem 6.1 tells us that

$$\mathrm{Eu}_{\mathbf{0}}\, f = (-1)^2(-1)\lambda^1_f(\mathbf{0}) = -1.$$

With a bit of work, one can use Theorem 5.5 to verify this calculation; we leave this as an exercise for the reader.

References

[1] D. A. H. Ament, J. J. Nuño-Ballesteros, B. Oréfice-Okamoto, and J. N. Tomazella, *The Euler obstruction of a function on a determinantal variety and on a curve*, Bull. Braz. Math. Soc. (N.S.) **47** (2016), no. 3, 955–970, DOI 10.1007/s00574-016-0198-y. MR3549078

[2] J.-P. Brasselet, Lê Dũng Tráng, and J. Seade, *Euler obstruction and indices of vector fields*, Topology **39** (2000), no. 6, 1193–1208, DOI 10.1016/S0040-9383(99)00009-9. MR1783853

[3] J.-P. Brasselet, D. Massey, A. J. Parameswaran, and J. Seade, *Euler obstruction and defects of functions on singular varieties*, J. London Math. Soc. (2) **70** (2004), no. 1, 59–76, DOI 10.1112/S0024610704005447. MR2064752

[4] J.-P. Brasselet, J. Seade, and T. Suwa, *Vector fields on singular varieties*, Lecture Notes in Mathematics, vol. 1987, Springer-Verlag, Berlin, 2009. MR2574165

[5] J.-L. Brylinski, A. S. Dubson, and M. Kashiwara, *Formule de l'indice pour modules holonomes et obstruction d'Euler locale* (French, with English summary), C. R. Acad. Sci. Paris Sér. I Math. **293** (1981), no. 12, 573–576. MR647684

[6] A. Dimca, *Sheaves in topology*, Universitext, Springer-Verlag, Berlin, 2004. MR2050072

[7] N. Dutertre and N. G. Grulha Jr., *Some notes on the Euler obstruction of a function*, J. Singul. **10** (2014), 82–91. MR3300287

[8] W. Ebeling and S. M. Gusein-Zade, *Radial index and Euler obstruction of a 1-form on a singular variety*, Geom. Dedicata **113** (2005), 231–241, DOI 10.1007/s10711-005-2184-1. MR2171307

[9] W. Ebeling and S. M. Gusein-Zade, *An equivariant version of the Euler obstruction*, Bull. Braz. Math. Soc. (N.S.) **48** (2017), no. 2, 199–208, DOI 10.1007/s00574-016-0022-8. MR3654142

[10] W. Fulton, *Intersection theory*, Ergebnisse der Mathematik und ihrer Grenzgebiete (3) [Results in Mathematics and Related Areas (3)], vol. 2, Springer-Verlag, Berlin, 1984. MR732620

[11] V. Ginsburg, *Characteristic varieties and vanishing cycles*, Invent. Math. **84** (1986), no. 2, 327–402, DOI 10.1007/BF01388811. MR833194

[12] G. González-Sprinberg, *L'obstruction locale d'Euler et le théorème de MacPherson* (French), The Euler-Poincaré characteristic (French), Astérisque, vol. 83, Soc. Math. France, Paris, 1981, pp. 7–32. MR629121

[13] M. Goresky and R. MacPherson, *Stratified Morse theory*, Ergebnisse der Mathematik und ihrer Grenzgebiete (3) [Results in Mathematics and Related Areas (3)], vol. 14, Springer-Verlag, Berlin, 1988. MR932724

[14] M. Kashiwara and P. Schapira, *Sheaves on manifolds*, Grundlehren der Mathematischen Wissenschaften [Fundamental Principles of Mathematical Sciences], vol. 292, Springer-Verlag, Berlin, 1990. With a chapter in French by Christian Houzel. MR1074006

[15] Lê Dũng Tráng, *Le concept de singularité isolée de fonction analytique* (French), Complex analytic singularities, Adv. Stud. Pure Math., vol. 8, North-Holland, Amsterdam, 1987, pp. 215–227, DOI 10.2969/aspm/00810215. MR894295

[16] Lê Dũng Tráng and B. Teissier, *Variétés polaires locales et classes de Chern des variétés singulières* (French), Ann. of Math. (2) **114** (1981), no. 3, 457–491, DOI 10.2307/1971299. MR634426

[17] R. D. MacPherson, *Chern classes for singular algebraic varieties*, Ann. of Math. (2) **100** (1974), 423–432, DOI 10.2307/1971080. MR0361141

[18] D. B. Massey, *The Lê varieties. I*, Invent. Math. **99** (1990), no. 2, 357–376, DOI 10.1007/BF01234423. MR1031905

[19] D. B. Massey, *The Lê varieties. II*, Invent. Math. **104** (1991), no. 1, 113–148, DOI 10.1007/BF01245068. MR1094048

[20] D. B. Massey, *Lê cycles and hypersurface singularities*, Lecture Notes in Mathematics, vol. 1615, Springer-Verlag, Berlin, 1995. MR1441075

[21] D. B. Massey, *Hypercohomology of Milnor fibres*, Topology **35** (1996), no. 4, 969–1003, DOI 10.1016/0040-9383(95)00054-2. MR1404920

[22] D. B. Massey, *Critical points of functions on singular spaces*, Topology Appl. **103** (2000), no. 1, 55–93, DOI 10.1016/S0166-8641(98)00161-8. MR1746918

[23] D. B. Massey, *A little microlocal Morse theory*, Math. Ann. **321** (2001), no. 2, 275–294, DOI 10.1007/s002080100227. MR1866489

[24] D. B. Massey, *The Sebastiani-Thom isomorphism in the derived category*, Compositio Math. **125** (2001), no. 3, 353–362, DOI 10.1023/A:1002608716514. MR1818986

[25] D. B. Massey, *Numerical control over complex analytic singularities*, Mem. Amer. Math. Soc. **163** (2003), no. 778, xii+268, DOI 10.1090/memo/0778. MR1962934

[26] D. B. Massey, *Singularities and enriched cycles*, Pacific J. Math. **215** (2004), no. 1, 35–84, DOI 10.2140/pjm.2004.215.35. MR2060494

[27] C. Sabbah, *Quelques remarques sur la géométrie des espaces conormaux* (French), Astérisque **130** (1985), 161–192. Differential systems and singularities (Luminy, 1983). MR804052

[28] J. Schürmann, *A general intersection formula for Lagrangian cycles*, Compos. Math. **140** (2004), no. 4, 1037–1052, DOI 10.1112/S0010437X04000272. MR2059229

[29] J. Schürmann, *Topology of singular spaces and constructible sheaves*, Instytut Matematyczny Polskiej Akademii Nauk. Monografie Matematyczne (New Series) [Mathematics Institute of the Polish Academy of Sciences. Mathematical Monographs (New Series)], vol. 63, Birkhäuser Verlag, Basel, 2003. MR2031639

[30] J. Seade, M. Tibăr, and A. Verjovsky, *Milnor numbers and Euler obstruction*, Bull. Braz. Math. Soc. (N.S.) **36** (2005), no. 2, 275–283, DOI 10.1007/s00574-005-0039-x. MR2152019

[31] B. Teissier, *Variétés polaires. II. Multiplicités polaires, sections planes, et conditions de Whitney* (French), Algebraic geometry (La Rábida, 1981), Lecture Notes in Math., vol. 961, Springer, Berlin, 1982, pp. 314–491, DOI 10.1007/BFb0071291. MR708342

Department of Mathematics, Northeastern University, Boston, MA 02115
Email address: `d.massey@northeastern.edu`

Contemporary Mathematics
Volume **742**, 2020
https://doi.org/10.1090/conm/742/14943

Hasse–Schmidt derivations versus classical derivations

Luis Narváez Macarro

Dedicated to Lê Dũng Tráng

Abstract. In this paper we survey the notion and basic results on multivariate Hasse–Schmidt derivations over arbitrary commutative algebras and we associate to such an object a family of classical derivations. We study the behavior of these derivations under the action of substitution maps and we prove that, in characteristic 0, the original multivariate Hasse–Schmidt derivation can be recovered from the associated family of classical derivations. Our constructions generalize a previous one by M. Mirzavaziri in the case of a base field of characteristic 0.

Introduction

Let k be a commutative ring and A a commutative k-algebra. A Hasse–Schmidt derivation of A over k of length $m \geq 0$ (or $m = \infty$), is a sequence $D = (D_0, D_1, \ldots, D_m)$ (or $D = (D_0, D_1, \ldots))$ of k-linear endomorphisms of A such that D_0 is the identity map and

$$D_\alpha(xy) = \sum_{\beta+\gamma=\alpha} D_\beta(x) D_\gamma(y), \quad \forall \alpha, \forall x, y \in A.$$

A such D can be seen as a power series $D = \sum_{\alpha=0}^{m} D_\alpha t^\alpha$ in the quotient ring $R[[t]]/\langle t^{m+1}\rangle$, with $R = \operatorname{End}_k(A)$ (the ring of endomorphisms of A as k-module). For $i \geq 1$, the ith component D_i turns out to be a k-linear differential operator of order $\leq i$ vanishing on 1, in particular D_1 is a k-derivation of A.

The notion of Hasse–Schmidt derivation was introduced in [**1**] in the case where k is a field of characteristic $p > 0$ and A a field of algebraic functions over k. This notion was used to understand, among others, Taylor expansions in this setting. But actually, Hasse–Schmidt derivations make sense in full generality.

If we are in characteristic 0 ($\mathbb{Q} \subset k$), then it is easy to produce examples of Hasse–Schmidt derivations: starting with a k-linear derivation $\delta : A \to A$ we consider its exponential:

$$e^{t\delta} = \sum_{\alpha=0}^{\infty} \frac{\delta^\alpha}{\alpha!} t^\alpha \in R[[t]], \quad R = \operatorname{End}_k(A).$$

2010 *Mathematics Subject Classification.* Primary: 14F10, 13N10; Secondary 13N15.

Key words and phrases. Hasse–Schmidt derivation, derivation, power series ring, substitution map.

Partially supported by MTM2016-75027-P, P12-FQM-2696 and FEDER.

It is clear that $e^{t\delta}$ is a Hasse–Schmidt derivation of A over k (of infinite length). This example also proves that, always under the characteristic 0 hypothesis, any k-linear derivation $\delta : A \to A$ appears as the 1-component of some Hasse–Schmidt derivation (of infinite length, and so, of any length $m \geq 1$) of A over k. This is what we call "to be ∞-integrable" (and so "m-integrable", for each $m \geq 1$) (see [**5**]). But if we are no more in characteristic 0, the situation becomes much more involved and integrable derivations deserve special attention (see [**4, 8, 13**] for several recent achievements in that direction).

As far as the author knows, two papers have been concerned with the description of Hasse–Schmidt derivations in terms of usual derivations, both in the case where k is a field of characteristic 0. In [**3**] it is proven that[1], if A is a (possibly non-commutative) k-algebra, then any Hasse–Schmidt derivation $D = (D_0 = \mathrm{Id}, D_1, D_2, \dots)$ of infinite length of A over k is determined by a unique sequence $\delta = (\delta_1, \delta_2, \dots)$ of classical derivations $\delta_i \in \mathrm{Der}_k(A)$. Namely, the expressions relating D and δ are:

$$\delta_n = \sum_{r=1}^{n} \frac{(-1)^{r+1}}{r} \sum_{\substack{n_1+\cdots+n_r=n \\ n_i>0}} D_{n_1} D_{n_2} \cdots D_{n_r}, \quad D_n = \sum_{r=1}^{n} \frac{1}{r!} \sum_{\substack{n_1+\cdots+n_r=n \\ n_i>0}} \delta_{n_1} \delta_{n_2} \cdots \delta_{n_r},$$

or in other words:

$$\sum_{n=0}^{\infty} D_n t^n = \exp\left(\sum_{n=1}^{\infty} \delta_n t^n \right).$$

A similar result is proven in [**7**]: any Hasse–Schmidt derivation $D = (D_0 = \mathrm{Id}, D_1, D_2, \dots)$ of infinite length of A over k determines, and is determined by a sequence $\delta = (\delta_1, \delta_2, \dots)$ of classical derivations given by the following recursive formula:

$$(n+1) D_{n+1} = \sum_{r=0}^{n} \delta_{r+1} D_{n-r}, \quad n \geq 0.$$

An interesting reinterpretation of both results can be found in [**2**].

The goal of this paper is twofold: to give a survey of multivariate Hasse–Schmidt derivations over a general commutative base ring k and a general commutative k-algebra A, as defined in [**9**]; and to generalize the construction in [**7**] to this setting.

One of our motivations is to understand the relationship between HS-modules, as defined in [**10**], and classical integrable connections. The paper [**11**] is devoted to prove that both notions are equivalent in characteristic 0, and the proof strongly depends on the constructions and results of the present paper.

A (p, Δ)-variate Hasse–Schmidt derivation of A over k is a family $D = (D_\alpha)_{\alpha \in \Delta}$ of k-linear endomorphisms of A such that D_0 is the identity map and:

$$D_\alpha(xy) = \sum_{\beta+\gamma=\alpha} D_\beta(x) D_\gamma(y), \quad \forall \alpha \in \Delta, \forall x, y \in A,$$

where $\Delta \subset \mathbb{N}^p$ is a non-empty *co-ideal*, i.e. a subset of $\mathbb{N}^p$ such that everytime $\alpha \in \Delta$ and $\alpha' \leq \alpha$ (i.e. $\alpha - \alpha' \in \mathbb{N}^p$) we have $\alpha' \in \Delta$. A simple but important idea is to think on Hasse–Schmidt derivations as series $D = \sum_{\alpha \in \Delta} D_\alpha \mathbf{s}^\alpha$ in the quotient

[1]This result has been "rediscovered" in [**12**] for A a commutative algebra over a field k of characteristic zero.

ring $R[[\mathbf{s}]]_\Delta$ of the power series ring $R[[\mathbf{s}]] = R[[s_1, \ldots, s_p]]$, $R = \mathrm{End}_k(A)$, by the two-sided monomial ideal generated by all $\mathbf{s}^\alpha$ with $\alpha \in \mathbb{N}^p \setminus \Delta$.

The set $\mathrm{HS}^p_k(A;\Delta)$ of (p,Δ)-variate Hasse–Schmidt derivations is a subgroup of the group of units $(R[[\mathbf{s}]]_\Delta)^\times$, and it also carries the action of substitution maps: given a substitution map $\varphi : A[[s_1, \ldots, s_p]]_\Delta \to A[[t_1, \ldots, t_q]]_\nabla$ and a (p,Δ)-variate Hasse–Schmidt derivation $D = \sum_{\alpha\in\Delta} D_\alpha \mathbf{s}^\alpha$ we obtain a new (q,∇)-variate Hasse–Schmidt derivation given by:

$$\varphi \bullet D := \sum_{\alpha\in\Delta} \varphi(\mathbf{s}^\alpha) D_\alpha.$$

This new structure is a key point in [**10**].

To generalize the construction in [**7**], we reinterpret the aforementioned recursive formula by means of the "logarithmic derivative type" maps:

$$\varepsilon^i : D \in \mathrm{HS}^p_k(A;\Delta) \longmapsto \varepsilon^i(D) := D^* \left(s_i \frac{\partial D}{\partial s_i} \right) \in R[[\mathbf{s}]]_\Delta, \quad i = 1, \ldots, p,$$

where D^* denotes the inverse of D. The starting point is to check that the coefficients of $\varepsilon(D)$ are always classical derivations, i.e. $\varepsilon(D) \in \mathrm{Der}_k(A)[[\mathbf{s}]]_\Delta$.

Let us comment on the content of the paper.

In section 1 we have gathered some notations and constructions on power series modules, powers series rings and substitution maps, most of them taken from [**9**], and we study the maps ε^i, and their conjugate $\overline{\varepsilon}^i$.

In section 2 we recall the notion and the basic properties of multivariate Hasse–Schmidt derivations and of the action of substitution maps on these objects.

Section 3 contains the main original results of this paper. First, we see how the ε^i or $\overline{\varepsilon}^i$ maps of section 1 allow us to associate to any multivariate Hasse–Schmidt derivation a power series whose coefficients are classical derivations, as explained before. When $\mathbb{Q} \subset k$ we obtain a characterization of multivariate Hasse–Schmidt derivations in terms of the ε^i (or $\overline{\varepsilon}^i$) maps, and we prove that any multivariate Hasse–Schmidt derivation can be constructed from a power series of classical derivations. To finish, we study the behavior of the ε^i maps of a multivariate Hasse–Schmidt derivation under the action of substitution maps.

1. Notations and preliminaries

1.1. Notations. Throughout the paper we will use the following notations:

-) k is a commutative ring and A a commutative k-algebra.

-) $\mathbf{s} = \{s_1, \ldots, s_p\}$, $\mathbf{t} = \{t_1, \ldots, t_q\}$, ... are sets of variables.

-) $\mathscr{U}^p(R;\Delta)$: see Notation 1.2.4.

-) $\mathbf{C}_e(\varphi, \alpha)$: see (1.11).

-) $\varphi \bullet r, r \bullet \varphi$: see 1.3.5.

-) φ^D: see 2.2.3.

1.2. Some constructions on power series rings and modules. Throughout this section, k will be a commutative ring, A a commutative k-algebra and R a ring, not-necessarily commutative.

Let $p \geq 0$ be an integer and let us call $\mathbf{s} = \{s_1, \ldots, s_p\}$ a set of p variables. The support of each $\alpha \in \mathbb{N}^p$ is defined as $\operatorname{supp} \alpha := \{i \mid \alpha_i \neq 0\}$. The monoid $\mathbb{N}^p$ is endowed with a natural partial ordering. Namely, for $\alpha, \beta \in \mathbb{N}^p$, we define:

$$\alpha \leq \beta \quad \stackrel{\text{def.}}{\Longleftrightarrow} \quad \exists \gamma \in \mathbb{N}^p \text{ such that } \beta = \alpha + \gamma \quad \Longleftrightarrow \quad \alpha_i \leq \beta_i \quad \forall i = 1 \ldots, p.$$

We denote $|\alpha| := \alpha_1 + \cdots + \alpha_p$.

Let $p \geq 1$ be an integer and $\mathbf{s} = \{s_1, \ldots, s_p\}$ a set of variables. If M is an abelian group and $M[[\mathbf{s}]]$ is the abelian group of power series with coefficients in M, the *support* of a series $m = \sum_\alpha m_\alpha \mathbf{s}^\alpha \in M[[\mathbf{s}]]$ is $\operatorname{supp}(m) := \{\alpha \in \mathbb{N}^p \mid m_\alpha \neq 0\} \subset \mathbb{N}^p$. We have $m = 0 \Leftrightarrow \operatorname{supp}(m) = \emptyset$.

The abelian group $M[[\mathbf{s}]]$ is clearly a $\mathbb{Z}[[\mathbf{s}]]$-module, which will be always endowed with the $\langle \mathbf{s} \rangle$-adic topology.

DEFINITION 1.2.1. We say that a subset $\Delta \subset \mathbb{N}^p$ is an *ideal* (resp. a *co-ideal*) of $\mathbb{N}^p$ if everytime $\alpha \in \Delta$ and $\alpha \leq \alpha'$ (resp. $\alpha' \leq \alpha$), then $\alpha' \in \Delta$.

It is clear that $\Delta \subset \mathbb{N}^p$ is an ideal if and only if its complement Δ^c is a co-ideal, and that the union and the intersection of any family of ideals (resp. of co-ideals) of $\mathbb{N}^p$ is again an ideal (resp. a co-ideal) of $\mathbb{N}^p$. Examples of ideals (resp. of co-ideals) of $\mathbb{N}^p$ are the $\beta + \mathbb{N}^p$ (resp. the $\{\alpha \in \mathbb{N}^p \mid \alpha \leq \beta\}$) with $\beta \in \mathbb{N}^p$. The $\{\alpha \in \mathbb{N}^p \mid |\alpha| \leq m\}$ with $m \geq 0$ are also co-ideals. Notice that a co-ideal $\Delta \subset \mathbb{N}^p$ is non-empty if and only if $\{0\} \subset \Delta$.

1.2.2 Let M be an abelian group. For each co-ideal $\Delta \subset \mathbb{N}^p$, we denote by Δ_M the closed sub-$\mathbb{Z}[[\mathbf{s}]]$-bimodule of $M[[\mathbf{s}]]$ whose elements are the formal power series $\sum_{\alpha \in \mathbb{N}^p} m_\alpha \mathbf{s}^\alpha$ such that $m_\alpha = 0$ whenever $\alpha \in \Delta$, and $M[[\mathbf{s}]]_\Delta := M[[\mathbf{s}]]/\Delta_M$. The elements in $M[[\mathbf{s}]]_\Delta$ are power series of the form $\sum_{\alpha \in \Delta} m_\alpha \mathbf{s}^\alpha$, $m_\alpha \in M$. If $f : M \to M'$ is a homomorphism of abelian groups, we will denote by $\overline{f} : M[[\mathbf{s}]]_\Delta \to M'[[\mathbf{s}]]_\Delta$ the $\mathbb{Z}[[\mathbf{s}]]_\Delta$-linear map defined as $\overline{f}\left(\sum_{\alpha \in \Delta} m_\alpha \mathbf{s}^\alpha\right) = \sum_{\alpha \in \Delta} f(m_\alpha) \mathbf{s}^\alpha$.

If R is a ring, then Δ_R is a closed two-sided ideal of $R[[\mathbf{s}]]$ and so $R[[\mathbf{s}]]_\Delta$ is a topological ring, which we always consider endowed with the $\langle \mathbf{s} \rangle$-adic topology (= to the quotient topology). Similarly, if M is an $(A; A)$-bimodule (central over k), then $M[[\mathbf{s}]]_\Delta$ is an $(A[[\mathbf{s}]_\Delta; A[[\mathbf{s}]]_\Delta)$-bimodule (central over $k[[\mathbf{s}]]_\Delta$).

For $\Delta' \subset \Delta$ non-empty co-ideals of $\mathbb{N}^p$, we have natural $\mathbb{Z}[[\mathbf{s}]]$-linear projections $\tau_{\Delta\Delta'} : M[[\mathbf{s}]]_\Delta \longrightarrow M[[\mathbf{s}]]_{\Delta'}$,that we call *truncations*:

$$\tau_{\Delta\Delta'} : \sum_{\alpha \in \Delta} m_\alpha \mathbf{s}^\alpha \in M[[\mathbf{s}]]_{\Delta'} \longmapsto \sum_{\alpha \in \Delta'} m_\alpha \mathbf{s}^\alpha \in M[[\mathbf{s}]]_\Delta.$$

If M is a ring (resp. an $(A; A)$-bimodule), then the truncations $\tau_{\Delta\Delta'}$ are ring homomorphisms (resp. $(A[[\mathbf{s}]]_\Delta; A[[\mathbf{s}]]_\Delta)$-linear maps). For $\Delta' = \{0\}$ we have $M[[\mathbf{s}]]_{\Delta'} = M$ and the kernel of $\tau_{\Delta\{0\}}$ will be denoted by $M[[\mathbf{s}]]_{\Delta,+}$. We have a bicontinuous isomorphism:

$$M[[\mathbf{s}]]_\Delta = \varprojlim M[[\mathbf{s}]]_{\Delta'}$$

where Δ' runs over all finite co-ideals contained in Δ.

DEFINITION 1.2.3. A k-algebra over A is a (not-necessarily commutative) k-algebra R endowed with a map of k-algebras $\iota : A \to R$. A map between two k-algebras $\iota : A \to R$ and $\iota' : A \to R'$ over A is a map $g : R \to R'$ of k-algebras such that $\iota' = g \circ \iota$.

It is clear that if R is a k-algebra over A, then $R[[\mathbf{s}]]_\Delta$ is a $k[[\mathbf{s}]]_\Delta$-algebra over $A[[\mathbf{s}]]_\Delta$.

NOTATION 1.2.4. Let R be a ring, $p \geq 1$ and $\Delta \subset \mathbb{N}^p$ a non-empty co-ideal. We denote by $\mathscr{U}^p(R;\Delta)$ the multiplicative sub-group of the units of $R[[\mathbf{s}]]_\Delta$ whose 0-degree coefficient is 1. The multiplicative inverse of a unit $r \in R[[\mathbf{s}]]_\Delta$ will be denoted by r^*. Clearly, $\mathscr{U}^p(R;\Delta)^{\text{opp}} = \mathscr{U}^p(R^{\text{opp}};\Delta)$. For $\Delta \subset \Delta'$ co-ideals we have $\tau_{\Delta'\Delta}(\mathscr{U}^p(R;\Delta')) \subset \mathscr{U}^p(R;\Delta)$ and the truncation map $\tau_{\Delta'\Delta} : \mathscr{U}^p(R;\Delta') \to \mathscr{U}^p(R;\Delta)$ is a group homomorphisms. Clearly, we have:

$$\mathscr{U}^p(R;\Delta) = \varprojlim_{\substack{\Delta' \subset \Delta \\ \sharp\Delta' < \infty}} \mathscr{U}^p(R;\Delta'). \tag{1.1}$$

If $p = 1$ and $\Delta = \{i \in \mathbb{N} \mid i \leq m\}$ we will simply denote $\mathscr{U}(R;m) := \mathscr{U}^1(R;\Delta)$.

For any ring homomorphism $f : R \to R'$, the induced ring homomorphism $\overline{f} : R[[\mathbf{s}]]_\Delta \to R'[[\mathbf{s}]]_\Delta$ sends $\mathscr{U}^p(R;\Delta)$ into $\mathscr{U}^p(R';\Delta)$ and so it induces natural group homomorphisms $\mathscr{U}^p(R;\Delta) \to \mathscr{U}^p(R';\Delta)$.

We recall the following easy result (cf. Lemma 2 in [**9**]).

LEMMA 1.2.5. *Let R be a ring and $\Delta \subset \mathbb{N}^p$ a non-empty co-ideal. The units in $R[[\mathbf{s}]]_\Delta$ are those power series $r = \sum r_\alpha \mathbf{s}^\alpha$ such that r_0 is a unit in R. Moreover, in the special case where $r_0 = 1$, the inverse $r^* = \sum r^*_\alpha \mathbf{s}^\alpha$ of r is given by $r^*_0 = 1$ and*

$$r^*_\alpha = \sum_{d=1}^{|\alpha|} (-1)^d \sum_{\alpha^\bullet \in \mathscr{P}(\alpha,d)} r_{\alpha^1} \cdots r_{\alpha^d} \quad \textit{for } \alpha \neq 0,$$

where $\mathscr{P}(\alpha,d)$ is the set of d-uples $\alpha^\bullet = (\alpha^1,\dots,\alpha^d)$ with $\alpha^i \in \mathbb{N}^{(\mathbf{s})}$, $\alpha^i \neq 0$, and $\alpha^1 + \cdots + \alpha^d = \alpha$.

1.2.6 Let E, F be A-modules. For each $r = \sum_\beta r_\beta \mathbf{s}^\beta \in \operatorname{Hom}_k(E,F)[[\mathbf{s}]]_\Delta$ we denote by $\widetilde{r} : E[[\mathbf{s}]]_\Delta \to F[[\mathbf{s}]]_\Delta$ the map defined by:

$$\widetilde{r}\left(\sum_{\alpha\in\Delta} e_\alpha \mathbf{s}^\alpha\right) := \sum_{\alpha\in\Delta}\left(\sum_{\beta+\gamma=\alpha} r_\beta(e_\gamma)\right)\mathbf{s}^\alpha,$$

which is obviously a $k[[\mathbf{s}]]_\Delta$-linear map. It is clear that the map:

$$r \in \operatorname{Hom}_k(E,F)[[\mathbf{s}]]_\Delta \longmapsto \widetilde{r} \in \operatorname{Hom}_{k[[\mathbf{s}]]_\Delta}(E[[\mathbf{s}]]_\Delta, F[[\mathbf{s}]]_\Delta) \tag{1.2}$$

is $(A[[\mathbf{s}]]_\Delta; A[[\mathbf{s}]]_\Delta)$-linear.

If $f : E[[\mathbf{s}]]_\Delta \to F[[\mathbf{s}]]_\Delta$ is a $k[[\mathbf{s}]]_\Delta$-linear map, let us denote by $f_\alpha : E \to F$, $\alpha \in \Delta$, the k-linear maps defined by:

$$f(e) = \sum_{\alpha\in\Delta} f_\alpha(e)\mathbf{s}^\alpha, \quad \forall e \in E.$$

If $g : E \to F[[\mathbf{s}]]_\Delta$ is a k-linear map, we denote by $g^e : E[[\mathbf{s}]]_\Delta \to F[[\mathbf{s}]]_\Delta$ the unique $k[[\mathbf{s}]]_\Delta$-linear map extending g to $E[[\mathbf{s}]]_\Delta = k[[\mathbf{s}]]_\Delta \widehat{\otimes}_k E$. It is given by:

$$g^e\left(\sum_\alpha e_\alpha \mathbf{s}^\alpha\right) := \sum_\alpha g(e_\alpha)\mathbf{s}^\alpha. \tag{1.3}$$

We have a $k[[\mathbf{s}]]_\Delta$-bilinear and $A[[\mathbf{s}]]_\Delta$-balanced map:

$$\langle -,-\rangle : (r,e) \in \operatorname{Hom}_k(E,F)[[\mathbf{s}]]_\Delta \times E[[\mathbf{s}]]_\Delta \longmapsto \langle r,e\rangle := \widetilde{r}(e) \in F[[\mathbf{s}]]_\Delta.$$

The following assertions are clear (see [**9**, Lemma 3]):

1) The map (1.2) is an isomorphism of $(A[[\mathbf{s}]]_\Delta; A[[\mathbf{s}]]_\Delta)$-bimodules. When $E = F$ it is an isomorphism of $k[[\mathbf{s}]]_\Delta$-algebras over $A[[\mathbf{s}]]_\Delta$.
2) The restriction map:
$$f \in \operatorname{Hom}_{k[[\mathbf{s}]]_\Delta}(E[[\mathbf{s}]]_\Delta, F[[\mathbf{s}]]_\Delta) \mapsto f|_E \in \operatorname{Hom}_k(E, F[[\mathbf{s}]]_\Delta),$$
is an isomorphism of $(A[[\mathbf{s}]]_\Delta; A)$-bimodules.

Let us call $R = \operatorname{End}_k(E)$. As a consequence of the above properties, the composition of the maps:

$$R[[\mathbf{s}]]_\Delta \xrightarrow{r\mapsto\widetilde{r}} \operatorname{End}_{k[[\mathbf{s}]]_\Delta}(E[[\mathbf{s}]]_\Delta) \xrightarrow{f\mapsto f|_E} \operatorname{Hom}_k(E, E[[\mathbf{s}]]_\Delta) \tag{1.4}$$

is an isomorphism of $(A[[\mathbf{s}]]_\Delta; A)$-bimodules, and so $\operatorname{Hom}_k(E, E[[\mathbf{s}]]_\Delta)$ inherits a natural structure of $k[[\mathbf{s}]]_\Delta$-algebra over $A[[\mathbf{s}]]_\Delta$. Namely, if $g,h : E \to E[[\mathbf{s}]]_\Delta$ are k-linear maps with:

$$g(e) = \sum_{\alpha\in\Delta} g_\alpha(e)\mathbf{s}^\alpha,\ h(e) = \sum_{\alpha\in\Delta} h_\alpha(e)\mathbf{s}^\alpha, \quad \forall e\in E, \quad g_\alpha, h_\alpha \in \operatorname{Hom}_k(E,E),$$

then the product $hg \in \operatorname{Hom}_k(E, E[[\mathbf{s}]]_\Delta)$ is given by:

$$(hg)(e) = \sum_{\alpha\in\Delta}\left(\sum_{\beta+\gamma=\alpha} (h_\beta\circ g_\gamma)(e)\right)\mathbf{s}^\alpha. \tag{1.5}$$

NOTATION 1.2.7. We denote:

$$\operatorname{Hom}_k^\circ(E, E[[\mathbf{s}]]_\Delta) := \{f \in \operatorname{Hom}_k(E, E[[\mathbf{s}]]_\Delta) \mid f(e) \equiv e \bmod \langle \mathbf{s}\rangle E[[\mathbf{s}]]_\Delta\ \forall e\in E\},$$

$$\operatorname{Aut}^\circ_{k[[\mathbf{s}]]_\Delta}(E[[\mathbf{s}]]_\Delta) :=$$
$$\left\{f \in \operatorname{Aut}_{k[[\mathbf{s}]]_\Delta}(E[[\mathbf{s}]]_\Delta) \mid f(e) \equiv e_0 \bmod \langle \mathbf{s}\rangle E[[\mathbf{s}]]_\Delta\ \forall e\in E[[\mathbf{s}]]_\Delta\right\}.$$

Let us notice that a $f \in \operatorname{Hom}_k(E, E[[\mathbf{s}]]_\Delta)$, given by $f(e) = \sum_{\alpha\in\Delta} f_\alpha(e)\mathbf{s}^\alpha$, belongs to $\operatorname{Hom}_k^\circ(E, E[[\mathbf{s}]]_\Delta)$ if and only if $f_0 = \operatorname{Id}_E$.

The isomorphism in (1.4) gives rise to a group isomorphism:

$$r \in \mathscr{U}^p(\operatorname{End}_k(E);\Delta) \xrightarrow{\sim} \widetilde{r} \in \operatorname{Aut}^\circ_{k[[\mathbf{s}]]_\Delta}(E[[\mathbf{s}]]_\Delta) \tag{1.6}$$

and to a bijection:

$$f \in \operatorname{Aut}^\circ_{k[[\mathbf{s}]]_\Delta}(E[[\mathbf{s}]]_\Delta) \xrightarrow{\sim} f|_E \in \operatorname{Hom}_k^\circ(E, E[[\mathbf{s}]]_\Delta). \tag{1.7}$$

So, $\operatorname{Hom}_k^\circ(E, E[[\mathbf{s}]]_\Delta)$ is naturally a group with the product described in (1.5).

If R is a (not necessarily commutative) k-algebra and $\Delta \subset \mathbb{N}^p$ is a co-ideal, any continuous k-linear map $h : k[[\mathbf{s}]]_\Delta \to k[[\mathbf{s}]]_\Delta$ induces a natural continuous left and right R-linear map:

$$h_R := \operatorname{Id}_R \widehat{\otimes}_k h : R[[\mathbf{s}]]_\Delta = R\widehat{\otimes}_k k[[\mathbf{s}]]_\Delta \longrightarrow R[[\mathbf{s}]]_\Delta = R\widehat{\otimes}_k k[[\mathbf{s}]]_\Delta$$

given by:

$$h_R\left(\sum_\alpha r_\alpha \mathbf{s}^\alpha\right) = \sum_\alpha r_\alpha h(\mathbf{s}^\alpha).$$

If $\mathfrak{d} : k[[\mathbf{s}]]_\Delta \to k[[\mathbf{s}]]_\Delta$ is k-derivation, it is continuous and $\mathfrak{d}_R : R[[\mathbf{s}]]_\Delta \to R[[\mathbf{s}]]_\Delta$ is a $(R;R)$-linear derivation, i.e. $\mathfrak{d}_R(sr) = s\mathfrak{d}_R(r)$, $\mathfrak{d}_R(rs) = \mathfrak{d}_R(r)s$, $\mathfrak{d}_R(rr') = \mathfrak{d}_R(r)r' + r\mathfrak{d}_R(r')$ for all $s \in R$ and for all $r, r' \in R[[\mathbf{s}]]_\Delta$.

The set of all $(R;R)$-linear derivations of $R[[\mathbf{s}]]_\Delta$ is a $k[[\mathbf{s}]]_\Delta$-Lie algebra and will be denoted by $\operatorname{Der}_R(R[[\mathbf{s}]]_\Delta)$. Moreover, the map:

$$\mathfrak{d} \in \operatorname{Der}_k(k[[\mathbf{s}]]_\Delta) \longmapsto \mathfrak{d}_R \in \operatorname{Der}_R(R[[\mathbf{s}]]_\Delta)$$

is clearly a map of $k[[\mathbf{s}]]_\Delta$-Lie algebras.

The following definition provides a particular family of k-derivations.

DEFINITION 1.2.8. For each $i = 1, \dots, p$, the *ith partial Euler k-derivation* is $\chi^i = s_i \frac{\partial}{\partial s_i} : k[[\mathbf{s}]] \to k[[\mathbf{s}]]$. It induces a k-derivation on each $k[[\mathbf{s}]]_\Delta$, which will be also denoted by χ^i.

The *Euler k-derivation* $\boldsymbol{\chi} : k[[\mathbf{s}]] \to k[[\mathbf{s}]]$ is defined as:

$$\boldsymbol{\chi} = \sum_{i=1}^p \chi^i, \quad \boldsymbol{\chi}\left(\sum_\alpha c_\alpha \mathbf{s}^\alpha\right) = \sum_\alpha |\alpha| c_\alpha \mathbf{s}^\alpha.$$

It induces a k-derivation on each $k[[\mathbf{s}]]_\Delta$, which will be also denoted by χ.

The proof of the following lemma is easy and it is left to the reader.

LEMMA 1.2.9. *Let E be an A-module and $r = \sum_\beta r_\beta \mathbf{s}^\beta \in \operatorname{Hom}_k(A,E)[[\mathbf{s}]]_\Delta$ a formal power series with coefficients in $\operatorname{Hom}_k(A,E)$. The following properties are equivalent:*

(1) $r \in \operatorname{Der}_k(A,E)[[\mathbf{s}]]_\Delta$.
(2) *For any $a \in A[[\mathbf{s}]]_\Delta$ we have $[r,a] = \widetilde{r}(a)$.*
(3) $\widetilde{r} \in \operatorname{Der}_{k[[\mathbf{s}]]_\Delta}(A[[\mathbf{s}]]_\Delta, E[[\mathbf{s}]]_\Delta)$.
(4) $\widetilde{r}|_A \in \operatorname{Der}_k(A, E[[\mathbf{s}]]_\Delta)$.

In particular, for each $r \in \operatorname{Der}_k(A)[[\mathbf{s}]]_\Delta$, we have that $\widetilde{r} \in \operatorname{Der}_{k[[\mathbf{s}]]_\Delta}(A[[\mathbf{s}]]_\Delta)$ (see 1.2.6) and that the $A[[\mathbf{s}]]_\Delta$-linear map

$$r \in \operatorname{Der}_k(A)[[\mathbf{s}]]_\Delta \longmapsto \widetilde{r} \in \operatorname{Der}_{k[[\mathbf{s}]]_\Delta}(A[[\mathbf{s}]]_\Delta) \tag{1.8}$$

is an isomorphism of $A[[\mathbf{s}]]_\Delta$-modules. Moreover, $\operatorname{Der}_k(A)[[\mathbf{s}]]_\Delta$ is a Lie algebra over $k[[\mathbf{s}]]_\Delta$, where the Lie bracket of $\delta = \sum_\alpha \delta_\alpha \mathbf{s}^\alpha, \varepsilon = \sum_\alpha \varepsilon_\alpha \mathbf{s}^\alpha \in \operatorname{Der}_k(A)[[\mathbf{s}]]_\Delta$ is given by:

$$[\delta, \varepsilon] = \delta\varepsilon - \varepsilon\delta = \sum_\alpha \left(\sum_{\beta+\gamma=\alpha} [\delta_\beta, \varepsilon_\gamma]\right) \mathbf{s}^\alpha,$$

and the map (1.8) is also an isomorphism of $k[[\mathbf{s}]]_\Delta$-Lie algebras.

LEMMA 1.2.10. *Let $\mathfrak{d} : k[[\mathbf{s}]]_\Delta \to k[[\mathbf{s}]]_\Delta$ be a k-derivation and $R = \operatorname{End}_k(A)$. Then, for each $r \in R[[\mathbf{s}]]_\Delta$ we have $\widetilde{\mathfrak{d}_R(r)} = [\mathfrak{d}_A, \widetilde{r}]$.*

PROOF. We have to prove that $\mathfrak{d}_A(\langle r,a\rangle) = \langle \mathfrak{d}_R(r),a\rangle + \langle r,\mathfrak{d}_A(a)\rangle$ for all $a \in A[[\mathbf{s}]]_\Delta$. By continuity, it is enough to prove the identity for $r = r_\alpha \mathbf{s}^\alpha$, $a = a_\beta \mathbf{s}^\beta$ with $\alpha,\beta \in \Delta$, $r_\alpha \in R$, $a_\beta \in A$:

$$\mathfrak{d}_A(\langle r,a\rangle) = \mathfrak{d}_A(\widetilde{r}(a)) = \mathfrak{d}_A(r_\alpha(a_\beta)\mathbf{s}^\alpha\mathbf{s}^\beta) = r_\alpha(a_\beta)\mathfrak{d}(\mathbf{s}^\alpha)\mathbf{s}^\beta + r_\alpha(a_\beta)\mathbf{s}^\alpha\mathfrak{d}(\mathbf{s}^\beta) =$$
$$\widetilde{\mathfrak{d}_R(r)}(a) + \widetilde{r}(\mathfrak{d}_A(a)) = \langle \mathfrak{d}_R(r),a\rangle + \langle r,\mathfrak{d}_A(a)\rangle.$$

□

DEFINITION 1.2.11. For any k-derivation $\mathfrak{d} : k[[\mathbf{s}]]_\Delta \to k[[\mathbf{s}]]_\Delta$ and any $r \in \mathscr{U}^p(R;\Delta)$ we define:

$$\varepsilon^{\mathfrak{d}}(r) := r^*\mathfrak{d}_R(r), \quad \overline{\varepsilon}^{\mathfrak{d}}(r) := \mathfrak{d}_R(r)r^*,$$

and we will write:

$$\varepsilon^{\mathfrak{d}}(r) = \sum_\alpha \varepsilon^{\mathfrak{d}}_\alpha(r)\mathbf{s}^\alpha, \quad \overline{\varepsilon}^{\mathfrak{d}}(r) = \sum_\alpha \overline{\varepsilon}^{\mathfrak{d}}_\alpha(r)\mathbf{s}^\alpha.$$

We will simply denote:

-) $\varepsilon^i(r) := \varepsilon^{\mathfrak{d}}(r)$, $\overline{\varepsilon}^i(r) := \overline{\varepsilon}^{\mathfrak{d}}(r)$ if $\mathfrak{d} = \chi^i$ (the ith partial Euler derivation), $i = 1,\dots,p$.
-) $\varepsilon(r) := \varepsilon^{\mathfrak{d}}(r)$, $\overline{\varepsilon}(r) := \overline{\varepsilon}^{\mathfrak{d}}(r)$ if $\mathfrak{d} = \boldsymbol{\chi}$ is the Euler derivation.

Observe that $\overline{\varepsilon}^{\mathfrak{d}}(r) = r\,\varepsilon^{\mathfrak{d}}(r)\,r^*$ and, for any co-ideal $\Delta' \subset \Delta$, we have $\tau_{\Delta\Delta'}(\varepsilon^{\mathfrak{d}}(r)) = \varepsilon^{\mathfrak{d}}(\tau_{\Delta\Delta'}(r))$, $\tau_{\Delta\Delta'}(\overline{\varepsilon}^{\mathfrak{d}}(r)) = \overline{\varepsilon}^{\mathfrak{d}}(\tau_{\Delta\Delta'}(r))$. Moreover, if E is an A-module and $R = \operatorname{End}_k(A)$, then

$$\widetilde{\varepsilon^{\mathfrak{d}}(r)} = \widetilde{r}^{-1}[\mathfrak{d}_A,\widetilde{r}] = \widetilde{r}^{-1}\mathfrak{d}_A\widetilde{r} - \mathfrak{d}_A, \quad \widetilde{\overline{\varepsilon}^{\mathfrak{d}}(r)} = [\mathfrak{d}_A,\widetilde{r}]\widetilde{r}^{-1} = \mathfrak{d}_A - \widetilde{r}\mathfrak{d}_A\widetilde{r}^{-1}.$$

The proof of the following lemma is straightforward.

LEMMA 1.2.12. *For each $r \in \mathscr{U}^p_k(R;\Delta)$, the maps:*

$$\mathfrak{d} \in \operatorname{Der}_k(k[[\mathbf{s}]]_\Delta) \longmapsto \varepsilon^{\mathfrak{d}}(r) \in R[[\mathbf{s}]]_\Delta, \quad \mathfrak{d} \in \operatorname{Der}_k(k[[\mathbf{s}]]_\Delta) \longmapsto \overline{\varepsilon}^{\mathfrak{d}}(r) \in R[[\mathbf{s}]]_\Delta$$

are $k[[\mathbf{s}]]_\Delta$-linear.

In particular:

$$\varepsilon(r) = \sum_{i=1}^p \varepsilon^i(r), \quad \overline{\varepsilon}(r) = \sum_{i=1}^p \overline{\varepsilon}^i(r).$$

LEMMA 1.2.13. *Let $\mathfrak{d},\mathfrak{d}' : k[[\mathbf{s}]]_\Delta \to k[[\mathbf{s}]]_\Delta$ be k-derivations and $r,r' \in \mathscr{U}^p_k(R;\Delta)$. Then, the following identities hold:*

(i) $\overline{\varepsilon}^{\mathfrak{d}}(1) = \varepsilon^{\mathfrak{d}}(1) = 0$, $\varepsilon^{\mathfrak{d}}(r'r) = \varepsilon^{\mathfrak{d}}(r) + r^*\varepsilon^{\mathfrak{d}}(r')\,r$, $\overline{\varepsilon}^{\mathfrak{d}}(r\,r') = \overline{\varepsilon}^{\mathfrak{d}}(r) + r\,\overline{\varepsilon}^{\mathfrak{d}}(r')\,r^*$.
(ii) $\varepsilon^{\mathfrak{d}}(r^*) = -r\,\varepsilon^{\mathfrak{d}}(r)\,r^* = -\overline{\varepsilon}^{\mathfrak{d}}(r)$.
(iii) $\varepsilon^{[\mathfrak{d},\mathfrak{d}']}(r) = \left[\varepsilon^{\mathfrak{d}}(r),\varepsilon^{\mathfrak{d}'}(r)\right] + \mathfrak{d}_R\left(\varepsilon^{\mathfrak{d}'}(r)\right) - \mathfrak{d}'_R\left(\varepsilon^{\mathfrak{d}}(r)\right)$.

PROOF. The proof of (i) is straightforward. For (ii) and (iii) one uses that $\mathfrak{d}_R(r^*) = -r^*\,\mathfrak{d}_R(r)\,r^*$. □

1.2.14 For each $r \in \mathscr{U}^p(R;\Delta)$ and each $i = 1,\dots,p$ we have:

$$\varepsilon^i(r) = r^*\chi^i_R(r) = \left(\sum_\alpha r^*_\alpha \mathbf{s}^\alpha\right)\left(\sum_\alpha \alpha_i r_\alpha \mathbf{s}^\alpha\right) = \sum_\alpha \left(\sum_{\beta+\gamma=\alpha} \gamma_i r^*_\beta r_\gamma\right)\mathbf{s}^\alpha,$$

$$\varepsilon(r) = r^*\boldsymbol{\chi}_R(r) = \left(\sum_\alpha r^*_\alpha \mathbf{s}^\alpha\right)\left(\sum_\alpha |\alpha| r_\alpha \mathbf{s}^\alpha\right) = \sum_\alpha \left(\sum_{\beta+\gamma=\alpha} |\gamma| r^*_\beta r_\gamma\right)\mathbf{s}^\alpha,$$

and so, by using Lemma 1.2.5, we obtain:

$$\varepsilon^i(r) = \sum_{\substack{\alpha\in\Delta\\ \alpha_i>0}} \left(\sum_{d=1}^{|\alpha|}(-1)^{d-1}\left(\sum_{\alpha^\bullet\in\mathscr{P}(\alpha,d)} \alpha^d_i r_{\alpha^1}\cdots r_{\alpha^d}\right)\right)\mathbf{s}^\alpha,$$

$$\varepsilon(r) = \sum_{\substack{\alpha\in\Delta\\ |\alpha|>0}} \left(\sum_{d=1}^{|\alpha|}(-1)^{d-1}\left(\sum_{\alpha^\bullet\in\mathscr{P}(\alpha,d)} |\alpha^d| r_{\alpha^1}\cdots r_{\alpha^d}\right)\right)\mathbf{s}^\alpha.$$

In a similar way we obtain:

$$\overline{\varepsilon}^i(r) = \sum_{\substack{\alpha\in\Delta\\ \alpha_i>0}} \left(\sum_{d=1}^{|\alpha|}(-1)^{d-1}\left(\sum_{\alpha^\bullet\in\mathscr{P}(\alpha,d)} \alpha^1_i r_{\alpha^1}\cdots r_{\alpha^d}\right)\right)\mathbf{s}^\alpha,$$

$$\overline{\varepsilon}(r) = \sum_{\substack{\alpha\in\Delta\\ |\alpha|>0}} \left(\sum_{d=1}^{|\alpha|}(-1)^{d-1}\left(\sum_{\alpha^\bullet\in\mathscr{P}(\alpha,d)} |\alpha^1| r_{\alpha^1}\cdots r_{\alpha^d}\right)\right)\mathbf{s}^\alpha.$$

In particular, we have $\varepsilon^i_\alpha(r) = \overline{\varepsilon}^i_\alpha(r) = 0$ whenever $\alpha_i = 0$, i.e. whenever $i \notin \operatorname{supp}\alpha$, and $\varepsilon_0(r) = \overline{\varepsilon}_0(r) = 0$:

$$\varepsilon^i(r) = \sum_{i\in\operatorname{supp}\alpha} \varepsilon^i_\alpha(r)\mathbf{s}^\alpha, \quad \overline{\varepsilon}^i(r) = \sum_{i\in\operatorname{supp}\alpha} \overline{\varepsilon}^i_\alpha(r)\mathbf{s}^\alpha,$$

$$\varepsilon(r) = \sum_{|\alpha|>0} \varepsilon_\alpha(r)\mathbf{s}^\alpha, \quad \overline{\varepsilon}(r) = \sum_{|\alpha|>0} \overline{\varepsilon}_\alpha(r)\mathbf{s}^\alpha,$$

and $\varepsilon^i(r), \overline{\varepsilon}^i(r), \varepsilon(r), \overline{\varepsilon}(r) \in R[[\mathbf{s}]]_{\Delta,+}$ (see (1.2.2)). The following recursive identities hold:

$$\alpha_i r_\alpha = \sum_{\substack{\beta+\gamma=\alpha\\ \gamma_i>0}} r_\beta\, \varepsilon^i_\gamma(r) = \sum_{\substack{\beta+\gamma=\alpha\\ \gamma_i>0}} \overline{\varepsilon}^i_\gamma(r)\, r_\beta,$$

$$\varepsilon^i_\alpha(r) = \alpha_i r_\alpha - \sum_{\substack{\beta+\gamma=\alpha\\ |\beta|,\gamma_i>0}} r_\beta\, \varepsilon^t_\gamma(r), \quad \overline{\varepsilon}^t_\alpha(r) = \alpha_t r_\alpha - \sum_{\substack{\beta+\gamma=\alpha\\ |\beta|,\gamma_i>0}} \overline{\varepsilon}^i_\gamma(r)\, r_\beta,$$

for all $\alpha \in \Delta$ with $\alpha_i > 0$, and:

$$|\alpha| r_\alpha = \sum_{\substack{\beta+\gamma=\alpha\\ |\gamma|>0}} r_\beta\, \varepsilon_\gamma(r) = \sum_{\substack{\beta+\gamma=\alpha\\ |\gamma|>0}} \overline{\varepsilon}_\gamma(r)\, r_\beta, \tag{1.9}$$

$$\varepsilon_\alpha(r) = |\alpha| r_\alpha - \sum_{\substack{\beta+\gamma=\alpha\\ |\beta|,|\gamma|>0}} r_\beta\, \varepsilon_\gamma(r), \quad \overline{\varepsilon}_\alpha(r) = |\alpha| r_\alpha - \sum_{\substack{\beta+\gamma=\alpha\\ |\beta|,|\gamma|>0}} \overline{\varepsilon}_\gamma(r)\, r_\beta,$$

for all $\alpha \in \Delta$.

REMARK 1.2.15. After (1.9), our definition of $\overline{\varepsilon}$ generalizes the construction in [**7**].

LEMMA 1.2.16. *For any $r \in \mathscr{U}^p(R;\Delta)$ and any $i,j = 1,\dots,p$ the following identity holds:*

$$\chi_R^j\left(\varepsilon^i(r)\right) - \chi_R^i\left(\varepsilon^j(r)\right) = [\varepsilon^i(r), \varepsilon^j(r)].$$

PROOF. Since $[\chi^i,\chi^j] = 0$, it is a consequence of Lemma 1.2.13, (iii). □

NOTATION 1.2.17. Under the above conditions, we will denote by $\Lambda^p(R;\Delta)$ the subset of $(R[[\mathbf{s}]]_{\Delta,+})^p$ whose elements are the families $\{\delta^i\}_{1\leq i\leq p}$ satisfying the following properties:

(a) If $\delta^i = \sum_{|\alpha|>0} \delta^i_\alpha \mathbf{s}^\alpha$, we have $\delta^i_\alpha = 0$ whenever $\alpha_i = 0$.

(b) For all $i,j = 1,\dots,p$ we have $\chi_R^j\left(\delta^i\right) - \chi_R^i\left(\delta^j\right) = [\delta^i,\delta^j]$.

Let us notice that property (b) may be explicitly written as:

$$\alpha_j \delta^i_\alpha - \alpha_i \delta^j_\alpha = \sum_{\substack{\beta+\gamma=\alpha\\ \beta_i,\gamma_j>0}} [\delta^i_\beta, \delta^j_\gamma] \tag{1.10}$$

for all $i,j = 1,\dots,p$ and for all $\alpha \in \Delta$ with $\alpha_i,\alpha_u > 0$. Let us also consider the map:

$$\mathbf{\Sigma} : \{\delta^i\} \in \Lambda^p(R;\Delta) \longmapsto \sum_{i=1}^p \delta^i \in R[[\mathbf{s}]]_{\Delta,+}.$$

After Lemma 1.2.16, we can consider the map:

$$\boldsymbol{\varepsilon} : D \in \mathscr{U}^p(R;\Delta) \longmapsto \{\varepsilon^i(r)\}_{1\leq i\leq p} \in \Lambda^p(R;\Delta)$$

and we obviously have $\varepsilon = \mathbf{\Sigma} \circ \boldsymbol{\varepsilon}$.

PROPOSITION 1.2.18. *Assume that $\mathbb{Q} \subset k$. Then, the three maps in the following commutative diagram:*

$$\begin{array}{ccc}
\mathscr{U}^p(R;\Delta) & \xrightarrow{\ \boldsymbol{\varepsilon}\ } & \Lambda^p(R;\Delta) \\
 & \searrow^{\varepsilon} & \big\downarrow \mathbf{\Sigma} \\
 & & R[[\mathbf{s}]]_{\Delta,+}
\end{array}$$

are bijective.

PROOF. The injectivity of ε is a straightforward consequence of (1.9). Let us prove the surjectivity of ε. Let $\overline{r} = \sum_\alpha \overline{r}_\alpha \mathbf{s}^\alpha$ be any element in $R[[\mathbf{s}]]_{\Delta,+}$. Since $\mathbb{Q} \subset k$, the differential equation

$$\boldsymbol{\chi}(Y) = Y\overline{r}, \quad Y \in R[[\mathbf{s}]]_\Delta,$$

has a unique solution $r \in R[[\mathbf{s}]]_\Delta$ with initial condition $r_0 = 1$, i.e. $r \in \mathscr{U}^p(R;\Delta)$. It is given recursively by:

$$|\alpha| r_\alpha = \sum_{\substack{\beta+\gamma=\alpha\\ |\gamma|>0}} r_\beta \overline{r}_\gamma, \quad \alpha \in \Delta, |\alpha| > 0,$$

and so $\varepsilon(r) = \overline{r}$. To finish, the only missing point is the injectivity of $\mathbf{\Sigma}$. Let $\{\delta^i\},\{\eta^i\} \in \Lambda^p(R;\Delta)$ be with $\sum_i \delta^i = \sum_i \eta^i$. It is clear that $\delta^i_\alpha = \eta^i_\alpha$ whenever $|\alpha| = 1$. Assume that $\delta^i_\beta = \eta^i_\beta$ for all $i = 1,\dots,p$ whenever $|\beta| < m$ and consider $\alpha \in \Delta$ with $|\alpha| = m$. By using (1.10) and the induction hypothesis we obtain:

$$\alpha_j \delta^i_\alpha - \alpha_i \delta^j_\alpha = \sum_{\substack{\beta+\gamma=\alpha\\ \beta_i,\gamma_j>0}} [\delta^i_\beta, \delta^j_\gamma] = \sum_{\substack{\beta+\gamma=\alpha\\ \beta_i,\gamma_j>0}} [\eta^i_\beta, \eta^j_\gamma] = \alpha_j \eta^i_\alpha - \alpha_i \eta^j_\alpha \quad \forall i,j \in \operatorname{supp}\alpha.$$

The above system of linear equations with rational coefficients joint with the linear equation:

$$\sum_{i\in\operatorname{supp}\alpha} \delta^i_\alpha = \sum_{i\in\operatorname{supp}\alpha} \eta^i_\alpha,$$

gives rise to a non singular system and we deduce that $\delta^i_\alpha = \eta^i_\alpha$ for all $i \in \operatorname{supp}\alpha$, and so $\delta^i_\alpha = \eta^i_\alpha$ for all $i = 1, \dots, p$. □

Notice that Lemma 1.2.16 and Proposition 1.2.18 can also be stated with the $\overline{\varepsilon}^i$ and $\overline{\varepsilon}$ instead of the ε^i and ε.

1.3. Substitution maps. In this section we give a summary of sections 2 and 3 of [**9**]. Let k be a commutative ring, A a commutative k-algebra, $\mathbf{s} = \{s_1,\dots,s_p\}, \mathbf{t} = \{t_1,\dots,t_q\}$ two sets of variables and $\Delta \subset \mathbb{N}^p, \nabla \subset \mathbb{N}^q$ non-empty co-ideals.

DEFINITION 1.3.1. An A-algebra map $\varphi : A[[\mathbf{s}]]_\Delta \to A[[\mathbf{t}]]_\nabla$ will be called a *substitution map* whenever $\operatorname{ord}(\varphi(s_i)) \geq 1$ for all $i = 1,\dots,p$. A such map is continuous and uniquely determined by the family $c = \{\varphi(s_i), i = 1,\dots,p\}$.

The *trivial* substitution map $A[[\mathbf{s}]]_\Delta \to A[[\mathbf{t}]]_\nabla$ is the one sending any s_i to 0. It will be denoted by $\mathbf{0}$.

The composition of substitution maps is obviously a substitution map. Any substitution map $\varphi : A[[\mathbf{s}]]_\Delta \to A[[\mathbf{t}]]_\nabla$ determines and is determined by a family:

$$\{\mathbf{C}_e(\varphi,\alpha), e \in \nabla, \alpha \in \Delta, |\alpha| \leq |e|\} \subset A, \quad \text{with } \mathbf{C}_0(\varphi, 0) = 1,$$

such that:

$$\varphi\left(\sum_{\alpha\in\Delta} a_\alpha \mathbf{s}^\alpha\right) = \sum_{e\in\nabla}\left(\sum_{\substack{\alpha\in\Delta\\ |\alpha|\leq|e|}} \mathbf{C}_e(\varphi,\alpha) a_\alpha\right)\mathbf{t}^e. \tag{1.11}$$

In section 3, 2., of [**9**] the reader can find the explicit expression of the $\mathbf{C}_e(\varphi,\alpha)$ in terms of the $\varphi(s_i)$. The following lemma is clear.

LEMMA 1.3.2. *If $\Delta' \subset \Delta \subset \mathbb{N}^p$ are non-empty co-ideals, the truncation $\tau_{\Delta\Delta'}$: $A[[\mathbf{s}]]_\Delta \to A[[\mathbf{s}]]_{\Delta'}$ is clearly a substitution map and $\mathbf{C}_\beta(\tau_{\Delta\Delta'},\alpha) = \delta_{\alpha\beta}$ for all $\alpha \in \Delta$ and for all $\beta \in \Delta'$ with $|\alpha| \leq |\beta|$.*

DEFINITION 1.3.3. We say that a substitution map $\varphi : A[[\mathbf{s}]]_\Delta \to A[[\mathbf{t}]]_\nabla$ has *constant coefficients* if $\varphi(s_i) \in k[[\mathbf{t}]]_\nabla$ for all $i = 1,\dots,p$. This is equivalent to saying that $\mathbf{C}_e(\varphi,\alpha) \in k$ for all $e \in \nabla$ and for all $\alpha \in \Delta$ with $|\alpha| \leq |e|$. Substitution maps which constant coefficients are induced by substitution maps $k[[\mathbf{s}]]_\Delta \to k[[\mathbf{t}]]_\nabla$.

1.3.4 Let M be an $(A;A)$-bimodule.

Any substitution map map $\varphi : A[[\mathbf{s}]]_\Delta \to A[[\mathbf{t}]]_\nabla$ induces $(A;A)$-linear maps:

$$\varphi_M := \varphi\widehat{\otimes}\operatorname{Id}_M : M[[\mathbf{s}]]_\Delta \equiv A[[\mathbf{s}]]_\Delta\widehat{\otimes}_A M \longrightarrow M[[\mathbf{t}]]_\nabla \equiv A[[\mathbf{t}]]_\nabla\widehat{\otimes}_A M$$

and

$${}_M\varphi := \operatorname{Id}_M\widehat{\otimes}\varphi : M[[\mathbf{s}]]_\Delta \equiv M\widehat{\otimes}_A A[[\mathbf{s}]]_\Delta \longrightarrow M[[\mathbf{t}]]_\nabla \equiv M\widehat{\otimes}_A A[[\mathbf{t}]]_\nabla.$$

We have:

$$\varphi_M\left(\sum_{\alpha\in\Delta} m_\alpha \mathbf{s}^\alpha\right) = \sum_{\alpha\in\Delta}\varphi(\mathbf{s}^\alpha)m_\alpha = \sum_{e\in\nabla}\left(\sum_{\substack{\alpha\in\Delta\\|\alpha|\le|e|}} \mathbf{C}_e(\varphi,\alpha)m_\alpha\right)\mathbf{t}^e,$$

$$_M\varphi\left(\sum_{\alpha\in\Delta} m_\alpha \mathbf{s}^\alpha\right) = \sum_{\alpha\in\Delta} m_\alpha\varphi(\mathbf{s}^\alpha) = \sum_{e\in\nabla}\left(\sum_{\substack{\alpha\in\Delta\\|\alpha|\le|e|}} m_\alpha\mathbf{C}_e(\varphi,\alpha)\right)\mathbf{t}^e$$

for all $m \in M[[\mathbf{s}]]_\Delta$. If M is a trivial bimodule, then $\varphi_M = {}_M\varphi$. If $\varphi' : A[[\mathbf{t}]]_\nabla \to A[[\mathbf{u}]]_\Omega$ is another substitution map and $\varphi'' = \varphi\circ\varphi'$, we have $\varphi''_M = \varphi_M \circ \varphi'_M$, ${}_M\varphi'' = {}_M\varphi \circ {}_M\varphi'$.

For all $m \in M[[\mathbf{s}]]_\Delta$ and all $a \in A[[\mathbf{s}]]_\nabla$, we have:

$$\varphi_M(am) = \varphi(a)\varphi_M(m), \quad {}_M\varphi(ma) = {}_M\varphi(m)\varphi(a),$$

i.e. φ_M is $(\varphi;A)$-linear and ${}_M\varphi$ is $(A;\varphi)$-linear. Moreover, φ_M and ${}_M\varphi$ are compatible with the augmentations, i.e.:

(1.12)

$$\varphi_M(m) \equiv m_0 \bmod \langle\mathbf{t}\rangle M[[\mathbf{t}]]_\nabla, \quad {}_M\varphi(m) \equiv m_0 \bmod \langle\mathbf{t}\rangle M[[\mathbf{t}]]_\nabla, \quad m\in M[[\mathbf{s}]]_\Delta.$$

If φ is the trivial substitution map (i.e. $\varphi(s_i) = 0$ for all $s_i \in \mathbf{s}$), then $\varphi_M : M[[\mathbf{s}]]_\Delta \to M[[\mathbf{t}]]_\nabla$ and ${}_M\varphi : M[[\mathbf{s}]]_\Delta \to M[[\mathbf{t}]]_\nabla$ are also trivial, i.e. $\varphi_M(m) = {}_M\varphi(m) = m_0$, for all $m \in M[[\mathbf{s}]]_\nabla$.

1.3.5 The above constructions apply in particular to the case of any k-algebra R over A, for which we have two induced continuous maps: $\varphi_R = \varphi\widehat{\otimes}\operatorname{Id}_R : R[[\mathbf{s}]]_\Delta \to R[[\mathbf{t}]]_\nabla$, which is $(A;R)$-linear, and ${}_R\varphi = \operatorname{Id}_R\widehat{\otimes}\varphi : R[[\mathbf{s}]]_\Delta \to R[[\mathbf{t}]]_\nabla$, which is $(R;A)$-linear. For $r \in R[[\mathbf{s}]]_\Delta$ we will denote $\varphi\bullet r := \varphi_R(r)$, $r\bullet\varphi := {}_R\varphi(r)$. Explicitly, if $r = \sum_\alpha r_\alpha \mathbf{s}^\alpha$ with $\alpha\in\Delta$, then:

$$\varphi\bullet r = \sum_{e\in\nabla}\left(\sum_{\substack{\alpha\in\Delta\\|\alpha|\le|e|}} \mathbf{C}_e(\varphi,\alpha)r_\alpha\right)\mathbf{t}^e, \quad r\bullet\varphi = \sum_{e\in\nabla}\left(\sum_{\substack{\alpha\in\Delta\\|\alpha|\le|e|}} r_\alpha\mathbf{C}_e(\varphi,\alpha)\right)\mathbf{t}^e. \tag{1.13}$$

From (1.12), we deduce that:

$$\varphi\bullet\mathscr{U}^p(R;\Delta) \subset \mathscr{U}^q(R;\nabla), \quad \mathscr{U}^p(R;\Delta)\bullet\varphi \subset \mathscr{U}^q(R;\nabla),$$

and $\varphi\bullet 1 = 1\bullet\varphi = 1$.

If φ is a substitution map with constant coefficients, then $\varphi_R = {}_R\varphi$ is a ring homomorphism over φ. In particular, $\varphi\bullet r = r\bullet\varphi$ and $\varphi\bullet(rr') = (\varphi\bullet r)(\varphi\bullet r')$.

If $\varphi = \mathbf{0} : A[[\mathbf{s}]]_\Delta \to A[[\mathbf{t}]]_\nabla$ is the trivial substitution map, then $\mathbf{0}\bullet r = r\bullet\mathbf{0} = r_0$ for all $r\in R[[\mathbf{s}]]_\Delta$. In particular, $\mathbf{0}\bullet r = r\bullet\mathbf{0} = 1$ for all $r \in \mathscr{U}^p(R;\Delta)$.

If $\mathbf{u} = \{u_1,\dots,u_r\}$ is another set of variables, $\Omega\subset\mathbb{N}^r$ is a non-empty co-ideal and $\psi : R[[\mathbf{t}]]_\nabla \to R[[\mathbf{u}]]_\Omega$ is another substitution map, one has:

$$\psi\bullet(\varphi\bullet r) = (\psi\circ\varphi)\bullet r, \quad (r\bullet\varphi)\bullet\psi = r\bullet(\psi\circ\varphi).$$

Since $(R[[\mathbf{s}]]_\Delta)^{\mathrm{opp}} = R^{\mathrm{opp}}[[\mathbf{s}]]_\Delta$, for any substitution map $\varphi : A[[\mathbf{s}]]_\Delta \to A[[\mathbf{t}]]_\nabla$ we have $(\varphi_R)^{\mathrm{opp}} = {}_{R^{\mathrm{opp}}}\varphi$ and $({}_R\varphi)^{\mathrm{opp}} = \varphi_{R^{\mathrm{opp}}}$.

For each substitution map $\varphi : A[[\mathbf{s}]]_\Delta \to A[[\mathbf{t}]]_\nabla$ we define the $(A;A)$-linear map:

$$\varphi_* : f \in \operatorname{Hom}_k(A, A[[\mathbf{s}]]_\Delta) \longmapsto \varphi_*(f) = \varphi \circ f \in \operatorname{Hom}_k(A, A[[\mathbf{t}]]_\nabla)$$

which induces another one $\overline{\varphi_*} : \operatorname{End}_{k[[\mathbf{s}]]_\Delta}(A[[\mathbf{s}]]_\Delta) \longrightarrow \operatorname{End}_{k[[\mathbf{t}]]_\nabla}(A[[\mathbf{t}]]_\nabla)$ given by:

$$\overline{\varphi_*}(f) := (\varphi_* (f|_A))^e = (\varphi \circ f|_A)^e \quad \forall f \in \operatorname{End}_{k[[\mathbf{s}]]_\Delta}(A[[\mathbf{s}]]_\Delta).$$

More generally, for any left A-modules E, F we have $(A;A)$-linear maps:

$$(\varphi_F)_* : f \in \operatorname{Hom}_k(E, F[[\mathbf{s}]]_\Delta) \longmapsto (\varphi_F)_*(f) = \varphi_F \circ f \in \operatorname{Hom}_k(E, F[[\mathbf{t}]]_\nabla),$$
$$\overline{(\varphi_F)_*} : \operatorname{Hom}_{k[[\mathbf{s}]]_\Delta}(E[[\mathbf{s}]]_\Delta, F[[\mathbf{s}]]_\Delta) \longrightarrow \operatorname{Hom}_{k[[\mathbf{t}]]_\nabla}(E[[\mathbf{t}]]_\nabla, F[[\mathbf{t}]]_\nabla),$$
$$\overline{(\varphi_F)_*}(f) := (\varphi_F \circ f|_E)^e .$$

Let us consider the $(A;A)$-bimodule $M = \operatorname{Hom}_k(E,F)$. For each $m \in M[[\mathbf{s}]]_\Delta$ and for each $e \in E$ we have $\widetilde{\varphi_M(m)}(e) = \varphi_F(\widetilde{m}(e))$, i.e.:

$$\widetilde{\varphi_M(m)}|_E = \varphi_F \circ (\widetilde{m}|_E)\,, \tag{1.14}$$

or more graphically, the following diagram is commutative (see (1.4)):

$$\begin{array}{ccccc} M[[\mathbf{s}]]_\Delta & \xrightarrow[m\mapsto\widetilde{m}]{\sim} & \operatorname{Hom}_{k[[\mathbf{s}]]_\Delta}(E[[\mathbf{s}]]_\Delta, F[[\mathbf{s}]]_\Delta) & \xrightarrow[\text{restr.}]{\sim} & \operatorname{Hom}_k(E, F[[\mathbf{s}]]_\Delta) \\ {\scriptstyle\varphi_M}\downarrow & & \downarrow{\scriptstyle\overline{(\varphi_F)_*}} & & {\scriptstyle(\varphi_F)_*}\downarrow \\ M[[\mathbf{t}]]_\nabla & \xrightarrow[m\mapsto\widetilde{m}]{\sim} & \operatorname{Hom}_{k[[\mathbf{t}]]_\nabla}(E[[\mathbf{t}]]_\nabla, F[[\mathbf{t}]]_\nabla) & \xrightarrow[\text{restr.}]{\sim} & \operatorname{Hom}_k(E, F[[\mathbf{t}]]_\nabla). \end{array} \tag{1.15}$$

In order to simplify notations, we will also write:

$$\varphi \bullet f := \overline{(\varphi_F)_*}(f) \quad \forall f \in \operatorname{Hom}_{k[[\mathbf{s}]]_\Delta}(E[[\mathbf{s}]]_\Delta, F[[\mathbf{s}]]_\Delta),$$

and so we have $\widetilde{\varphi\bullet m} = \varphi\bullet\widetilde{m}$ for all $m \in M[[\mathbf{s}]]_\Delta$. Let us notice that $(\varphi\bullet f)(e) = (\varphi_F \circ f)(e)$ for all $e \in E$, i.e.:

$$\boxed{(\varphi\bullet f)|_E = (\varphi_F \circ f)|_E = \varphi_F \circ (f|_E), \text{ but in general } \varphi\bullet f \neq \varphi_F \circ f.} \tag{1.16}$$

If $\varphi = \mathbf{0}$ is the trivial substitution map, then for each k-linear map $f = \sum_\alpha f_\alpha \mathbf{s}^\alpha : E \to E[[\mathbf{s}]]_\Delta$ (resp. $f = \sum_\alpha f_\alpha \mathbf{s}^\alpha \in \operatorname{End}_k(E)[[\mathbf{s}]]_\Delta \equiv \operatorname{End}_{k[[\mathbf{s}]]_\Delta}(E[[\mathbf{s}]]_\Delta)$), we have $\mathbf{0}\bullet f = f\bullet\mathbf{0} = f_0 \in \operatorname{End}_k(E) \subset \operatorname{Hom}_k(E, E[[\mathbf{s}]]_\Delta)$ (resp. $\mathbf{0}\bullet f = f\bullet\mathbf{0} = f_0^e = \overline{f_0} \in \operatorname{End}_{k[[\mathbf{s}]]_\Delta}(E[[\mathbf{s}]]_\Delta)$).

If $\varphi : A[[\mathbf{s}]]_\Delta \to A[[\mathbf{t}]]_\nabla$ is a substitution map, we have:

$$\varphi\bullet(af) = \varphi(a)\,(\varphi\bullet f)\,,\ (fa)\bullet\varphi = (f\bullet\varphi)\,\varphi(a)$$

for all $a \in A[[\mathbf{s}]]_\Delta$ and for all $f \in \operatorname{Hom}_k(E, E[[\mathbf{s}]]_\Delta)$ (or $f \in \operatorname{End}_{k[[\mathbf{s}]]_\Delta}(E[[\mathbf{s}]]_\Delta)$). Moreover:

$$(\varphi_E)_*(\operatorname{Hom}_k^\circ(E, M[[\mathbf{s}]]_\Delta)) \subset \operatorname{Hom}_k^\circ(E, E[[\mathbf{t}]]_\nabla),$$
$$\varphi\bullet\left(\operatorname{Aut}^\circ_{k[[\mathbf{s}]]_\Delta}(E[[\mathbf{s}]]_\Delta)\right) \subset \operatorname{Aut}^\circ_{k[[\mathbf{t}]]_\nabla}(E[[\mathbf{t}]]_\nabla),$$

and so we have a commutative diagram:

$$\begin{array}{ccccc} \mathscr{U}^p(R;\Delta) & \xrightarrow[r\mapsto\widetilde{r}]{\sim} & \operatorname{Aut}^\circ_{k[[\mathbf{s}]]_\Delta}(E[[\mathbf{s}]]_\Delta) & \xrightarrow[\text{restr.}]{\sim} & \operatorname{Hom}_k^\circ(E, E[[\mathbf{s}]]_\Delta) \\ {\scriptstyle\varphi\bullet(-)}\downarrow & & \downarrow{\scriptstyle\varphi\bullet(-)} & & \downarrow{\scriptstyle(\varphi_E)_*} \\ \mathscr{U}^q(R;\nabla) & \xrightarrow[r\mapsto\widetilde{r}]{\sim} & \operatorname{Aut}^\circ_{k[[\mathbf{t}]]_\nabla}(E[[\mathbf{t}]]_\nabla) & \xrightarrow[\text{restr.}]{\sim} & \operatorname{Hom}_k(E, F[[\mathbf{t}]]_\nabla). \end{array} \tag{1.17}$$

Now we are going to see how the $\varepsilon^i(r), \overline{\varepsilon}^i(r), \varepsilon(r), \overline{\varepsilon}(r)$ (see 1.2.14) can be expressed in terms of the action of substitution maps.

Let us consider the power series ring $A[[\mathbf{s},\tau]] = A[[\mathbf{s}]]\widehat{\otimes}_A A[[\tau]]$, and for each $i = 1,\dots,p$ we denote $\sigma^i : A[[\mathbf{s}]] \to A[[\mathbf{s},\tau]]$ the substitution map (with constant coefficients) defined by:

$$\sigma^i(s_j) = \begin{cases} s_i + s_i\tau & \text{if } j = i \\ s_j & \text{if } j \neq i. \end{cases}$$

Let us also denote $\sigma : A[[\mathbf{s}]] \to A[[\mathbf{s},\tau]]$ the substitution map (with constant coefficients) defined by:

$$\sigma(s_i) = s_i + s_i\tau \quad \forall i = 1,\dots,p,$$

and $\iota : A[[\mathbf{s}]] \to A[[\mathbf{s},\tau]]$ the substitution map induced by the inclusion $\mathbf{s} \hookrightarrow \mathbf{s}\sqcup\{\tau\}$. We often consider ι as an inclusion $A[[\mathbf{s}]] \hookrightarrow A[[\mathbf{s},\tau]]$.

It is clear that for each non-empty co-ideal $\Delta \subset \mathbb{N}^p$, the substitution maps $\sigma^i, \sigma, \iota : A[[\mathbf{s}]] \to A[[\mathbf{s},\tau]]$ induce new substitution maps $A[[\mathbf{s}]]_\Delta \to A[[\mathbf{s},\tau]]_{\Delta\times\{0,1\}}$, which will be also denoted by the same letters. Moreover, as a consequence of Taylor's expansion we have:

$$\sigma^i(a) = a + \chi^i_A(a)\tau, \quad \sigma(a) = a + \chi_A(a)\tau$$

where $\chi^i = s_i\frac{\partial}{\partial s_i}$ and $\chi = \sum_i \chi^i$ (see Definition1.2.8).

The proof of the following lemma is clear.

LEMMA 1.3.6. *The map $\xi : R[[\mathbf{s}]]_{\Delta,+} \to \mathscr{U}^{p+1}(R;\Delta\times\{0,1\})$ defined as:*

$$\xi\left(\sum_{\alpha\in\Delta,|\alpha|>0} r_\alpha \mathbf{s}^\alpha\right) = 1 + \sum_{\alpha\in\Delta,|\alpha|>0} r_\alpha \mathbf{s}^\alpha\tau$$

is a group homomorphism.

Let us notice that the map ξ above is injective and its image is the set of $r \in \mathscr{U}^{p+1}(R;\Delta\times\{0,1\})$ such that $\operatorname{supp} r \subset \{(0,0)\} \cup ((\Delta\setminus\{0\})\times\{1\})$.

PROPOSITION 1.3.7. *For each $r \in \mathscr{U}^p(R;\Delta)$, the following properties hold:*

(1) $r^*(\sigma^i\bullet r) = \xi(\varepsilon^i(r))$, $(\sigma^i\bullet r)r^* = \xi(\overline{\varepsilon}^i(r))$.
(2) $r^*(\sigma\bullet r) = \xi(\varepsilon(r))$, $(\sigma\bullet r)r^* = \xi(\overline{\varepsilon}(r))$.

PROOF. It is a straightforward consequence of Taylor's expansion formula:

$$\sigma^i\bullet r = r + \chi^i_R(r)\tau, \quad \sigma\bullet r = r + \chi_R(r)\tau.$$

□

Let us notice that in the above proposition, the action $\iota\bullet(-) : R[[\mathbf{s}]]_\Delta \to R[[\mathbf{s},\tau]]_{\Delta\times\{0,1\}}$ is simply considered as an inclusion.

2. Multivariate Hasse–Schmidt derivations

2.1. Basic definitions. In this section we recall some notions and results of the theory of Hasse–Schmidt derivations [**1, 6**] as developed in [**9**].

From now on k will be a commutative ring, A a commutative k-algebra, $\mathbf{s} = \{s_1, \dots, s_p\}$ a set of variables and $\Delta \subset \mathbb{N}^p$ a non-empty co-ideal.

DEFINITION 2.1.1. A (p, Δ)*-variate Hasse–Schmidt derivation*, or a (p, Δ)*-variate HS-derivation* for short, of A over k is a family $D = (D_\alpha)_{\alpha \in \Delta}$ of k-linear maps $D_\alpha : A \longrightarrow A$, satisfying the following Leibniz type identities:

$$D_0 = \mathrm{Id}_A, \quad D_\alpha(xy) = \sum_{\beta+\gamma=\alpha} D_\beta(x) D_\gamma(y)$$

for all $x, y \in A$ and for all $\alpha \in \Delta$. We denote by $\mathrm{HS}^p_k(A; \Delta)$ the set of all (p, Δ)-variate HS-derivations of A over k. For $p = 1$, a 1-variate HS-derivation will be simply called a *Hasse–Schmidt derivation* (a HS-derivation for short), or a *higher derivation*[2], and we will simply write $\mathrm{HS}_k(A; m) := \mathrm{HS}^1_k(A; \Delta)$ for $\Delta = \{q \in \mathbb{N} \mid q \leq m\}$.[3]

Any (p, Δ)-variate HS-derivation D of A over k can be understood as a power series:

$$\sum_{\alpha \in \Delta} D_\alpha \mathbf{s}^\alpha \in R[[\mathbf{s}]]_\Delta, \quad R = \mathrm{End}_k(A),$$

and so we consider $\mathrm{HS}^p_k(A; \Delta) \subset R[[\mathbf{s}]]_\Delta$. Actually $\mathrm{HS}^p_k(A; \Delta)$ is a (multiplicative) sub-group of $\mathscr{U}^p(R; \Delta)$. The group operation in $\mathrm{HS}^p_k(A; \Delta)$ is explicitly given by:

$$(D, E) \in \mathrm{HS}^p_k(A; \Delta) \times \mathrm{HS}^p_k(A; \Delta) \longmapsto D \circ E \in \mathrm{HS}^p_k(A; \Delta)$$

with:

$$(D \circ E)_\alpha = \sum_{\beta+\gamma=\alpha} D_\beta \circ E_\gamma,$$

and the identity element of $\mathrm{HS}^p_k(A; \Delta)$ is $\mathbb{I}$ with $\mathbb{I}_0 = \mathrm{Id}$ and $\mathbb{I}_\alpha = 0$ for all $\alpha \neq 0$. The inverse of a $D \in \mathrm{HS}^p_k(A; \Delta)$ will be denoted by D^*.

For $\Delta' \subset \Delta \subset \mathbb{N}^p$ non-empty co-ideals, we have truncations:

$$\tau_{\Delta\Delta'} : \mathrm{HS}^p_k(A; \Delta) \longrightarrow \mathrm{HS}^p_k(A; \Delta'),$$

which obviously are group homomorphisms. Since any $D \in \mathrm{HS}^p_k(A; \Delta)$ is determined by its finite truncations, we have a natural group isomorphism

$$\mathrm{HS}^p_k(A) = \varprojlim_{\substack{\Delta' \subset \Delta \\ \sharp\Delta' < \infty}} \mathrm{HS}^p_k(A; \Delta'). \tag{2.1}$$

The proof of the following proposition is straightforward and it is left to the reader (see Notation 1.2.4 and 1.2.6).

PROPOSITION 2.1.2. *Let us denote* $R = \mathrm{End}_k(A)$ *and let* $D = \sum_\alpha D_\alpha \mathbf{s}^\alpha \in R[[\mathbf{s}]]_\Delta$ *be a power series. The following properties are equivalent:*

(a) D *is a* $(\mathbf{s}, \Delta)$*-variate HS-derivation of* A *over* k.
(b) *The map* $\widetilde{D} : A[[\mathbf{s}]]_\Delta \to A[[\mathbf{s}]]_\Delta$ *is a (continuous)* $k[[\mathbf{s}]]_\Delta$*-algebra homomorphism compatible with the natural augmentation* $A[[\mathbf{s}]]_\Delta \to A$.

---

[2]This terminology is used for instance in [**6**].

[3]These HS-derivations are called of length m in [**8**].

(c) $D \in \mathscr{U}^p(R;\Delta)$ *and for all* $a \in A[[\mathbf{s}]]_\Delta$ *we have* $Da = \widetilde{D}(a)D$.
(d) $D \in \mathscr{U}^p(R;\Delta)$ *and for all* $a \in A$ *we have* $Da = \widetilde{D}(a)D$.

Moreover, in such a case $\widetilde{D}$ *is a* $k[[\mathbf{s}]]_\Delta$*-algebra automorphism of* $A[[\mathbf{s}]]_\Delta$.

NOTATION 2.1.3. Let us denote:

$$\operatorname{Hom}^\circ_{k-\mathrm{alg}}(A, A[[\mathbf{s}]]_\Delta) := \{f \in \operatorname{Hom}_{k-\mathrm{alg}}(A, A[[\mathbf{s}]]_\Delta),\ f(a) \equiv a \bmod \langle \mathbf{s}\rangle\ \forall a \in A\},$$
$$\operatorname{Aut}^\circ_{k[[\mathbf{s}]]_\Delta-\mathrm{alg}}(A[[\mathbf{s}]]_\Delta) :=$$
$$\left\{f \in \operatorname{Aut}_{k[[\mathbf{s}]]_\Delta-\mathrm{alg}}(A[[\mathbf{s}]]_\Delta) \mid f(a) \equiv a_0 \bmod \langle \mathbf{s}\rangle\ \forall a \in A[[\mathbf{s}]]_\Delta\right\}.$$

It is clear that $\operatorname{Hom}^\circ_{k-\mathrm{alg}}(A, A[[\mathbf{s}]]_\Delta) \subset \operatorname{Hom}^\circ_k(A, A[[\mathbf{s}]]_\Delta)$ and

$$\operatorname{Aut}^\circ_{k[[\mathbf{s}]]_\Delta-\mathrm{alg}}(A[[\mathbf{s}]]_\Delta) \subset \operatorname{Aut}^\circ_{k[[\mathbf{s}]]_\Delta}(A[[\mathbf{s}]]_\Delta)$$

(see Notation 1.2.7) are subgroups, and we have group isomorphisms (see (1.7) and (1.6)):

$$\operatorname{HS}^p_k(A;\Delta) \xrightarrow[\simeq]{D\mapsto\widetilde{D}} \operatorname{Aut}^\circ_{k[[\mathbf{s}]]_\Delta-\mathrm{alg}}(A[[\mathbf{s}]]_\Delta) \xrightarrow[\simeq]{\text{restr.}} \operatorname{Hom}^\circ_{k-\mathrm{alg}}(A, A[[\mathbf{s}]]_\Delta). \tag{2.2}$$

The composition of the above isomorphisms is given by:
(2.3)

$$D \in \operatorname{HS}^p_k(A;\Delta) \overset{\sim}{\longmapsto} \Phi_D := \left[a \in A \mapsto \sum_{\alpha\in\Delta} D_\alpha(a)\mathbf{s}^\alpha\right] \in \operatorname{Hom}^\circ_{k-\mathrm{alg}}(A, A[[\mathbf{s}]]_\Delta).$$

For each HS-derivation $D \in \operatorname{HS}^p_k(A;\Delta)$ we have $\widetilde{D} = (\Phi_D)^e$, i.e.:

$$\widetilde{D}\left(\sum_{\alpha\in\Delta} a_\alpha \mathbf{s}^\alpha\right) = \sum_{\alpha\in\Delta} \Phi_D(a_\alpha)\mathbf{s}^\alpha$$

for all $\sum_\alpha a_\alpha \mathbf{s}^\alpha \in A[[\mathbf{s}]]_\Delta$, and for any $E \in \operatorname{HS}^p_k(A;\Delta)$ we have $\Phi_{D\circ E} = \widetilde{D}\circ\Phi_E$. If $\Delta' \subset \Delta$ is another non-empty co-ideal and we denote by $\pi_{\Delta\Delta'} : A[[\mathbf{s}]]_\Delta \to A[[\mathbf{s}]]_{\Delta'}$ the projection (or truncation), one has $\Phi_{\tau_{\Delta\Delta'}(D)} = \pi_{\Delta\Delta'}\circ\Phi_D$.

2.2. The action of substitution maps on HS-derivations. Now, we recall the action of substitution maps on HS-derivations [**9**, §6]. Let $\mathbf{s} = \{s_1,\dots,s_p\}$, $\mathbf{t} = \{t_1,\dots,t_p\}$ be sets of variables, $\Delta \subset \mathbb{N}^p$, $\nabla \subset \mathbb{N}^q$ non-empty co-ideals and let us write $R = \operatorname{End}_k(A)$.

Let us recall Proposition 10 in [**9**].

PROPOSITION 2.2.1. *For any substitution map* $\varphi : A[[\mathbf{s}]]_\Delta \to A[[\mathbf{t}]]_\nabla$, *we have:*
1) $\varphi_*\left(\operatorname{Hom}^\circ_{k-\mathrm{alg}}(A, A[[\mathbf{s}]]_\Delta)\right) \subset \operatorname{Hom}^\circ_{k-\mathrm{alg}}(A, A[[\mathbf{t}]]_\nabla)$,
2) $\varphi\bullet \operatorname{HS}^p_k(A;\Delta) \subset \operatorname{HS}^q_k(A;\nabla)$,
3) $\varphi\bullet \operatorname{Aut}^\circ_{k[[\mathbf{s}]]_\Delta-\mathrm{alg}}(A[[\mathbf{s}]]_\Delta) \subset \operatorname{Aut}^\circ_{k[[\mathbf{t}]]_\nabla-\mathrm{alg}}(A[[\mathbf{t}]]_\nabla)$.

Then we have a commutative diagram:

$$\begin{array}{ccccc} \operatorname{Hom}^\circ_{k-\mathrm{alg}}(A, A[[\mathbf{s}]]_\Delta) & \xleftarrow[\Phi_D \leftarrow\!\shortmid D]{\sim} & \operatorname{HS}^p_k(A;\Delta) & \xrightarrow{\sim} & \operatorname{Aut}^\circ_{k[[\mathbf{s}]]_\Delta-\mathrm{alg}}(A[[\mathbf{s}]]_\Delta)n \\ \downarrow{\scriptstyle\varphi_*} & & \downarrow{\scriptstyle\varphi\bullet(-)} & & \downarrow{\scriptstyle\varphi\bullet(-)} \\ \operatorname{Hom}^\circ_{k-\mathrm{alg}}(A, A[[\mathbf{t}]]_\nabla) & \xleftarrow[\Phi_D \leftarrow\!\shortmid D]{\sim} & \operatorname{HS}^q_k(A;\nabla) & \xrightarrow{\sim} & \operatorname{Aut}^\circ_{k[[\mathbf{t}]]_\nabla-\mathrm{alg}}(A[[\mathbf{t}]]_\nabla). \end{array} \tag{2.4}$$

In particular, for any HS-derivation $D \in \mathrm{HS}_k^p(A;\Delta)$ we have $\varphi\bullet D \in \mathrm{HS}_k^q(A;\nabla)$ (see 1.3.5). Moreover $\Phi_{\varphi\bullet D} = \varphi \circ \Phi_D$.

It is clear that for any co-ideals $\Delta' \subset \Delta$ and $\nabla' \subset \nabla$ with $\varphi(\Delta'_A/\Delta_A) \subset \nabla'_A/\nabla_A$ we have:

$$\tau_{\nabla\nabla'}(\varphi\bullet D) = \varphi'\bullet\tau_{\Delta\Delta'}(D), \tag{2.5}$$

where $\varphi' : A[[\mathbf{s}]]_{\Delta'} \to A[[\mathbf{t}]]_{\nabla'}$ is the substitution map induced by φ.

2.2.2 Let $\mathbf{u} = \{u_1,\ldots,u_r\}$ be another set of variables, $\Omega \subset \mathbb{N}^r$ a non-empty co-ideal, $\varphi : A[[\mathbf{s}]]_\Delta \to A[[\mathbf{t}]]_\nabla$, $\psi : A[[\mathbf{t}]]_\nabla \to A[[\mathbf{u}]]_\Omega$ substitution maps and $D, D' \in \mathrm{HS}_k^p(A;\Delta)$ HS-derivations. From 1.3.5 we deduce the following properties:

-) If we denote $E := \varphi\bullet D \in \mathrm{HS}_k^q(A;\nabla)$, we have

$$E_0 = \mathrm{Id}, \quad E_e = \sum_{\substack{\alpha\in\Delta \\ |\alpha|\le|e|}} \mathbf{C}_e(\varphi,\alpha) D_\alpha, \quad \forall e \in \nabla. \tag{2.6}$$

-) If $\varphi = \mathbf{0}$ is the trivial substitution map or if $D = \mathbb{I}$, then $\varphi\bullet D = \mathbb{I}$.

-) If φ has constant coefficients, then $\varphi\bullet(D\circ D') = (\varphi\bullet D)\circ(\varphi\bullet D')$ and $(\varphi\bullet D)^* = \varphi\bullet D^*$ (the general case is treated in Proposition 2.2.3).

-) $\psi\bullet(\varphi\bullet D) = (\psi\circ\varphi)\bullet D$.

The following result is proven in Propositions 11 and 12 of [**9**].

PROPOSITION 2.2.3. *Let $\varphi : A[[\mathbf{s}]]_\Delta \to A[[\mathbf{t}]]_\nabla$ be a substitution map. Then, the following assertions hold:*

(i) *For each $D \in \mathrm{HS}_k^p(A;\Delta)$ there is a unique substitution map $\varphi^D : A[[\mathbf{s}]]_\Delta \to A[[\mathbf{t}]]_\nabla$ such that $\left(\widetilde{\varphi\bullet D}\right)\circ\varphi^D = \varphi\circ\widetilde{D}$. Moreover, $(\varphi\bullet D)^* = \varphi^D\bullet D^*$, $\varphi^{\mathbb{I}} = \varphi$ and:*

$$\mathbf{C}_e(\varphi, f+\nu) = \sum_{\substack{\beta+\gamma=e \\ |f+g|\le|\beta|,|\nu|\le|\gamma|}} \mathbf{C}_\beta(\varphi, f+g) D_g(\mathbf{C}_\gamma(\varphi^D,\nu))$$

for all $e \in \Delta$ and for all $f,\nu \in \nabla$ with $|f+\nu| \le |e|$.

(ii) *For each $D, E \in \mathrm{HS}_k^p(A;\nabla)$, we have $\varphi\bullet(D\circ E) = (\varphi\bullet D)\circ(\varphi^D\bullet E)$ and $\left(\varphi^D\right)^E = \varphi^{D\circ E}$. In particular, $\left(\varphi^D\right)^{D^*} = \varphi$.*

(iii) *If ψ is another composable substitution map, then $(\varphi\circ\psi)^D = \varphi^{\psi\bullet D}\circ\psi^D$.*

(iv) *If φ has constant coefficients then $\varphi^D = \varphi$.*

3. Main results

3.1. The derivations associated with a Hasse–Schmidt derivation. In this section k will be a commutative ring, A a commutative k-algebra, $R = \mathrm{End}_k(A)$, $\mathbf{s} = \{s_1,\ldots,s_p\}$ a set of variables and $\Delta \subset \mathbb{N}^p$ a non-empty co-ideal.

LEMMA 3.1.1. *Let $\mathfrak{d} : k[[\mathbf{s}]]_\Delta \to k[[\mathbf{s}]]_\Delta$ be a k-derivation and $D \in \mathrm{HS}_k^p(A;\Delta)$ a HS-derivation. Then, for each $a \in A[[\mathbf{s}]]_\Delta$ we have $\mathfrak{d}_R(D)a = \widetilde{\mathfrak{d}_R(D)}(a)D + \widetilde{D}(a)\mathfrak{d}_R(D)$.*

PROOF. By using Lemma 1.2.10 we have:

$$\mathfrak{d}_R(D)a = \mathfrak{d}_R(Da) - D\mathfrak{d}_A(a) = \mathfrak{d}_R(\widetilde{D}(a)D) - \widetilde{D}(\mathfrak{d}_A(a))D =$$

$$\mathfrak{d}_A(\widetilde{D}(a))D + \widetilde{D}(a)\mathfrak{d}_R(D) - \widetilde{D}(\mathfrak{d}_A(a))D = \widetilde{\mathfrak{d}_R(D)}(a)D + \widetilde{D}(a)\mathfrak{d}_R(D).$$

□

From now on, we will denote to simplify $\mathfrak{d}_R = \mathfrak{d}$ and $\mathfrak{d}_A = \mathfrak{d}$.

PROPOSITION 3.1.2. *Let $\mathfrak{d} : k[[\mathbf{s}]]_\Delta \to k[[\mathbf{s}]]_\Delta$ be a k-derivation. Then, for any $D \in \mathrm{HS}^p_k(A;\Delta)$ we have $\varepsilon^{\mathfrak{d}}(D), \overline{\varepsilon}^{\mathfrak{d}}(D) \in \mathrm{Der}_k(A)[[\mathbf{s}]]_{\Delta,+} = \mathrm{Der}_k(A)[[\mathbf{s}]]_\Delta \cap R[[\mathbf{s}]]_{\Delta,+}$.*

PROOF. Remember that $\overline{\varepsilon}^{\mathfrak{d}}(D) = \mathfrak{d}(D)D^*$ and $\varepsilon^{\mathfrak{d}}(D) = D^*\mathfrak{d}(D)$ (Definition 1.2.11). We will use Lemma 1.2.9 and Lemma 3.1.1. For any $a \in A[[\mathbf{s}]]_\Delta$ we have:

$$(\mathfrak{d}(D)D^*)\,a = \mathfrak{d}(D)\widetilde{D^*}(a)D^* = \left[\widetilde{\mathfrak{d}(D)}(\widetilde{D^*}(a))D + \widetilde{D}(\widetilde{D^*}(a))\mathfrak{d}(D)\right] D^* =$$

$$\widetilde{\mathfrak{d}(D)}(\widetilde{D^*}(a)) + a\mathfrak{d}(D)D^* = (\widetilde{\mathfrak{d}(D)D^*})(a) + a\,(\mathfrak{d}(D)D^*),$$

and so $[\mathfrak{d}(D)D^*, a] = (\widetilde{\mathfrak{d}(D)D^*})(a)$ and $\mathfrak{d}(D)D^* \in \mathrm{Der}_k(A)[[\mathbf{s}]]_\Delta$. The proof for $\varepsilon^{\mathfrak{d}}(D)$ is completely similar. □

EXAMPLE 3.1.3. If $D \in \mathrm{HS}_k(A;m)$ is a 1-variate HS-derivation of length m, then:

$$\varepsilon_1(D) = D_1,\ \varepsilon_2(D) = 2D_2 - D_1^2,\ \varepsilon_3(D) = 3D_3 - 2D_1D_2 - D_2D_1 + D_1^3, \dots$$

Let us recall that the map $\xi : R[[\mathbf{s}]]_{\Delta,+} \to \mathscr{U}^{p+1}(R;\Delta\times\{0,1\})$ has been defined in Lemma 1.3.6. The proof of the following lemma is clear.

LEMMA 3.1.4. *For each $\delta \in \mathrm{Der}_k(A)[[\mathbf{s}]]_{\Delta,+}$ we have $\xi(\delta) \in \mathrm{HS}^{p+1}_k(A;\Delta\times\{0,1\})$.*

So we have a group homomorphism $\xi : \mathrm{Der}_k(A)[[\mathbf{s}]]_{\Delta,+} \to \mathrm{HS}^{p+1}_k(A;\Delta\times\{0,1\})$ whose image is the set of $D \in \mathrm{HS}^{p+1}_k(A;\Delta\times\{0,1\})$ such that $\operatorname{supp} D \subset \{(0,0)\} \cup ((\Delta\setminus\{0\})\times\{1\})$.

The following proposition provides a characterization of HS-derivations in characteristic 0.

PROPOSITION 3.1.5. *Assume that $\mathbb{Q} \subset k$, $R = \mathrm{End}_k(A)$ and $D \in \mathscr{U}^p(R;\Delta)$. The following properties are equivalent:*

(a) $D \in \mathrm{HS}^p_k(A;\Delta)$.
(b) $\varepsilon^{\mathfrak{d}}(D) \in \mathrm{Der}_k(A)[[\mathbf{s}]]_\Delta$ *for all k-derivations $\mathfrak{d} : k[[\mathbf{s}]]_\Delta \to k[[\mathbf{s}]]_\Delta$.*
(c) $\varepsilon(D) \in \mathrm{Der}_k(A)[[\mathbf{s}]]_\Delta$.

PROOF. The implication (a) ⇒ (b) comes from Proposition 3.1.2 and (b) ⇒ (c) is obvious. Let us prove (c) ⇒ (a). Write $\delta = \varepsilon(D)$, i.e. $\boldsymbol{\chi}_R(D) = D\,\delta$. After Proposition 2.1.2, we need to prove that $D\,a = \widetilde{D}(a)D$ for all $a \in A$, and since $D\,a - \widetilde{D}(a)D$ belongs to the augmentation ideal of $R[[\mathbf{s}]]_\Delta$ and $\mathbb{Q} \subset k$, it is enough to prove that $\boldsymbol{\chi}_R\left(D\,a - \widetilde{D}(a)D\right) = 0$. By using that $\boldsymbol{\chi}_A(a) = 0$ and Lemma 1.2.10

we have:

$$\begin{gathered}\chi_R\left(D\,a - \widetilde{D}(a)D\right) = \chi_R(D)\,a + D\,\chi_A(a) - \chi_A(\widetilde{D}(a))D - \widetilde{D}(a)\chi_R(D) = \\ D\,\delta\,a - \widetilde{\chi_R(D)}(a)D - \widetilde{D}(\chi_A(a))D - \widetilde{D}(a)D\,\delta = \\ D\,a\,\delta + D\,\widetilde{\delta}(a) - \widetilde{\chi_R(D)}(a)D - \widetilde{D}(a)D\,\delta = \\ \widetilde{D}(a)D\,\delta + \widetilde{D}(\widetilde{\delta}(a))D - \widetilde{(D\,\delta)}(a)D - \widetilde{D}(a)D\,\delta = 0.\end{gathered}$$

□

THEOREM 3.1.6. *Assume that $\mathbb{Q} \subset k$. Then, all the maps in the following commutative diagram:*

$$\begin{array}{ccc} \mathrm{HS}^p_k(A;\Delta) & \xrightarrow{\ \varepsilon\ } & \Lambda^p(R;\Delta) \bigcap (\mathrm{Der}_k(A)[[\mathbf{s}]]_{\Delta,+})^p \\ & \searrow^{\varepsilon} & \downarrow \boldsymbol{\Sigma} \\ & & \mathrm{Der}_k(A)[[\mathbf{s}]]_{\Delta,+} \end{array}$$

are bijective.

PROOF. It is a consequence of Proposition 1.2.18 and Proposition 3.1.5. □

A similar result holds for $\overline{\varepsilon}$ instead of ε.

REMARK 3.1.7. Let us notice that, in Theorem 3.1.6, $\mathrm{HS}^p_k(A;\Delta)$ carries a group structure (non-commutative in general) and an action of substitution maps, and on the other hand $\mathrm{Der}_k(A)[[\mathbf{s}]]_{\Delta,+}$ carries an $A[[\mathbf{s}]]_\Delta$-module structure and a $k[[\mathbf{s}]]_\Delta$-Lie algebra structure, but the bijection $\varepsilon : \mathrm{HS}^p_k(A;\Delta) \xrightarrow{\sim} \mathrm{Der}_k(A)[[\mathbf{s}]]_{\Delta,+}$ is not compatible with these structures. The formulas expressing the behavior of ε with respect to the group operation on HS-derivations or the behavior of ε^{-1} with respect to the addition of power series with coefficients in $\mathrm{Der}_k(A)$, turn out to be complicated and have a similar flavor to Baker-Campbell-Hausdorff formula.

3.2. The behavior under the action of substitution maps.

DEFINITION 3.2.1. Let S be a k-algebra over A, $r \in \mathscr{U}^p(S;\Delta)$, $D \in \mathrm{HS}^p_k(A;\Delta)$, $r' \in S[[\mathbf{s}]]_\Delta$ and $\delta \in \mathrm{Der}_k(A)[[\mathbf{s}]]_\Delta$. We say that

-) r is a *D-element* if $r\,a = \widetilde{D}(a)\,r$ for all $a \in A[[\mathbf{s}]]_\Delta$.
-) r' is a *δ-element* if $r'\,a = ar' + \widetilde{\delta}(a)\,1_S$ for all $a \in A[[\mathbf{s}]]_\Delta$.

It is clear that $D \in \mathrm{HS}^p_k(A;\Delta) \subset \mathscr{U}^p(\mathrm{End}_k(A);\Delta)$ is a D-element. For $D = \mathbb{I}$ the identity HS-derivation, a $r \in \mathscr{U}^p(S;\Delta)$ is an *$\mathbb{I}$-element* if and only if r commutes with all $a \in A[[\mathbf{s}]]_\Delta$. If $E \in \mathrm{HS}^p_k(A;\Delta)$ is another HS-derivation, $r \in \mathscr{U}^p(S;\Delta)$ is a D-element and $s \in \mathscr{U}^p(S;\Delta)$ is an E-element, then rs is a $(D \circ E)$-element.

The following lemma provides a characterization of D-elements. Its proof is easy and it is left to the reader.

LEMMA 3.2.2. *With the above notations, for each $r = \sum_\alpha r_\alpha \mathbf{s}^\alpha \in \mathscr{U}^p(S;\Delta)$ the following properties are equivalent:*

-) *r is a D-element.*
-) *$br = r\,\widetilde{D^*}(b)$ for all $b \in A[[\mathbf{s}]]_\Delta$.*
-) *r^* is a D^*-element.*

-) *If* $r=\sum_\alpha r_\alpha \mathbf{s}^\alpha$, *we have* $r_\alpha\, a = \sum_{\beta+\gamma=\alpha} D_\beta(a) r_\gamma$ *for all* $a\in A$ *and for all* $\alpha\in\Delta$.

-) $r\, a = \widetilde{D}(a) r$ *for all* $a\in A$.

The following proposition reproduces Proposition 2.2.6 of [**10**].

PROPOSITION 3.2.3. *Let* S *be a* k*-algebra over* A, $D\in \operatorname{HS}^p_k(A;\Delta)$, $\varphi: A[[\mathbf{s}]]_\Delta \to A[\mathbf{u}]]_\nabla$ *a substitution map and* $r\in \mathscr{U}^p(S;\Delta)$ *a* D*-element. Then the following identities hold:*

(a) $\varphi_S(r)$ *is a* $(\varphi\bullet D)$*-element.*

(b) $\varphi_S(rr') = \varphi_S(r)\varphi^D_S(r')$ *for all* $r'\in R[[\mathbf{s}]]_\Delta$. *In particular,* $\varphi_S(r)^* = \varphi^D_S(r^*)$.

Moreover, if E *is an* A*-module and* $S=\operatorname{End}_k(E)$, *then the following identity holds:*

(c) $\langle \varphi\bullet r, \varphi^D_E(e)\rangle = \varphi_E(\langle r,e\rangle)$ *for all* $e\in E[[\mathbf{s}]]_\Delta$. *In other words:* $\varphi_E\circ\widetilde{r} = (\varphi\bullet\widetilde{r})\circ\varphi^D_E$.

Lemma 1.2.10 and Proposition 3.1.2 can be generalized in the following way.

PROPOSITION 3.2.4. *Let* S *be a* k*-algebra over* A, $D\in\operatorname{HS}^p_k(A;\Delta)$, $r\in\mathscr{U}^p(S;\Delta)$ *a* D*-element and* $\mathfrak{d}: k[[\mathbf{s}]]_\Delta\to k[[\mathbf{s}]]_\Delta$ *a* k*-derivation. Then, the following properties hold:*

(1) $\mathfrak{d}(r)\, a = \langle \mathfrak{d}(D), a\rangle\, r + \langle D,a\rangle\, \mathfrak{d}(r)$ *for all* $a\in A$

(2) $\varepsilon^{\mathfrak{d}}(r)$ *is a* $\varepsilon^{\mathfrak{d}}(D)$*-element and* $\overline{\varepsilon}^{\mathfrak{d}}(r)$ *is a* $\overline{\varepsilon}^{\mathfrak{d}}(D)$*-element.*

PROOF. (1) Since $\delta(a)=0$ we have:

$$\mathfrak{d}(r)\, a = \delta(r\, a) = \mathfrak{d}\left(\widetilde{D}(a) r\right) = \mathfrak{d}\left(\langle D,a\rangle\, r\right) = \langle \mathfrak{d}(D), a\rangle\, r + \langle D,a\rangle\, \mathfrak{d}(r).$$

(2) For all $a\in A$ we have:

$$\varepsilon^{\mathfrak{d}}(r)\, a = r^*\, \mathfrak{d}(r)\, a \overset{(1)}{=} r^*\left(\widetilde{\mathfrak{d}(D)}(a)\, r + \widetilde{D}(a)\, \mathfrak{d}(r)\right) =$$

$$\widetilde{D^*}(\widetilde{\mathfrak{d}(D)}(a)) r^*\, r + \widetilde{D^*}(\widetilde{D}(a)) r^* \mathfrak{d}(r) = \widetilde{\varepsilon^{\mathfrak{d}}(D)}(a)\, 1_S + a\, \varepsilon^{\mathfrak{d}}(r).$$

The proof for $\overline{\varepsilon}^{\mathfrak{d}}(r)$ is similar. □

Let us consider two sets of variables $\mathbf{s}=\{s_1,\dots,s_p\}$ and $\mathbf{u}=\{u_1,\dots,u_q\}$, and let us denote by $\{v^1,\dots,v^p\}$ the canonical basis of $\mathbb{N}^p$: $v^i_l=\delta_{il}$.

THEOREM 3.2.5. *For each non-empty co-ideals* $\Delta\subset\mathbb{N}^p, \Omega\subset\mathbb{N}^q$, *each substitution map* $\varphi: A[[\mathbf{s}]]_\Delta\to A[[\mathbf{u}]]_\Omega$ *and each HS-derivation* $D\in\operatorname{HS}^p_k(A;\Delta)$, *there exists a family*

$$\left\{\mathbf{N}^{j,i}_{e,h} \mid 1\le j\le q, 1\le i\le p, e\in\Omega, h\in\Delta, |h|\le|e|\right\}\subset A$$

such that for any k*-algebra* S *over* A *and any* D*-element* $r\in\mathscr{U}^p(S;\Delta)$, *we have:*

$$\varepsilon^j_e(\varphi\bullet r) = \sum_{\substack{0<|h|\le|e|\\ i\in\operatorname{supp} h}} \mathbf{N}^{j,i}_{e,h}\varepsilon^i_h(r) \quad \forall e\in\Omega, \forall j=1,\dots,q. \tag{3.1}$$

Moreover, $\mathbf{N}^{j,i}_{e,h} = \sum_{f,g,\beta} g_j \mathbf{C}_f(\varphi^D, \beta+h-v^i) D^*_\beta(\mathbf{C}_g(\varphi, v^i))$, *where* $f,g\in\Omega$, $\beta\in\Delta$, $f+g=e$, $|\beta+h|-1\le|f|$ *and* $g_j>0$, *whenever* $e_j, h_i>0$, *and* $\mathbf{N}^{j,i}_{e,h}=0$ *otherwise.*

PROOF. Let us write $r = \sum_{\alpha\in\Delta} r_\alpha \mathbf{s}^\alpha$. For each $\alpha \in \Delta$ and each $j = 1,\dots,q$ we have:

$$\chi^j(\varphi(\mathbf{s}^\alpha)) = \chi^j\left(\prod_{i=1}^{p} \varphi(s_i)^{\alpha_i}\right) = \sum_{\alpha_i\neq 0} \alpha_i \varphi(s_i)^{\alpha_i-1}\left(\prod_{\substack{1\leq l\leq p\\ l\neq i}} \varphi(s_l)^{\alpha_l}\right)\chi^j(\varphi(s_i)) =$$

$$\sum_{\alpha_i\neq 0}\alpha_i\varphi\left(\mathbf{s}^{\alpha-v^i}\right)\chi^j(\varphi(s_i)) = \sum_{\alpha_i\neq 0}\alpha_i\left(\sum_{|e|\geq|\alpha|-1}\mathbf{C}_e(\varphi,\alpha-v^i)\mathbf{u}^e\right)\left(\sum_{e_j>0} e_j\mathbf{C}_e(\varphi,v^i)\mathbf{u}^e\right) =$$

$$\sum_{\alpha_i\neq 0}\alpha_i\left(\sum_{\substack{|e|\geq|\alpha|\\ e_j>0}}\left(\sum_{\substack{e'+e''=e\\ |e'|\geq|\alpha|-1\\ e''_j>0}} e''_j\mathbf{C}_{e'}(\varphi,\alpha-v^i)\mathbf{C}_{e''}(\varphi,v^i)\right)\mathbf{u}^e\right) = \sum_{\alpha_i\neq 0}\alpha_i\left(\sum_{\substack{|e|\geq|\alpha|\\ e_j>0}}\mathbf{M}^{j,i}_{\alpha,e}\mathbf{u}^e\right)$$

with:

$$\mathbf{M}^{j,i}_{\alpha,e} := \sum_{\substack{e'+e''=e\\ |e'|\geq|\alpha|-1\\ e''_j>0}} e''_j\mathbf{C}_{e'}(\varphi,\alpha-v^i)\mathbf{C}_{e''}(\varphi,v^i)$$

for $i\in\operatorname{supp}\alpha$ and $e\in\Omega$ with $e_j>0$ and $|e|\geq|\alpha|$. If either $\alpha_i=0$ or $e_j=0$, we set $\mathbf{M}^{j,i}_{\alpha,e}=0$. So, for each $j=1,\dots,q$ we have:

$$\varepsilon^j(\varphi\bullet r) = (\varphi\bullet r)^*\chi^j(\varphi\bullet r) = \cdots = (\varphi^D\bullet r^*)\left(\sum_{\alpha\in\Delta}\chi^j(\varphi(\mathbf{s}^\alpha))r_\alpha\right) =$$

$$\left(\sum_{e\in\Omega}\left(\sum_{|\alpha|\leq|e|}\mathbf{C}_e(\varphi^D,\alpha)r^*_\alpha\right)\mathbf{u}^e\right)\left(\sum_{e_j>0}\left(\sum_{\substack{|\alpha|\leq|e|\\ \alpha_i\neq 0}}\alpha_i\mathbf{M}^{j,i}_{\alpha,e}r_\alpha\right)\mathbf{u}^e\right) =$$

$$\sum_{e_j>0}\left(\sum_{\substack{f+g=e\\ |\mu|\leq|f|,|\nu|\leq|g|\\ g_j>0,i\in\operatorname{supp}\nu}}\nu_t\mathbf{C}_f(\varphi^D,\mu)r^*_\mu\mathbf{M}^{j,i}_{\nu,g}r_\nu\right)\mathbf{u}^e \overset{(\star)}{=}$$

$$\sum_{e_j>0}\left(\sum_{\substack{f+g=e\\ |\beta+\gamma|\leq|f|,0<|\nu|\leq|g|\\ g_j>0,i\in\operatorname{supp}\nu}}\nu_i\mathbf{C}_f(\varphi^D,\beta+\gamma)D^*_\beta(\mathbf{M}^{j,i}_{\nu,g})r^*_\gamma r_\nu\right)\mathbf{u}^e =$$

$$\sum_{e_j>0}\left(\sum_{\substack{0<|h|\leq|e|\\ i\in\operatorname{supp}h}}\sum_{\substack{\gamma+\nu=h\\ i\in\operatorname{supp}\nu}}\left(\sum_{\substack{f+g=e\\ |\beta+\gamma|\leq|f|\\ |\nu|\leq|g|,g_j>0}}\mathbf{C}_f(\varphi^D,\beta+\gamma)D^*_\beta(\mathbf{M}^{j,i}_{\nu,g})\right)\nu_i r^*_\gamma r_\nu\right)\mathbf{u}^e =$$

$$\sum_{e_j>0}\left(\sum_{\substack{0<|h|\leq|e|\\ i\in\operatorname{supp}h}}\sum_{\substack{\gamma+\nu=h\\ i\in\operatorname{supp}\nu}} N^{j,i}_{e,\nu,\gamma}\nu_i r^*_\gamma r_\nu\right)\mathbf{u}^e,$$

where equality $(\star)$ comes from the fact that r^* is a D^*-element (see Lemma 3.2.2) and

$$N^{j,i}_{e,\nu,\gamma} = \begin{cases} \displaystyle\sum_{\substack{f+g=e\\ |\beta+\gamma|\le|f|\\ |\nu|\le|g|, g_j>0}} \mathbf{C}_f(\varphi^D,\beta+\gamma)D^*_\beta(\mathbf{M}^{j,i}_{\nu,g}) & \text{if } \nu_i>0, e_j>0, |e|\ge|\nu+\gamma|,\\ 0 & \text{otherwise.}\end{cases}$$

But, for $h=\nu+\gamma$, we have:

$$N^{j,i}_{e,\nu,\gamma} = \sum_{\substack{f+g=e\\ |\beta+\gamma|\le|f|\\ |\nu|\le|g|, g_j>0}} \mathbf{C}_f(\varphi^D,\beta+\gamma)D^*_\beta(\mathbf{M}^{j,i}_{\nu,g}) =$$

$$\sum_{\substack{f+g=e\\ |\beta+\gamma|\le|f|\\ |\nu|\le|g|, g_j>0}} \mathbf{C}_f(\varphi^D,\beta+\gamma)\, D^*_\beta \left(\sum_{\substack{g'+g''=g\\ |g'|\ge|\nu|-1\\ g''_j>0}} g''_j \mathbf{C}_{g'}(\varphi,\nu-v^i)\mathbf{C}_{g''}(\varphi,v^i)\right) =$$

$$\sum_{\substack{f+g'+g''=e\\ |\beta'+\beta''+\gamma|\le|f|\\ |\nu|-1\le|g'|, g''_j>0}} g''_u \mathbf{C}_f(\varphi^D,\beta'+\beta''+\gamma)\, D^*_{\beta'}\left(\mathbf{C}_{g'}(\varphi,\nu-v^i)\right)\, D^*_{\beta''}\left(\mathbf{C}_{g''}(\varphi,v^i)\right) =$$

$$\sum_{\substack{\rho+g''=e\\ |\beta''+h|-1\le|\rho|\\ g''_j>0}} g''_j \left(\sum_{\substack{f+g'=\rho\\ |\beta''+\gamma+\beta'|\le|f|\\ |\nu|-1\le|g'|}} \mathbf{C}_f(\varphi^D,\beta''+\gamma+\beta')\, D^*_{\beta'}\left(\mathbf{C}_{g'}(\varphi,\nu-v^i)\right)\right) D^*_{\beta''}\left(\mathbf{C}_{g''}(\varphi,v^i)\right) \overset{(\star)}{=}$$

$$\sum_{\substack{\rho+g''=e\\ |\beta''+h|-1\le|\rho|\\ g''_j>0}} g''_j\, \mathbf{C}_\rho(\varphi^D,\beta''+\gamma+\nu-v^i)\, D^*_{\beta''}\left(\mathbf{C}_{g''}(\varphi,v^i)\right),$$

where equality $(\star)$ comes from Proposition 2.2.3, and so $N^{j,i}_{e,\nu,\gamma}$ only depends on $h=\gamma+\nu$. By setting:

$$\mathbf{N}^{j,i}_{e,h} := \sum_{\substack{f+g=e\\ |\beta+h|-1\le|f|\\ g_j>0}} g_j \mathbf{C}_f(\varphi^D,\beta+h-v^i)D^*_\beta(\mathbf{C}_g(\varphi,v^i))$$

for $e_j,h_i>0, |e|\ge|h|$ and $\mathbf{N}^{j,i}_{e,h}:=0$ otherwise, we have $N^{j,i}_{e,\nu,\gamma}=\mathbf{N}^{j,i}_{e,\gamma+\nu}$, and so:

$$\varepsilon^j(\varphi\bullet r) = \cdots = \sum_{e_j>0}\left(\sum_{\substack{0<|h|\le|e|\\ o\in\operatorname{supp} h}} \sum_{\substack{\gamma+\nu=h\\ i\in\operatorname{supp}\nu}} N^{j,i}_{e,\nu,\gamma}\nu_i r^*_\gamma r_\nu\right)\mathbf{u}^e =$$

$$\sum_{e_j>0}\left(\sum_{\substack{0<|h|\le|e|\\ i\in\operatorname{supp} h}} \sum_{\substack{\gamma+\nu=h\\ i\in\operatorname{supp}\nu}} \mathbf{N}^{j,i}_{e,\gamma+\nu}\nu_i r^*_\gamma r_\nu\right)\mathbf{u}^e = \sum_{e_j>0}\left(\sum_{\substack{0<|h|\le|e|\\ i\in\operatorname{supp} h}} \mathbf{N}^{j,i}_{e,h} \sum_{\substack{\gamma+\nu=h\\ i\in\operatorname{supp}\nu}} \nu_i r^*_\gamma r_\nu\right)\mathbf{u}^e =$$

$$\sum_{e_j>0}\left(\sum_{\substack{0<|h|\le|e|\\ i\in\operatorname{supp} h}} \mathbf{N}^{j,i}_{e,h}\varepsilon^i_h(r)\right)\mathbf{u}^e. \qquad \square$$

COROLLARY 3.2.6. *Under the hypotheses of Theorem 3.2.5, we have:*

$$\varepsilon_e^j(\varphi \bullet D) = \sum_{\substack{0<|h|\le|e| \\ i\in\operatorname{supp} h}} \mathbf{N}_{e,h}^{j,i} \varepsilon_h^i(D) \quad \forall e \in \Omega, \forall j = 1, \ldots, q.$$

References

[1] F. K. Schmidt and H. Hasse, *Noch eine Begründung der Theorie der höheren Differentialquotienten in einem algebraischen Funktionenkörper einer Unbestimmten. (Nach einer brieflichen Mitteilung von F.K. Schmidt in Jena)* (German), J. Reine Angew. Math. **177** (1937), 215–237, DOI 10.1515/crll.1937.177.215. MR1581557

[2] M. Hazewinkel, *Symmetric functions, noncommutative symmetric functions and quasisymmetric functions. II*, Acta Appl. Math. **85** (2005), no. 1-3, 319–340, DOI 10.1007/s10440-004-5635-z. MR2128927

[3] N. Heerema, *Higher derivations and automorphisms of complete local rings*, Bull. Amer. Math. Soc. **76** (1970), 1212–1225, DOI 10.1090/S0002-9904-1970-12609-X. MR266916

[4] D. Hoffmann and P. Kowalski, *Integrating Hasse-Schmidt derivations*, J. Pure Appl. Algebra **219** (2015), no. 4, 875–896, DOI 10.1016/j.jpaa.2014.05.024. MR3282115

[5] H. Matsumura, *Integrable derivations*, Nagoya Math. J. **87** (1982), 227–245. MR676593

[6] H. Matsumura, *Commutative ring theory*, Cambridge Studies in Advanced Mathematics, vol. 8, Cambridge University Press, Cambridge, 1986. Translated from the Japanese by M. Reid. MR879273

[7] M. Mirzavaziri, *Characterization of higher derivations on algebras*, Comm. Algebra **38** (2010), no. 3, 981–987, DOI 10.1080/00927870902828751. MR2650383

[8] L. Narváez Macarro, *On the modules of m-integrable derivations in non-zero characteristic*, Adv. Math. **229** (2012), no. 5, 2712–2740, DOI 10.1016/j.aim.2012.01.015. MR2889143

[9] L. Narváez Macarro, *On Hasse-Schmidt derivations: the action of substitution maps*, Singularities, algebraic geometry, commutative algebra, and related topics, Springer, Cham, 2018, pp. 219–262, DOI 10.1007/978-3-319-96827-8_10. MR3839798

[10] L. Narváez Macarro, *Rings of differential operators as enveloping algebras of Hasse-Schmidt derivations*, J. Pure Appl. Algebra **224** (2020), no. 1, 320–361, DOI 10.1016/j.jpaa.2019.05.009. MR3986424

[11] L. Narváez Macarro. Hasse–Schmidt modules versus integrable connections. `https://arxiv.org/abs/1903.08985`; arXiv:1903.08985

[12] S. A. Saymeh, *On Hasse-Schmidt higher derivations*, Osaka J. Math. **23** (1986), no. 2, 503–508. MR856904

[13] M. P. Tirado Hernández. Integrable derivations in the sense of Hasse–Schmidt for some binomial plane curves, Rend. Sem. Mat. Univ. Padova, to appear. (`arXiv:1807.10502`).

LUIS NARVÁEZ MACARRO, DEPARTAMENTO DE ÁLGEBRA & INSTITUTO DE MATEMÁTICAS (IMUS), FACULTAD DE MATEMÁTICAS, UNIVERSIDAD DE SEVILLA, CALLE TARFIA S/N, 41012 SEVILLA, SPAIN

Email address: `narvaez@us.es`

Contemporary Mathematics
Volume **742**, 2020
https://doi.org/10.1090/conm/742/14944

On the connectivity of Milnor fiber for mixed functions

Mutsuo Oka

Dedicated to Professor Lê Dũng Tráng for his 70th birthday

Abstract. In this note, we prove the connectivity of the Milnor fiber for a mixed polynomial $f(\mathbf{z}, \bar{\mathbf{z}})$, assuming the existence of a sequence of smooth points of $f^{-1}(0)$ converging to the origin. This result gives also a another proof for the connectivity of the Milnor fiber of a non-reduced complex analytic function which is proved by A. Dimca

1. Introduction

Let $f(\mathbf{z}, \bar{\mathbf{z}}) = \sum_{\nu,\mu} c_{\nu,\mu}\mathbf{z}^\nu\bar{\mathbf{z}}^\mu$ be a mixed polynomial of n-variables $\mathbf{z} = (z_1, \dots, z_n) \in \mathbb{C}^n$ which has a singularity at the origin. We say that f satisfies *Hamm-Lê condition* at the origin if there exists a positive number r_0 such that for any $0 < r_1 \le r_0$, there exists a positive number $\delta(r_1)$ such that the hypersurface $f^{-1}(\eta) \cap B_{r_0}^{2n}$ is non-empty, non-singular and it intersects transversely with the sphere S_r^{2n-1} for any r, η with $r_1 \le r \le r_0$ and $0 \ne |\eta| \le \delta(r_1)$. Here B_r^{2n} is the ball $\{\mathbf{z} \in \mathbb{C}^n \mid \|\mathbf{z}\| \le r\}$ of radius r and S_r^{2n-1} is the boundary sphere of B_r^{2n}. We call such a positive number r_0 a stable Milnor radius. If f is a holomorphic function, Hamm-Lê condition is always satisfied (Hamm-Lê [**H-L**]). For a strongly non-degenerate mixed function, this condition is also satisfied provided $f(\mathbf{z}, \bar{\mathbf{z}})$ is either convenient ([**O2**]) or locally uniformly tame along vanishing coordinate subspaces ([**O3**]). The following assertion is immediate from Hamm-Lê condition and Ehresman's fibration theorem ([**W**]).

Lemma 1.1 ([**H-L, O3**]). *Assume that $f(\mathbf{z}, \bar{\mathbf{z}})$ be a mixed function of n-variables with $n \ge 2$ which satisfies the Hamm-Lê condition. Choose $r_0 > 0$ as above and take arbitrary positive numbers r, r_1, with $0 < r_1 \le r \le r_0$ and take a positive number δ with $\delta \le \delta(r_1)$. Consider the following tubular set and its boundary*

$$E(r,\delta) := \{\mathbf{z} \in B_r^{2n} \mid |f(\mathbf{z}, \bar{\mathbf{z}})| \le \delta\}$$
$$\partial E(r,\delta) := \{\mathbf{z} \in B_r^{2n} \mid |f(\mathbf{z}, \bar{\mathbf{z}})| = \delta\}.$$

Then the mappings

$$f : E(r,\delta) \setminus f^{-1}(0) \to D_\delta \setminus \{0\} \text{ and}$$
$$f : \partial E(r,\delta) \to S_\delta^1$$

2010 *Mathematics Subject Classification.* 14J17, 32S55.
Key words and phrases. Milnor fiber, connectivity.

are locally trivial fibrations where $D_\delta := \{\zeta \mid |\zeta| \le \delta\}$ *and* $S_\delta^1 = \partial D_\delta$ *(the boundary circle). They are homotopically equivalent and their isomorphism classes do not depend on the choice of* r *and* δ.

This is called *the tubular Milnor fibration* of f at the origin and the fiber $f^{-1}(\delta) \cap B_r$ is called the Milnor fiber at the origin. It is known that the tubular Milnor fibration is equivalent to the spherical Milnor fibration (see [**M, O2**])

$$\varphi : S_r \setminus K \to S^1, \quad \varphi(z) = f(\mathbf{z}, \bar{\mathbf{z}})/|\mathbf{f}(\mathbf{z}, \bar{\mathbf{z}})|$$

2. Statement of the result

2.1. Connectivity of Milnor fibers of holomorphic functions. We first recall basic facts about the connectivity of the Milnor fibers of holomorphic functions. Assume that $f : (U, \mathbf{0}) \to (\mathbf{C}, 0)$ be a holomorphic function with a singularity at the origin where U is an open neighborhood of the origin. A fundamental result for the connectivity is

THEOREM 2.1 (Milnor [**M**]). *The Milnor fiber* F *has a homotopy type of an* $(n-1)$*-dimensional CW-complex. If further the origin is an isolated singularity,* F *is* $(n-2)$*-connected.*

Let $\varphi : S_r^{2n-1} \setminus K \to S^1$ be the spherical Milnor fibration. The proof depends on Morse theory. Milnor proved that the fiber has a homotopy type of an $n-1$ dimensional CW-complex, by showing that the index of a suitable Morse function is less than or equal to $n-1$. Secondly, if the origin is an isolated singularity, $\bar{F}$ is a manifold with boundary and the inclusion $F \subset \bar{F}$ is a homotopy equivalence in the sphere S_r^{2n-1} and the vanishing $H_j(F) = 0$ for $0 < j < n-1$ follows from the Alexander duality and the homotopy equivalence $F_\pi \subset S_r^{2n-1} \setminus \bar{F}$ where $F_\pi = \varphi^{-1}(-1)$. The simply connectedness of F for $n \ge 3$ follows from handle body argument.

Kato and Matsumoto further generalized this assertion as follows.

THEOREM 2.2 (Kato and Matsumoto [**K-M**]). *Assume that* s *is the dimension of the critical points locus at the origin. Then* F *is* $(n-s-2)$*-connected*

Their proof depends on the above result of Milnor and an inductive argument using Whitney stratification.

2.2. Mixed functions. For mixed functions, no similar connectivity statement is known. A main reason is that the tangent space of a mixed hypersurface has no complex structure and Morse function argument does not help so much as in the holomorphic case.

The following is our result which is a first step for the connectivity of the Milnor fiber of a mixed function. For the proof, we use an elementary but completely different viewpoint which is nothing to do with Morse functions. We use a one-parameter group of diffeomorphisms which contains the monodromy map.

THEOREM 2.3. *Assume that* $f(\mathbf{z}, \bar{\mathbf{z}})$ *is a mixed function of* n*-complex variables* $\mathbf{z} = (z_1, \dots, z_n)$ *with* $n \ge 2$ *which satisfies Hamm-Lê condition at the origin and take a stable Milnor radius* r_0 *and assume that there exists a mixed smooth point* $\mathbf{w} \in f^{-1}(0)$ *with* $\|\mathbf{w}\| < r_0$ *and the sphere of radius* $\|\mathbf{w}\|$ *intersect* $f^{-1}(0)$ *transversely at* $\mathbf{w}$*. We assume also that* $\dim_{\mathbb{R}} f^{-1}(0) = 2n-2$*. Then the Milnor fiber is connected.*

REMARK 2.4. Suppose that f is a holomorphic function and s is the dimension of the critical points locus of f at the origin. If f is non-reduced, then $s = n-1$ and Theorem 2.2 says nothing about the connectivity.

2.3. Proof of Theorem 2.3. Let r_0 be a positive number which satisfies the Hamm-Lê condition. Take a mixed smooth point $\mathbf{w} \in f^{-1}(0)$ with $\|\mathbf{w}\| < r_0$. Put $r := \|\mathbf{w}\|$. Fix positive numbers r_1 with $r_1 < r$ and $\delta \le \delta(r_1)$ and we consider the Milnor fibration

$$(\star) \quad f : \partial E(r_0, \delta) \to S^1_\delta.$$

Let $F_\theta := f^{-1}(\delta e^{i\theta}) \cap B_{r_0}$ for $\theta \in \mathbb{R}$ be the Milnor fiber over $\delta e^{i\theta}$. We assume also the sphere with radius r, S^{2n-1}_r intersects transversely with the hypersurface $f^{-1}(0)$ at the smooth point $\mathbf{w}$. For the proof, we use a certain one-parameter family of diffeomorphisms associated with the tubular Milnor fibration $(\star)$. Recall that a one-parameter family of the characteristic diffeomorphisms h_θ, $\theta \in \mathbb{R}$ are constructed by integrating a given horizontal vector field $\mathcal{V}$. Here a vector field $\mathcal{V}$ is called *a horizontal vector field* on $\partial E(r_0, \delta)$ if it satisfies the following property:

$$Tf_p(\mathcal{V}(p)) = \frac{\partial}{\partial \theta}(f(p)), \quad p \in \partial E(r_0, \delta)$$

where $Tf_p : T_p\partial E(r_0, \delta) \to T_{f(p)}S^1_\delta$ is the tangential map and furthermore $\mathcal{V}$ is tangent to $S_{r_0} \cap \partial E(r_0, \delta)$ on the boundary and $\frac{\partial}{\partial\theta}$ is the unit angular vector field along S^1_δ. In other word, $\frac{\partial}{\partial\theta}$ is the tangent vector of the curve $t \mapsto \delta e^{it}$. As $\partial E(r_0, \delta)$ is a compact manifold with boundary, there is an integral $\varphi : \partial E(r_0, \delta) \times \mathbb{R} \to \partial E(r_0, \delta)$ such that $\varphi(p, 0) = p$ and $\varphi(p, t)$, $-\infty < t < \infty$ is the integral of $\mathcal{V}$ starting at p for $t = 0$. Let $h_\theta : \partial E(r_0, \delta) \to \partial E(r_0, \delta)$, $\theta \in \mathbb{R}$ be the corresponding one parameter family of characteristic diffeomorphisms, which are defined by $h_\theta(p) = \varphi(p, \theta)$. Note that $\{h_\theta\}$ satisfy the property

$$h_\theta(F_\eta) = F_{\eta+\theta}, \quad \eta, \theta \in \mathbf{R}.$$

They also satisfy the equalities

$$h_0 = \mathrm{id}, \quad h_\theta \circ h_\xi = h_{\theta+\xi}, \quad \theta, \xi \in \mathbb{R}.$$

Using one-parameter family $\{h_\theta, \theta \in \mathbb{R}\}$, the monodromy map $h : F \to F$ is given by the restriction $h_{2\pi}|F$ with $F = F_0$.

Now we construct $\mathcal{V}$ more carefully. We take a local real-analytic chart U with local coordinates $(x_1, y_1, \dots, x_n, y_n)$ centered at the given smooth point $\mathbf{w} \in f^{-1}(0)$ such that $f(x_1, y_1, \dots, x_n, y_n) = x_n + iy_n$. We take δ sufficiently small and we take a normal disc D of $f^{-1}(0)$ centered at $\mathbf{w}$ so that $|f(\mathbf{z})| = \delta$ for any $\mathbf{z} \in \partial D$ and thus $\partial D \subset \partial E(r_0, \delta)$. In the above coordinates, we can assume $D = \{(0, \dots, 0, x_n, y_n) \,|\, x_n^2 + y_n^2 \le \delta^2\}$. By the transversality assumption, we may assume that $\mathcal{V}$ is tangent to ∂D which implies that $h_{2\pi}(\mathbf{z}) = \mathbf{z}$ for $\mathbf{z} \in \partial D$. Note that the family of tubular neighborhoods $\{E(r, \delta) | \delta < \delta(r)\}$ and the family of disk neighborhoods $\{B^{2n}_s, s > 0\}$ are cofinal neighborhood systems of the origin. Thus $E(r, \delta)$ is contractible and $E \setminus f^{-1}(0)$ is connected by the assumption $\mathrm{codim}_{\mathbb{R}}\, \mathrm{f}^{-1}(0) = 2$, $\partial E(r, \delta)$ is also connected, as $E(r_0, \delta) \setminus f^{-1}(0)$ and $\partial E(r_0, \delta)$ are homotopy equivalent.

Now we are ready to prove the connectivity of the Milnor fiber F_0. Fix a point $p \in \partial D \cap F_0$ and take an arbitrary point $q \in F_0$. As $\partial E(r, \delta)$ is connected, we can find a path $\sigma : [0, 1] \to \partial E(r_0, \delta)$ such that $\sigma(0) = q$ and $\sigma(1) = p$. We will show that $q \in F_0$ can be joined to p by a path in the fiber F_0, modifying the path σ.

Consider first the closed loop $f\circ\sigma : (I,\{0,1\})\to(S^1_\delta,\delta)$ and let m be the rotation number, that is

$$m=\frac{1}{2\pi}\int_0^1 d\arg f(\sigma(\theta)).$$

Let $\omega : I\to\partial D$ be the clockwise rotation m-times along the boundary of D starting at p. Consider the path σ' which is given as the composition of paths $\sigma' := \sigma\cdot\omega : I\to\partial E(r,\delta)$ and define a function $\psi:[0,1]\to\mathbb{R}$ by

$$\psi(t)=\int_0^t d\arg f(\sigma'(t))dt.$$

Note that $\psi(t)\equiv\arg f(\sigma'(t))$ modulo 2π and $\psi(0)=\psi(1)=0$. This follows from the observation that the rotation number of $f\circ\sigma'$ is zero by the definition of ω. Now we can deform σ' using characteristic diffeomorphisms h_t, $t\in\mathbb{R}$ into a path in the fiber F_0 as follows. Define a modified path

$$\hat\sigma(t) := h_{-\psi(t)}(\sigma'(t)),\quad 0\le t\le 1.$$

Then we have the equality

$$\arg f(\hat\sigma(t)) = -\psi(t)+\arg f(\sigma'(t))\equiv 0,\quad 0\le t\le 1.$$

Thus the deformed path $\hat\sigma$ is entirely included in F_0. By the construction, $\hat\sigma(0)=q$ and $\hat\sigma(1)=p$ and $\hat\sigma$ is the path which connect q and p in F_0. □

2.4. Generalization of Main Theorem for holomorphic functions. The following is known by A. Dimca [**D**]. We give another direct proof of this assertion, using a similar argument as in the proof of Theorem 2.3.

THEOREM 2.5. (Prpoposition 2.3, [**D**]) *Assume that f is a germ of holomorphic function at the origin of $\mathbb{C}^n$, $n\ge 2$ with $f(\mathbf{0})=0$ and assume that f is factored as $f_1^{n_1}\cdots f_r^{n_r}$ where $f_1,\dots,f_r$ are irreducible in $\mathcal{O}_n$ and mutually coprime. Put $n_0=\gcd(n_1,\dots,n_r)$. Then Milnor fiber of f at the origin is connected if and only if $n_0=1$.*

Here $\mathcal{O}_n$ is the ring of germs of holomorphic functions at the origin.

PROOF. Let F be the Milnor fiber of f. We can write $f=g^{n_0}$ with $g=f_1^{n_1/n_0}\cdots f_r^{n_r/n_0}$ and F is diffeomorphic to disjoint sum of n_0 copies of the Milnor fibers of g. Thus F is not connected if $n_0\ge 2$. Assume $n_0=1$. Take a stable radius r_0 which satisfies Hamm-Lê condition. For each $1\le i\le r$, we consider the reduced irreducible component $V_i=\{f_i=0\}$ and take a non-singular point $p_i\in V_i\setminus\bigcup_{j\ne i}V_j$ with $\|p_i\|<r_0$. Choose a positive number $r<r_0$ with $r\le\|p_i\|, i=1,\dots,r$ and we consider Milnor fibration

$$f:\partial E(r_0,\delta)\to S^1_\delta,\quad \delta\le\delta(r)$$

We assume that in a sufficiently small neighborhood U_i of p_i, any analytic branch of f^{1/n_i} is a well-defined single-valued function. Taking a one branch $\widetilde{f_i}$ and we take $\widetilde{f_i}$ as the last coordinate function of a complex analytic coordinate system $\mathbf{z}_i=(z_{i1},\dots,z_{in})$ in U_i i.e., $\widetilde{f_i}(\mathbf{z}_i)=z_{in}$. Consider a normal disk at p_i which is defined by $D_i=\{(0,\dots,0,z_{in})\,|\,|z_{in}|\le\delta^{1/n_i}\}$ where we assume that δ is sufficiently small so that $D_i\subset U_i$. Note that $\widetilde{f_i}$ is locally written as $\widetilde{f_i}=f_i\cdot u_i$ where u_i is a unit in U_i and $f(\partial D_i)=S^1_\delta$ and $\partial D_i\subset\partial E(r,\delta)$. Construct a one parameter family of characteristic diffeomorphisms h_t, $t\in\mathbb{R}$ as before such that ∂D_i is stable by h_t

and $h_t|\partial D_i$ is the rotation of angle t/n_i, under the identification $\widetilde{f_i} : \partial D_i \to S^1_{\delta^{1/n_i}}$. (Note that $f : \partial D_i \to S^1_\delta$ is the n_i rotation.) $F \cap \partial D_i$ can be identified with n_i-th roots of δ and we put them as $F \cap \partial D_i = \{p_{i,0}, \dots, p_{i,n_i-1}\}$ so that the monodromy $h := h_{2\pi}$ acts simply as a cyclic permutation $p_{i,j} \mapsto p_{i,j+1}$. Using a similar discussion as in the proof of Theorem 2.3, for a given $q \in F$, we can connect q to some point in $F \cap D_j$ for any j. To see this, we first take a path σ so that $\sigma(0) = q$ and $\sigma(1) = p_{j,0}$, then in the argument of the proof of Theorem 2.3, we simply replace ω by m/n_j rotation in the clockwise direction along ∂D_j and do the same argument, where m is the rotation number of $f \circ \sigma$. Note that ω need not be a closed loop but the image of ω by f is a closed loop and it gives $-m$ rotation. Thus the proof is reduced to show that the points $\{p_{1,0}, \dots, p_{1,n_1-1}\}$ are in the same connected component of F.

In particular taking $q = p_{1,0}$, we can find a path ℓ_j in F which connects $p_{1,0}$ to some p_{j,ν_j} for any j, $j = 2, \dots, r$. Let $\ell_j^a = h^a(\ell_j)$ the image of ℓ_j by h^a. Then it connects $p_{1,a}$ to p_{j,ν_j+a} in F. Now we consider the image of ℓ_j^a, $a = 0, \dots, n_1 - 1$ under h^{n_j} which fixes p_{j,ν_j+a} but the other end $p_{1,a}$ of ℓ_j^a goes to p_{1,n_j+a}. As F is stable by the monodromy map $h = h_{2\pi}$, this image is also a path in F. Thus $p_{1,a}$ and $p_{1,n_j+a} = h^{n_j}(p_{1,a})$ are connected by the path $\ell_j^a \cdot (h^{n_j} \circ (\ell_j^a)^{-1})$ in F for any a. See Figure 1 for $a = 0$. As we have assumed $\gcd(n_1, \dots, n_r) = 1$, there exist integers $a_1, \dots, a_r$ so that we can write $1 = \sum_i^r a_i n_i$. Then

$$p_{1,1} = h(p_{i,0}) = h^{\sum_{i=1}^r a_i n_i}(p_{1,0}) = h^{a_r n_r}(\cdots(h^{a_2 n_2}(p_{1,0})\cdots)).$$

Put $p_{1,\mu_2} := h^{a_2 n_2}(p_{1,0})$ and $p_{1,\mu_{j+1}} := h^{n_{j+1}a_{j+1}}(p_{1,\mu_j})$ for $j = 2, \dots, r-1$. Then points $p_{1,\mu_2}, \dots, p_{1,\mu_r} = p_{1,1}$ are in the same connected component of $p_{1,0}$ in F. Thus $p_{1,1}$ and $p_{1,0}$ are in the same component. Repeating the same argument, we conclude that $p_{1,0}, \dots, p_{1,n_1-1}$ are all in the same connected component of F. Combining the above observation, this proves that F is connected.

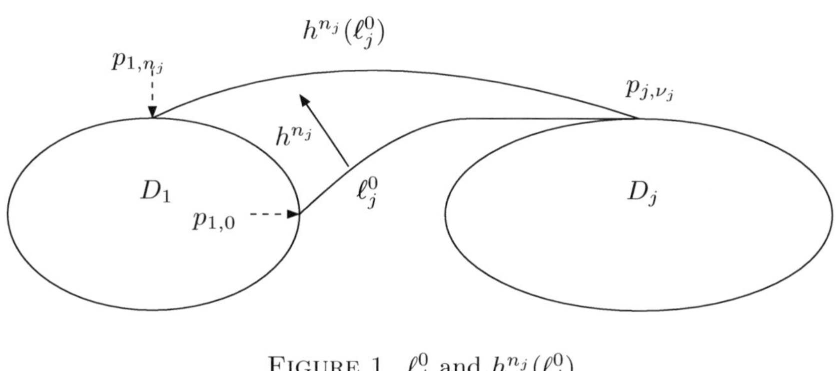

FIGURE 1. ℓ_j^0 and $h^{n_j}(\ell_j^0)$

□

REMARK 2.6. The special case $f = z_1^{a_1} \cdots z_k^{a_k}$ has been considered in Example (3.7), [**O1**]. For mixed function case, the above argument does not work. There does not exist any correspondence between irreducible components of $f^{-1}(0)$ as real algebraic sets and the irreducible factors of $f(\mathbf{z}, \bar{\mathbf{z}})$ as a polynomial in $\mathbb{C}[\mathbf{z}, \bar{\mathbf{z}}]$ or a germ of an analytic function of $2n$ variables. For example, let $f(z, w, \bar{z}, \bar{w})$ be the strongly homogeneous mixed polynomial of two variables which come from the Rhie's Lens equation $\varphi_n = 0$,

$$\varphi_n(z) := \bar{z} - \frac{z^{n-2}}{z^{n-1} - a^{n-1}} - \frac{\varepsilon}{z}, \; n \geq 2$$

where ε, a are sufficiently small positive numbers $0 < \varepsilon \ll a \ll 1$. f is obtained as the homogenization of the numerator of φ_n. Then f is irreducible but $f^{-1}(0)$ has $5n-5$ line components in $\mathbb{C}^2$([**O4**, **R**]).

3. Application

3.1. Convenient strongly non-degenerate mixed functions. Let $f(\mathbf{z}, \bar{\mathbf{z}})$ be a strongly non-degenerate convenient mixed polynomial. Then the Hamm-Lê condition is satisfied (Lemma 28, [**O2**]). Furthermore the mixed hypersurface $V = f^{-1}(0)$ has an isolated singularity at the origin (Corollary 20, [**O2**]). Thus the assumption of Theorem 2.3 is satisfied and we have

THEOREM 3.1. *Let $f(\mathbf{z}, \bar{\mathbf{z}})$ be a strongly non-degenerate convenient mixed function and $n \geq 2$. Then the Milnor fiber is connected.*

3.2. Non-convenient mixed polynomials. Let $f(\mathbf{z}, \bar{\mathbf{z}})$ be a strongly non-degenerate mixed polynomial which is not convenient. In general, such a mixed hypersurface defined by f has a non-isolated singular locus. Let I be a non-empty subset of $\{1, \dots, n\}$. We use the following notations.

$$\mathbb{C}^I = \{\mathbf{z} \,|\, z_j = 0, \, \forall j \notin I\},$$
$$\mathbb{C}^{*I} = \{\mathbf{z} \,|\, z_j = 0 \iff j \notin I\}.$$

We say $\mathbb{C}^I$ is *a vanishing coordinate subspace for* f if the restriction of f to the coordinate subspace $\mathbb{C}^I$, denoted by f^I, is identically zero. Otherwise, we say $\mathbb{C}^I$ *a non-vanishing coordinate subspace.* Put $V^\sharp$ be the union of $V \cap \mathbb{C}^{*I}$ for all I such that $\mathbb{C}^I$ is a non-vanishing coordinate subspace. We know that $V^\sharp$ is mixed non-singular as a germ at the origin (Theorem 19, [**O2**]). We say that f *is uniformly locally tame on the vanishing coordinate subspace* $\mathbb{C}^I$ if there exists a positive number $\varepsilon > 0$ and for any non-negative weight vector $P = (p_1, \dots, p_n) \in \mathbb{Z}^n_{\geq 0}$ with $I(P) = I$ where $I(P) := \{i \,|\, p_i = 0\}$, the face function f_P is strongly non-degenerate as a function of $\{z_j \,|\, j \notin I\}$ for any fixed $\mathbf{z}_I = (z_i)_{i \in I} \in \mathbb{C}^{*I}$ with $\sum_{i \in I} |z_i|^2 \leq \varepsilon$. A strongly non-degenerate mixed function which is uniformly locally tame along any vanishing coordinate subspaces satisfies Hamm-Lê condition (Proposition 11, [**O3**]). Thus we have

THEOREM 3.2. *Assume that $f(\mathbf{z}, \bar{\mathbf{z}})$ is a strongly non-degenerate mixed function which is uniformly locally tame along vanishing coordinate subspaces and $n \geq 2$. We assume also that $V^\sharp$ is a non-empty germ of mixed variety. Then the Milnor fiber is connected.*

A similar but weaker result is proved for a strongly mixed homogeneous polynomial ([**O5**]). As a next working problem, we also propose a conjecture about the fundamental groups of the Milnor fiber.

References

[D] A. Dimca, *Singularities and topology of hypersurfaces*, Universitext, Springer-Verlag, New York, 1992. MR1194180

[H-L] H. A. Hamm and Lê Dũng Tráng, *Un théorème de Zariski du type de Lefschetz*, Ann. Sci. École Norm. Sup. (4) **6** (1973), 317–355. MR0401755

[K-M] M. Kato and Y. Matsumoto, *On the connectivity of the Milnor fiber of a holomorphic function at a critical point*, Manifolds—Tokyo 1973 (Proc. Internat. Conf., Tokyo, 1973), Univ. Tokyo Press, Tokyo, 1975, pp. 131–136. MR0372880

[M] J. Milnor, *Singular points of complex hypersurfaces*, Annals of Mathematics Studies, No. 61, Princeton University Press, Princeton, N.J.; University of Tokyo Press, Tokyo, 1968. MR0239612

[O1] M. Oka, *Non-degenerate complete intersection singularity*, Actualités Mathématiques. [Current Mathematical Topics], Hermann, Paris, 1997. MR1483897

[O2] M. Oka, *Non-degenerate mixed functions*, Kodai Math. J. **33** (2010), no. 1, 1–62, DOI 10.2996/kmj/1270559157. MR2732230
38,

[O3] M. Oka, *On Milnor fibrations of mixed functions, a_f-condition and boundary stability*, Kodai Math. J. **38** (2015), no. 3, 581–603, DOI 10.2996/kmj/1446210596. MR3417523
in Singularities and Foliations. Geometry, Topology and Applications,

[O4] M. Oka, *On the roots of an extended Lens equation and an application*, Singularities and foliations. geometry, topology and applications, Springer Proc. Math. Stat., vol. 222, Springer, Cham, 2018, pp. 489–511. MR3779572

[O5] M. Oka, *Smooth mixed projective curves and a conjecture*, J. Singul. **18** (2018), 329–349. MR3899545

[R] S.H. Rhie. *n-point Gravitational Lenses with 5(n-1) Images.* arXiv:astro-ph/0305166, May 2003.

[W] J. A. Wolf, *Differentiable fibre spaces and mappings compatible with Riemannian metrics*, Michigan Math. J. **11** (1964), 65–70. MR0159285

Department of Mathematics, Tokyo University of Science, Kagurazaka 1-3, Shinjuku-ku, Tokyo 162-8601

Email address: oka@rs.kagu.tus.ac.jp

Contemporary Mathematics
Volume **742**, 2020
https://doi.org/10.1090/conm/742/14945

Some classes of homeomorphisms that preserve multiplicity and tangent cones

J. Edson Sampaio

Dedicated to Professor Lê Dũng Tráng on the occasion of his 70th birthday

Abstract. In this paper we present some applications of A'Campo-Lê's Theorem and we study some relations between Zariski's Questions A and B. It is presented some classes of homeomorphisms that preserve multiplicity and tangent cones of complex analytic sets. Moreover, we present a class of homeomorphisms that has the multiplicity as an invariant when we consider right equivalence and this class contains many known classes of homeomorphisms that preserve tangent cones. In particular, we present some effective approaches to Zariski's Question A. We show a version of these results looking at infinity. Additionally, we present some results related with Nash modification and Lipschitz Geometry.

1. Introduction

In 1973, N. A'Campo in [**1**] and Lê D. T. in [**17**] proved separately the following result.

Theorem 1.1 (A'Campo-Lê's Theorem). *Let $f, g\colon (\mathbb{C}^n,0) \to (\mathbb{C},0)$ be two complex analytic functions. Suppose that $V(f)$ is smooth at 0. If there is a homeomorphism $\varphi\colon (\mathbb{C}^n, V(f),0) \to (\mathbb{C}^n, V(g),0)$, then $V(g)$ is also smooth at 0.*

The main aim of the this paper is to give some applications of this theorem to Zariski multiplicity question and to Lipschitz Geometry (see Section 2). Thus, we are going to describe these applications.

Let $f\colon (\mathbb{C}^n,0) \to (\mathbb{C},0)$ be the germ of a reduced holomorphic function at the origin. We recall that *the multiplicity of* $V(f)$ at the origin, denoted by $m(V(f),0)$, is defined as follows: we write

$$f = f_m + f_{m+1} + \cdots + f_k + \cdots$$

where each f_k is a homogeneous polynomial of degree k and $f_m \neq 0$. Then, $m(V(f),0) := m$. In this case, the tangent cone of $V(f)$ at the origin is $C(V(f),0) = V(f_m)$.

2010 *Mathematics Subject Classification.* Primary MSC 14B05 - MSC 32S50.

Key words and phrases. Multiplicity - Tangent cones - Zariski's Problems.

The author was supported by the ERCEA 615655 NMST Consolidator Grant and also by the Basque Government through the BERC 2018-2021 program and by the Spanish Ministry of Science, Innovation and Universities: BCAM Severo Ochoa accreditation SEV-2017-0718.

In 1971, O. Zariski in [**29**] proposes many questions and the most known among them is the following.

Question A. Let $f, g\colon (\mathbb{C}^n, 0) \to (\mathbb{C}, 0)$ be two complex analytic functions. If there is a homeomorphism $\varphi\colon (\mathbb{C}^n, V(f), 0) \to (\mathbb{C}^n, V(g), 0)$, is it true that $m(V(f), 0) = m(V(g), 0)$?

In order to know more about this question, you can see [**9**].

Let us remark that since $V(f)$ is smooth at 0 if and only if $m(V(f), 0) = 1$, then A'Campo-Lê's Theorem gives a positive answer to Question A when $m(V(f), 0) = 1$.

Let $Bl_0(\mathbb{C}^n) = \{(x, [v]) \in \mathbb{C}^n \times \mathbb{C}P^{n-1}; x \wedge v = 0\}$ and $\beta\colon Bl_0(\mathbb{C}^n) \to \mathbb{C}^n$ be the projection onto $\mathbb{C}^n$, where $x \wedge v = 0$ means that there exists $\lambda \in \mathbb{C}$ such that $x = \lambda v$. If $f\colon (\mathbb{C}^n, 0) \to (\mathbb{C}, 0)$ is a complex analytic function, we define $Bl_0(V(f)) = \overline{\beta^{-1}(V(f) \setminus \{0\})}$ and $E_0(f) = Bl_0(V(f)) \cap (\{0\} \times \mathbb{C}P^{n-1})$. Remark that $E_0(f) = \{0\} \times \mathbb{P}C(V(f), 0)$, where $\mathbb{P}C(V(f), 0)$ is the projectivized tangent cone of $V(f)$.

Thus, another question asked in [**29**] was the following.

Question B. Let $f, g\colon (\mathbb{C}^n, 0) \to (\mathbb{C}, 0)$ be two complex analytic functions. If there is a homeomorphism $\varphi\colon (\mathbb{C}^n, V(f), 0) \to (\mathbb{C}^n, V(g), 0)$, is there a homeomorphism $h\colon E_0(f) \to E_0(g)$ such that for each $p \in E_0(f)$ the germs $(Bl_0(V(f)), p)$ and $(Bl_0(V(g)), h(p))$ are homeomorphic?

We would like to consider also the following versions of Questions A and B:

Let $f, g\colon (\mathbb{C}^n, 0) \to (\mathbb{C}, 0)$ be two reduced complex analytic functions. Suppose that there is a homeomorphism $\varphi\colon (\mathbb{C}^n, 0) \to (\mathbb{C}^n, 0)$ such that $f = g \circ \varphi$.

Question A'. Is it true that $m(V(f), 0) = m(V(g), 0)$?

Question B'. Suppose that $C(V(f), 0)$ or $C(V(g), 0)$ is a linear subspace. Is there a homeomorphism between $(C(V(f), 0), 0)$ and $(C(V(g), 0), 0)$ or between $E_0(f)$ and $E_0(g)$?

Let me remark that by using Corollary in [**15**] together with Corollary 2 in [**21**], we obtain that, in the case of functions with isolated singularities, Question A' is equivalent to Question A. Moreover, even in the case of isolated singularities, Question A is still an open problem. However, J. Fernández de Bobadilla in [**12**] showed that Question B has a negative answer and in [**2**] the authors presented complex analytic functions $f, g\colon (\mathbb{C}^3, 0) \to (\mathbb{C}, 0)$ with isolated singularities such that there is a homeomorphism $\varphi\colon (\mathbb{C}^3, 0) \to (\mathbb{C}^3, 0)$ satisfying $f = g \circ \varphi$, but there is no homeomorphism $\tilde{h}\colon (\mathbb{C}^3, 0) \to (\mathbb{C}^3, 0)$ such that $\tilde{h}(C(V(f), 0)) = C(V(g), 0)$. In [**2**], the authors used A'Campo-Lê's Theorem to obtain that a such $\tilde{h}$ preserves the singular points of the tangent cones and, then, arrived in a contradiction. However, a homeomorphism $h\colon (C(V(f), 0), 0) \to (C(V(g), 0), 0)$ does not need to preserve singular points when h is not a restriction of a homeomorphism $\tilde{h}\colon (\mathbb{C}^3, 0) \to (\mathbb{C}^3, 0)$. Here, we show that the example presented in [**2**] still gives a negative answer to Question B' when we do not require that $C(V(f), 0)$ or $C(V(g), 0)$ need to be a linear subspace (see Proposition 2.8), though that the Question B' has a positive answer when $n = 2$.

As a first application of A'Campo-Lê's Theorem, we show that Questions A' and B' are related. In fact, we prove by using A'Campo-Lê's Theorem the following.

THEOREM 1.2. *If Question B' has a positive answer then Question A' has a positive answer as well.*

Another problem related with the above questions is to know if the relative multiplicities are topological invariants. To be more precise, let us define which are the relative multiplicities. Let $f\colon (\mathbb{C}^n,0) \to (\mathbb{C},0)$ be a reduced complex analytic function and

$$f = f_m + f_{m+1} + \cdots + f_k + \cdots$$

is as before. We have the decomposition of f_m in irreducible polynomials

$$f_m = h_1^{k_1} \cdots h_r^{k_r}.$$

Thus, we define the relative multiplicity of $V(f)$ (along of $V(h_i)$) by $k_{V(f)}(V(h_i))\colon = k_i$. In particular, we obtain that $m(V(f),0) = \sum k_{V(f)}(V(h_i))m(V(h_i),0)$.
Let $f,g\colon (\mathbb{C}^n,0) \to (\mathbb{C},0)$ be two reduced complex analytic functions. Let $X_1,...,X_r$ (resp. $Y_1,...,Y_s$) be the irreducible components of $C(V(f),0)$ (resp. $C(V(g),0)$). Suppose that there is a homeomorphism $\varphi\colon (\mathbb{C}^n,0) \to (\mathbb{C}^n,0)$ such that $f = g \circ \varphi$.

Question B$_{Rel}$. Is there a bijection $\sigma\colon \{1,...,r\} \to \{1,...,s\}$ such that $k_{V(f)}(X_i) = k_{V(g)}(Y_{\sigma(i)})$, for all $i \in \{1,...,r\}$?

Question B'$_{Rel}$. Suppose that $C(V(f),0)$ or $C(V(g),0)$ is a linear subspace. Is it true that $r = s = 1$ and $k_{V(f)}(X_1) = k_{V(g)}(Y_1)$?
Question B$_{Rel}$ was also asked in January 2017 on Mathoverflow.net [**26**] and it already was answered in many particular cases, for example, the answer is known to be yes in the following special cases:

- if $n = 2$ in [**28**];
- if φ is a real analytic diffeomorphism in [**16**];
- if φ is a subanalytic bi-Lipschitz homeomorphism in [**27**];
- if φ is a bi-Lipschitz homeomorphism in [**10**];
- if φ is a blow-spherical homeomorphism in [**24**].

So, as a second application of A'Campo-Lê's Theorem, we show the following.

THEOREM 1.3. *If Question B'$_{Rel}$ has a positive answer then Question A' also has a positive answer.*

In Proposition 2.8, we show that Question B$_{Rel}$ has a negative answer.
We present also an application of A'Campo-Lê's Theorem to Lipschitz Geometry. Let us remind the following.

DEFINITION 1.4. Let $X \subset \mathbb{R}^n$ and $Y \subset \mathbb{R}^m$. A mapping $f\colon X \to Y$ is called **Lipschitz** if there exists $\lambda > 0$ such that is

$$\|f(x_1) - f(x_2)\| \le \lambda\|x_1 - x_2\|, \quad \text{for all } x_1, x_2 \in X.$$

A Lipschitz mapping $f\colon X \to Y$ is called **bi-Lipschitz** if its inverse mapping exists and is Lipschitz. When there exists an bi-Lipschitz mapping $f\colon X \to Y$ we say that X and Y are bi-Lipschitz homeomorphic.

In [**19**], the authors presented several notions related with the notion of Lipschitz manifold and among them, they had the following.

DEFINITION 1.5. A subset $M \subset \mathbb{R}^N$ is called a k-dimensional **Lipschitz submanifold** if for every $x_0 \in M$ there exist an open set $U \subset \mathbb{R}^N$ and a bi-Lipschitz mapping $\varphi\colon U \to \varphi(U) \subset \mathbb{R}^N$ such that $x_0 \in U$, $\varphi(M \cap U) = \{(x_1,...,x_N) \in \varphi(U);\, x_{k+1} = ... = x_N = 0\}$ and $\varphi(x_0) = 0$.

DEFINITION 1.6. A subset $M \subset \mathbb{R}^N$ is called a k-dimensional **parametric Lipschitz manifold** if for every $x_0 \in M$ there exist an open neighborhood $U \subset \mathbb{R}^N$ of x_0, an open set $W \subset \mathbb{R}^k$ and a bi-Lipschitz mapping $\varphi\colon W \to M \cap U$.

Moreover, the authors mention in ([**19**], page 11) that they could not prove or disprove that a parametric Lipschitz manifold is a Lipschitz submanifold. Here, by using A'Campo-Lê's Theorem, we disprove this.

PROPOSITION 1.7. *There are parametric Lipschitz manifolds that are not Lipschitz submanifolds.*

In [**6**], the authors presented germs at the origin of real semialgebraic sets (of dimension two) $S_1, S_2 \subset \mathbb{R}^3$ such that they are bi-Lipschitz homeomorphic but there is no bi-Lipschitz homeomorphism $\phi\colon (\mathbb{R}^3,0) \to (\mathbb{R}^3,0)$ such that $\phi(S_1) = S_2$.

Finally, in Section 3, we consider one version more of Question B. In fact, we consider a version of Question B by considering Nash modifications instead of the blowing up of the origin. Let us be more precise with this.

If $\mathbb{K}$ is $\mathbb{R}$ or $\mathbb{C}$ and $X \subset \mathbb{K}^n$ is an analytic subset, we denote by $\mathrm{Sing}(X)$ to be the singular points of X and by $\mathrm{Reg}(X) := X \setminus \mathrm{Sing}(X)$ to be the regular points of X.

DEFINITION 1.8. Let $X \subset \mathbb{K}^n$ be an analytic subset of pure dimension r and $G_r^n(\mathbb{K})$ be the Grassmannian of r-planes in $\mathbb{K}^n$. We define a map $\nu\colon \mathrm{Reg}(X) \to X \times G_r^n(\mathbb{K})$, by $\nu(x) := (x, T_xX)$, where T_xX is the tangent space of X at x. The closure of the image of ν, denoted by X'_{Nash}, is called the **Nash modification of** X and the mapping induced by first projection $\eta_X : X'_{Nash} \to X$ is called the **Nash mapping of** X.

Thus, we have the following.

Question B$_{Nash}$. Suppose that two germs of analytic sets $(X,0)$ and $(Y,0)$ are bi-Lipschitz homeomorphic. Is there a homeomorphism between X'_{Nash} and Y'_{Nash} sending $\eta_X^{-1}(\mathrm{Sing}(X))$ over $\eta_Y^{-1}(\mathrm{Sing}(Y))$?

By proving in Proposition 3.2 that the cone over a LNE subset of $\mathbb{S}^n$ is also LNE (see definition 3.1), we show the following.

PROPOSITION 1.9. *Question B_{Nash} has a negative answer when we consider real analytic sets.*

2. Some applications of A'Campo-Lê's Theorem

2.1. Invariance of the multiplicity by right equivalence. In this Subsection, among other things we prove that a positive answer to Question B' or B'$_{Rel}$ implies in a positive answer to Question A'.

Let me introduce a new class of homeomorphisms that essentially generalizes bi-Lipschitz equivalence, weak directional equivalence, differentiable equivalence and C^1 equivalence.

DEFINITION 2.1. Let $\varphi\colon (\mathbb{C}^n,0)\to(\mathbb{C}^n,0)$ be a homeomorphism. We say that φ preserves linear tangent cones if for any complex analytic functions $f,g\colon(\mathbb{C}^n,0)\to(\mathbb{C},0)$ such that $f=g\circ\varphi$ and $C(V(f),0)$ or $C(V(g),0)$ is a linear subspace of $\mathbb{C}^n$, we have that there exists a homeomorphism ϕ between $(C(V(f),0),0)$ and $(C(V(g),0),0)$ or between $E_0(f)$ and $E_0(g)$.

DEFINITION 2.2. For each $n\in\mathbb{N}$, let $\mathcal{H}_n$ be the collection of all germs of homeomorphisms from $(\mathbb{C}^n,0)$ to $(\mathbb{C}^n,0)$ and let $\mathcal{S}\colon\mathcal{H}\to\mathcal{H}$ be the mapping given by $\mathcal{S}(\varphi)=\varphi\times id_{\mathbb{C}}$, where $\mathcal{H}:=\bigcup\limits_{n\in\mathbb{N}}\mathcal{H}_n$. We say that $\mathcal{T}\subset\mathcal{H}$ is an invariant set of $\mathcal{S}$ if $\mathcal{S}(\mathcal{T})\subset\mathcal{T}$.

DEFINITION 2.3. Let $\mathcal{T}$ be a subset of $\mathcal{H}$. We say that $\mathcal{T}$ preserves linear tangent cones if each element $\varphi\in\mathcal{T}$ preserves linear tangent cones. We say that $\mathcal{T}$ preserves weakly linear tangent cones if for each element $\varphi\in\mathcal{T}$ there exists a positive integer m such that $\mathcal{S}^m(\varphi)$ preserves linear tangent cones.

EXAMPLE 2.4. If *Lip* is the collection, for each $n\in\mathbb{N}$, of all germs of bi-Lipschitz homeomorphisms $h\colon(\mathbb{C}^n,0)\to(\mathbb{C}^n,0)$, then by Theorem 3.2 in [**23**], it is easy to verify that *Lip* preserves linear tangent cones and is an invariant set of $\mathcal{S}$.

THEOREM 2.5. *Let $f,g\colon(\mathbb{C}^n,0)\to(\mathbb{C},0)$ be complex analytic functions and let $\mathcal{T}\subset\mathcal{H}$ be an invariant set of $\mathcal{S}$. Suppose that there exists $\varphi\in\mathcal{T}$ such that $f=g\circ\varphi$. If $\mathcal{T}$ preserves weakly linear tangent cones, then* $\mathrm{ord}_0 f=\mathrm{ord}_0 g$.

PROOF. Since $\mathcal{T}$ preserves weakly linear tangent cones, there exists a positive integer m such that $\tilde\varphi:=\mathcal{S}^m(\varphi)$ preserves linear tangent cones.

Claim. $\mathrm{ord}_0 f\geq\mathrm{ord}_0 g$.

Suppose that $\mathrm{ord}_0 f<\mathrm{ord}_0 g$. If $k=\mathrm{ord}_0 f$, then by A'Campo-Lê's Theorem, we have that $k>1$. Let us define $\tilde f,\tilde g\colon(\mathbb{C}^n\times\mathbb{C}^m,0)\to(\mathbb{C},0)$ by $\tilde f(x,t)=f(x)+t_m^k$ and $\tilde g(x,t)=g(x)+t_m^k$, where $t=(t_1,\dots,t_m)$. Then, $C(V(\tilde g),0)=\{t_m=0\}=\mathbb{C}^{n+m-1}\times\{0\}$ and $\tilde f=\tilde g\circ\tilde\varphi$. Moreover, since $f_k:=\mathbf{in}f\not\equiv0$ and $k>1$, we get that $f_k+t_m^k$ cannot be a power of a linear form and, in particular, $C(V(\tilde f),0)=V(f_k+t_m^k)$ is not a linear subspace.

Since $\tilde f=\tilde g\circ\tilde\varphi$ and $\tilde\varphi$ preserves linear tangent cones, we have two possible cases:

1. There exists a homeomorphism $\tilde\phi\colon(V(f_k+t_m^k),0)\to(\mathbb{C}^{n+m-1}\times\{0\},0)$.

Thus, by Prill's Theorem (Theorem in [**20**]), $V(f_k+t_m^k)$ is a linear subspace of $\mathbb{C}^{n+m}$, which is a contradiction. Then, $\mathrm{ord}_0 f\geq\mathrm{ord}_0 g$.

2. There exists a homeomorphism $\tilde\phi\colon E_0(\tilde f)\to E_0(\tilde g)$.

If $\dim E_0(\tilde f)\neq2$, by Corollary 2.12 in ([**8**], p. 145), $\mathbb{P}V(f_k+t_m^k)$ is a linear subspace of $\mathbb{C}P^{n+m-1}$, since $E_0(\tilde f)\cong\mathbb{P}V(f_k+t_m^k)$ and $E_0(\tilde g)\cong\mathbb{C}P^{n+m-2}$. In particular, $C(V(\tilde f),0)$ is a linear subspace of $\mathbb{C}^{n+m}$, which is a contradiction. If $\dim E_0(\tilde f)=2$ and, in particular, $n+m=4$, then there exists a positive integer r such that $\bar\varphi=\mathcal{S}^r(\tilde\varphi)$ preserves linear tangent cones, since $\mathcal{T}\subset\mathcal{H}$ is an invariant set of $\mathcal{S}$ and preserves weakly linear tangent cones. Thus, we define $\bar f,\bar g\colon(\mathbb{C}^4\times\mathbb{C}^r,0)\to(\mathbb{C},0)$ by $\bar f(z,s)=\tilde f(z)+s_r^{k+1}$ and $\bar g(z,s)=\tilde g(z)+s_r^{k+1}$, where $s=(s_1,\dots,s_r)$. Therefore, $\bar f=\bar g\circ\bar\varphi$. Moreover, $C(V(\bar g),0)$ is the linear subspace $\{t_m=0\}\subset\mathbb{C}^{r+4}$, $C(V(\bar f),0)=V(f_k+t_m^k)\subset\mathbb{C}^{r+4}$ and as before $C(V(\bar f),0)$ cannot be a linear subspace. However, since $\bar{\varphi}$ preserves linear tangent cones and $\dim E_0(\bar{f}) = 2+r \neq 2$, by the same reason as before, $C(V(\bar{f}),0)$ is a linear subspace of $\mathbb{C}^{n+m+r}$, which is a contradiction. Then, in any case, $\mathrm{ord}_0 f \geq \mathrm{ord}_0 g$, which finish the proof of Claim.

We need also to show that $\mathrm{ord}_0 f \leq \mathrm{ord}_0 g$. In order to do this, let us define $\mathcal{T}^{-1} = \{\psi^{-1};\, \psi \in \mathcal{T}\}$. Now, it is clear that $\mathcal{T}^{-1}$ is an invariant set of $\mathcal{S}$, $g = f\circ\varphi^{-1}$ and $\varphi^{-1} \in \mathcal{T}^{-1}$. Moreover, by definition, a homeomorphism ψ preserves linear tangent cones if and only if ψ^{-1} preserves linear tangent cones, as well. Therefore, by Claim, $\mathrm{ord}_0 g \leq \mathrm{ord}_0 f$, which finish the proof. $\square$

As a direct consequence, we obtain Theorem 1.2.

DEFINITION 2.6. We say that a homeomorphism $\varphi\colon (\mathbb{C}^n,0) \to (\mathbb{C}^n,0)$ preserves weakly relative multiplicities if for any two reduced complex analytic functions $f,g\colon (\mathbb{C}^n,0) \to (\mathbb{C},0)$ such that $C(V(f),0)$ or $C(V(g),0)$ is a linear subspace and $f = g\circ\varphi$, then $C(V(f),0)$ and $C(V(g),0)$ are irreducible and $k_{V(f)}(C(V(f),0)) = k_{V(g)}(C(V(g),0))$.

THEOREM 2.7. *Let $f,g\colon (\mathbb{C}^n,0) \to (\mathbb{C},0)$ be reduced complex analytic functions and let $\mathcal{T} \subset \mathcal{H}$ be an invariant set of $\mathcal{S}$. Suppose that any element of $\mathcal{T}$ preserves weakly relative multiplicities. If there exists $\varphi \in \mathcal{T}$ such that $f = g\circ\varphi$, then $m(V(f),0) = m(V(g),0)$.*

PROOF. Suppose that $m(V(f),0) = \mathrm{ord}_0 f < \mathrm{ord}_0 g = m(V(g),0)$. If $k = \mathrm{ord}(f)$, then, as before, by A'Campo-Lê's Theorem, we have that $k > 1$. Let f_k be the homogeneous polynomial formed by the monomials of f that have degree k. Let us define $\tilde{f},\tilde{g}\colon (\mathbb{C}^{n+1},0) \to (\mathbb{C},0)$ by $\tilde{f}(x,t) = f(x)+t^k$ and $\tilde{g}(x,t) = g(x)+t^k$. Then, $C(V(\tilde{g}),0) = \{t=0\} = \mathbb{C}^n\times\{0\}$. Moreover, since $f_k \not\equiv 0$ and $k>1$, we get that $f_k + t^k$ cannot be a power of a linear form and, in particular, $C(V(\tilde{f}),0) = V(f_k+t^k)$ is not a linear subspace. Now, let us define $\tilde{\varphi}\colon (\mathbb{C}^{n+1},0) \to (\mathbb{C}^{n+1},0)$ by $\tilde{\varphi}(x,t) = (\varphi(x),t)$. We have that $\tilde{\varphi} \in \mathcal{T}$. Then, $C(V(\tilde{f}),0)$ is irreducible and $k_{V(\tilde{f})}(C(V(\tilde{f}),0)) = k_{V(\tilde{g})}(C(V(\tilde{g}),0)) = k$, since $\tilde{\varphi}$ preserves weakly relative multiplicities. Therefore, we have that $k = \mathrm{ord}(\tilde{f}) = k\cdot m(C(V(\tilde{f}),0),0)$. Thus, $m(C(V(\tilde{f}),0),0) = 1$ and, then, $C(V(\tilde{f}),0)$ is smooth at 0, which is a contradiction, since $C(V(\tilde{f}),0)$ cannot be a linear subspace. Therefore, $\mathrm{ord}(f) \geq \mathrm{ord}(g)$ and by similar arguments as in the proof of Theorem 2.5, we obtain also $\mathrm{ord}(f) \leq \mathrm{ord}(g)$. $\square$

As a direct consequence, we obtain Theorem 1.3.

Let us finish this Subsection by showing that Question B_{Rel} has a negative answer and that Question B' also has a negative answer, when we do not require that $C(V(f),0)$ or $C(V(g),0)$ need to be a linear subspace. Here, as it already was remarked in the Introduction, we cannot use A'Campo-Lê's Theorem as it was used in [**2**].

PROPOSITION 2.8. *There exists a family $\{V_t\}$ of hypersurface singularities topologically equisingular such that $C(V(f_t),0)$ is not homeomorphic to $C(V(f_0),0)$ for any $t \neq 0$. Moreover, this same family gives a negative answer to Question B_{Rel}.*

PROOF. In the paper [**2**] the authors showed that the family $\{V_t\}$ of hypersurface singularities defined as zero locus of

$$f_t = z^{12} + zy^3x + ty^2x^3 + x^6 + y^5$$

is topologically equisingular. In fact, they showed that there exists a family of homeomorphisms $\{\varphi_t \colon (\mathbb{C}^3, 0) \to (\mathbb{C}^3, 0)\}$ satisfying $f_t = f_0 \circ \varphi_t$.

However, $\mathbf{in}(f_t) = zy^3x + ty^2x^3 + y^5 = y^2(xyz + tx^3 + y^3)$ for $t \neq 0$ and $\mathbf{in}(f_0) = y^3(xz+y^2)$. Thus, $m(C(V(f_t),0),0) = 4$ for $t \neq 0$ and $m(C(V(f_0),0),0) = 3$. Therefore, by Proposition 3.5 in [**13**], $C(V(f_t),0)$ is not homeomorphic to $C(V(f_0),0)$ for any $t \neq 0$, which finish the first part of the proof.

Moreover, as it was said by J. Fernández de Bobadilla in a private talk with the author, that family gives also a negative answer to Question B_{Rel}. In fact, it is clear to see that $k_{V(f_t)}(V(y)) = 2$ and $k_{V(f_t)}(V(xyz + tx^3 + y^3)) = 1$ for $t \neq 0$, but $k_{V(f_0)}(V(y)) = 3$ and $k_{V(f_0)}(V(xz + y^2)) = 1$.

Thus,

$$k_{V(f_0)}(V(y)) \neq k_{V(f_t)}(V(y))$$

and

$$k_{V(f_0)}(V(y)) \neq k_{V(f_t)}(V(xyz + tx^3 + y^3)),$$

showing that $V(f_t)$ and $V(f_0)$ have different relative multiplicities, when $t \neq 0$ and this gives a negative answer to Question B_{Rel}. □

2.2. Lipschitz submanifold vs. parametric Lipschitz manifold.

PROOF OF PROPOSITION 1.7. For each odd natural number k, let $f \colon \mathbb{C}^{k+1} \to \mathbb{C}$ be the polynomial given by $f_k(x_1, ..., x_k, z) = x_1^2 + ... + x_k^2 - z^3$. We have that $V(f_k)$ has a unique singularity at $0 \in \mathbb{C}^{k+1}$. Moreover, by Proposition in ([**18**], Theorem 2.10), there exist $\varepsilon > 0$ and a homeomorphism $h \colon (B_\varepsilon, V(f_k) \cap B_\varepsilon) \to (B_\varepsilon, CL_{V(f_k)})$ such that $h(0) = 0$ and $V(f_k) \cap \mathbb{S}_\varepsilon^{2k+1}$ is a compact smooth manifold, where

$$B_\varepsilon = \{v \in \mathbb{C}^{k+1}; \|v\| \leq \varepsilon\},$$

$$\mathbb{S}_\varepsilon^{2k+1} = \{v \in \mathbb{C}^{k+1}; \|v\| = \varepsilon\}$$

and

$$CL_{V(f_k)} = \{tv; t \in [0,1] \text{ and } v \in V(f_k) \cap \mathbb{S}_\varepsilon^{2k+1}\}.$$

However, it was proven in [**7**] that there is a diffeomorphism $\phi \colon \mathbb{S}^{2k-1} \to V(f_k) \cap \mathbb{S}_\varepsilon^{2k+1}$. Since $V(f_k) \cap \mathbb{S}_\varepsilon^{2k+1}$ and $\mathbb{S}^{2k-1}$ are compact smooth manifolds, we have that ϕ is bi-Lipschitz homeomorphism. Therefore, the mapping $\varphi \colon \mathbb{R}^{2k} \to C_k$ given by

$$\varphi(x) = \begin{cases} \|x\| \cdot \phi(\frac{x}{\|x\|}), & \text{if } x \neq 0 \\ 0, & \text{if } x = 0 \end{cases}$$

is a bi-Lipschitz homeomorphism as well, where $C_k = \{tv; t \geq 0 \text{ and } v \in V(f_k) \cap \mathbb{S}_\varepsilon^{2k+1}\}$. In particular, C_k is a parametric Lipschitz manifold.

Claim 1. C_k is not a Lipschitz submanifold.

Suppose that C_k is a Lipschitz submanifold. Then, since $0 \in C_k$, there exist an open neighborhood $U \subset \mathbb{C}^{k+1} \equiv \mathbb{R}^{2k+2}$ and a bi-Lipschitz homeomorphism $\phi \colon U \to \phi(U) \subset \mathbb{C}^{k+1}$ such that $\phi(0) = 0$ and $\phi(C_k \cap U) = V(z) \cap \phi(U)$, where $V(z) = \{(x_1, ..., x_k, z) \in \mathbb{C}^{k+1}; z = 0\}$. Therefore, we have a homeomorphism $\phi \colon (\mathbb{C}^{k+1}, V(z), 0) \to (\mathbb{C}^{k+1}, V(f_k), 0)$. Since $V(z)$ is smooth at 0, by A'Campo-Lê's Theorem, $V(f_k)$ is smooth at 0, which is a contradiction. Thus, Claim 1 is proven and we finish the proof. □

We would like to remark that we cannot obtain counterexamples, like in the proof of Proposition 1.7, being complex analytic sets, since it was proven by the author in [**23**, Theorem 4.2] (see also Theorem 3.1 in [**5**]) that if a complex analytic set X is a parametric Lipschitz manifold then X is smooth and, in particular, it is a Lipschitz submanifold.

2.3. Invariance of the degree by right equivalence. Let $f\colon \mathbb{C}^n \to \mathbb{C}$ be a complex polynomial with degree $d > 0$. We write

$$f = f_m + f_{m+1} + \cdots + f_{d-1} + f_d$$

where each f_k is a homogeneous polynomial of degree k. Then, the tangent cone of $V(f)$ at infinity is $C_\infty(V(f)) = V(f_d)$.

DEFINITION 2.9. For each $n \in \mathbb{N}$, let $\mathcal{H}_{n,\infty}$ be the collection of all germs of homeomorphisms at infinity from $\mathbb{C}^n$ to $\mathbb{C}^n$. Let $\mathcal{T}_{\mathcal{H},\infty} \subset \mathcal{H}_\infty := \bigcup\limits_{n\in\mathbb{N}} \mathcal{H}_{n,\infty}$ be the maximal subset with respect the inclusion satisfying:

i) If $\varphi \in \mathcal{T}_{\mathcal{H},\infty}$, then $\varphi \times \mathrm{id}_{\mathbb{C}} \in \mathcal{T}_{\mathcal{H},\infty}$;
ii) If $\varphi \in \mathcal{T}_{\mathcal{H},\infty}$ and $f, g\colon \mathbb{C}^n \to \mathbb{C}$ are complex polynomials such that $\varphi(V(f)) = V(g)$ (outside of a compact subset), then there exists a homeomorphism $\phi\colon C_\infty(V(f)) \to C_\infty(V(g))$.

EXAMPLE 2.10. If $Lip_{n,\infty}$ is the collection of all germs of bi-Lipschitz homeomorphisms at infinity from $\mathbb{C}^n$ to $\mathbb{C}^n$, then by Theorem 4.5 in [**11**], it is easy to verify that $Lip_{n,\infty} \subset \mathcal{T}_{\mathcal{H},\infty}$.

As an application of the proof of Theorem 2.5, we obtain a result about invariance of the degree.

COROLLARY 2.11. *Let $f, g\colon \mathbb{C}^n \to \mathbb{C}$ be complex polynomials. If there exists $\varphi \in \mathcal{T}_{\mathcal{H},\infty}$ such that $f = g \circ \varphi$ (outside of a compact subset), then* $\deg(f) = \deg(g)$.

PROOF. Suppose that $\deg(f) > \deg(g)$. If $k = \deg(f)$, then we have that $k > 1$, since g cannot be constant. Let f_k be the homogeneous polynomial formed by the monomials of f that have degree k. Let us define $\tilde{f}, \tilde{g}\colon (\mathbb{C}^{n+1}, 0) \to (\mathbb{C}, 0)$ by $\tilde{f}(x,t) = f(x) + t^k$ and $\tilde{g}(x,t) = g(x) + t^k$. Then, $C(V(\tilde{g}), 0) = \{t = 0\} = \mathbb{C}^n \times \{0\}$. Moreover, since $f_k \not\equiv 0$ and $k > 1$, we get that $f_k + t^k$ cannot be a power of a linear form and, in particular, $C(V(\tilde{f}), 0) = V(f_k + t^k)$ is not a linear subspace. Now, let us define $\tilde{\varphi}\colon (\mathbb{C}^{n+1}, 0) \to (\mathbb{C}^{n+1}, 0)$ by $\tilde{\varphi}(x,t) = (\varphi(x), t)$. By item i) of the definition 2.9, we have that $\tilde{\varphi} \in \mathcal{T}_{\mathcal{H},\infty}$. By item iii) of the definition 2.9, there exists a homeomorphism $\tilde{\phi}\colon (V(f_k + t^k), 0) \to (\mathbb{C}^n \times \{0\}, 0)$. Thus, by Prill's Theorem (Theorem in [**20**]), $V(f_k + t^k)$ is a linear subspace, which is a contradiction. Therefore, $\mathrm{ord}(f) \leq \mathrm{ord}(g)$ and by similar arguments as in the proof of Theorem 2.5, we obtain also $\mathrm{ord}(f) \geq \mathrm{ord}(g)$. □

We would like to say that in general it is hard the degree to be preserved for some equivalence, as we can see in next example.

EXAMPLE 2.12. Let $f, g\colon \mathbb{C}^n \to \mathbb{C}$ be two complex polynomials such that $f(x,y) = y - x^2$ and $g(x,y) = y$. Let $\varphi\colon \mathbb{C}^2 \to \mathbb{C}^2$ be the polynomial diffeomorphism given by $\varphi(x,y) = (x, y - x^2)$. Then $f = g \circ \varphi$. However, $\deg(f) = 2$ and $\deg(g) = 1$.

In particular, a polynomial diffeomorphism does not need to belong to $\mathcal{T}_{\mathcal{H},\infty}$.

Thus, by Example 2.10 and Corollary 2.11, we obtain the following particular case of Corollary 3.15 in [**25**].

COROLLARY 2.13. *Let $f, g\colon \mathbb{C}^n \to \mathbb{C}$ be two polynomials. If there exists $\varphi \in Lip_{n,\infty}$ such that $f = g \circ \varphi$ (outside of a compact subset) then* $\deg(f) = \deg(g)$.

3. Homeomorphism between two Nash Modifications

In this Section we show that Question B_{Nash} has a negative answer when we consider real analytic sets. In order to do this, we need of some preliminaries.

Let $X \subset \mathbb{R}^m$ be a path connected subset. Let us consider the following distance on X: given two points $x_1, x_2 \in X$, $d_X(x_1, x_2)$ is the infimum of the lengths of paths on X connecting x_1 to x_2. Let us observe that

$$\|x_1 - x_2\| \le d_X(x_1, x_2), \quad \forall\, x_1, x_2 \in X.$$

DEFINITION 3.1. A subset $X \subset \mathbb{R}^m$ is called **Lipschitz normally embedded (LNE)** if there exists a constant $k \ge 1$ such that

$$d_X(x_1, x_2) \le k\|x_1 - x_2\|, \quad \forall\, x_1, x_2 \in X.$$

PROPOSITION 3.2. *Let $L \subset \mathbb{S}^k$ be a set and* $\mathrm{Cone}(L) := \{tx;\, x \in L$ *and* $t \in [0, +\infty)\}$. *If L is a LNE set then* $\mathrm{Cone}(L)$ *is a LNE set, as well.*

Let us remark that Proposition 3.2 already was proven in ([**14**], Proposition 2.8 (a)). However, in his thesis ([**22**], Example 4.3.7), the author of this paper proved Proposition 3.2 for the case of a surface in $\mathbb{R}^3$ with isolated singularity and as it was remarked in [**14**], it is possible to generalize the proof in [**22**] to arbitrary dimensions and, thus, we present this generalization below.

PROOF OF PROPOSITION 3.2. Since L is LNE, there exists a constant $k \ge 1$ such that

$$d_L(x_1, x_2) \le k\|x_1 - x_2\|, \quad \forall\, x_1, x_2 \in L.$$

It is enough to show that there exists a positive constant C such that $d_V(x, y) \le C\|x - y\|$ for all $x, y \in V$. Given $x, y \in V$, we can suppose that $\|y\| \ge \|x\|$. If $x = 0$, then $d_V(0, y) = \|0 - y\| = \|y\|$, since V is a real cone with vertex at origin. If $x \ne 0$ and again using the fact that V is a real cone with vertex at origin, we have $\frac{x}{\|x\|}, \frac{y}{\|y\|} \in L$. Thus, given $\varepsilon > 0$, there exists a Lipschitz curve $\alpha\colon [0,1] \to L$ such that $\alpha(0) = \frac{x}{\|x\|}$, $\alpha(1) = \frac{y}{\|y\|}$ and

$$length(\alpha) = \int_0^1 \|\alpha'(t)\| dt \le d_L(\tfrac{x}{\|x\|}, \tfrac{y}{\|y\|}) + \varepsilon \left\| \tfrac{x}{\|x\|} - \tfrac{y}{\|y\|} \right\| \le (k + \varepsilon) \left\| \tfrac{x}{\|x\|} - \tfrac{y}{\|y\|} \right\|.$$

Then, we define $\gamma\colon [0,1]\to V$ given by $\gamma(t)=(t\|y\|+(1-t)\|x\|)\alpha(t)$. It is clear that γ is a Lipschitz curve satisfying $\gamma(0)=x$ and $\gamma(1)=y$. Hence,

$$\begin{aligned}
d_V(x,y) &\leq length(\gamma)=\int_0^1\|\gamma'(t)\|dt\\
&= \int_0^1\|(\|y\|-\|x\|)\alpha(t)+(t\|y\|+(1-t)\|x\|)\alpha'(t)\|dt\\
&\leq \int_0^1\big|\|y\|-\|x\|\big|\cdot\|\alpha(t)\|dt+\int_0^1(t\|y\|+(1-t)\|x\|)\|\alpha'(t)\|dt\\
&\leq \big|\|y\|-\|x\|\big|+\|y\|\int_0^1\|\alpha'(t)\|dt,\text{ since }\|y\|\geq\|x\|\\
&\leq \|x-y\|+\|y\|(k+\varepsilon)\left\|\tfrac{x}{\|x\|}-\tfrac{y}{\|y\|}\right\|\\
&\leq \|x-y\|+(k+\varepsilon)\|\tfrac{\|y\|}{\|x\|}x-y\|\\
&\leq \|x-y\|+(k+\varepsilon)\left\|\|y\|\tfrac{x}{\|x\|}-\|x\|\tfrac{x}{\|x\|}\right\|+(k+\varepsilon)\|x-y\|\\
&= (k+\varepsilon)\big|\|y\|-\|x\|\big|+(k+\varepsilon+1)\|x-y\|\\
&\leq (2(k+\varepsilon)+1)\|x-y\|.
\end{aligned}$$

Therefore, by taking the limit as $\varepsilon\to 0$, we obtain that $d_V(x,y)\leq C\|x-y\|$ for all $x,y\in V$, where $C=2k+1$. □

DEFINITION 3.3. Let $X\subset\mathbb{R}^n$ and $Y\subset\mathbb{R}^m$. A mapping $f\colon X\to Y$ is called **inner Lipschitz** if there exists $\lambda>0$ such that is

$$d_Y(f(x_1),f(x_2))\leq\lambda d_X(x_1,x_2),\quad\text{for all } x_1,x_2\in X.$$

A Lipschitz mapping $f\colon X\to Y$ is called **inner bi-Lipschitz** if its inverse mapping exists and is inner Lipschitz. When there exists an inner bi-Lipschitz mapping $f\colon X\to Y$ we say that X and Y are inner bi-Lipschitz homeomorphic.

PROOF OF PROPOSITION 1.9. Let us consider $V=\{(x,y,z)\in\mathbb{R}^3;\ z^{2019}=x^{2019}+y^{2019}\}$. Since the link of V, $L:=V\cap\mathbb{S}^2$, is a smooth submanifold of $\mathbb{R}^3$, connected and compact, then L is LNE. By Proposition 3.2, V is LNE.

Therefore, by Theorem 8.3 in [**3**], there exists $\beta\geq 1$ such that $(V,0)$ and $(X_\beta,0)$ are inner bi-Lipschitz homeomorphic, where $X_\beta=\{(x,y,z)\in\mathbb{R}^3;\ x^2+y^2=z^{2\beta}$ and $z\geq 0\}$. Since $(V,0)$ and $(X_\beta,0)$ are LNE sets, then $(V,0)$ and $(X_\beta,0)$ are also bi-Lipschitz homeomorphic. In addiction, X_β has zero density at the origin if and only if $\beta>1$. However, the tangent cone of V at origin is itself and, in particular, V does not have zero density at the origin, then $\beta=1$, since an inner bi-Lipschitz homeomorphism preserves zero density (see Proposition 2.3 in [**4**] and Théorème 3.8 in [**16**]). Moreover, $(X_1,0)$ and $(\mathbb{R}^2,0)$ are bi-Lipschitz homeomorphic, since X_1 is the graph of the Lipschitz mapping $f\colon(\mathbb{R}^2,0)\to(\mathbb{R},0)$ given by $f(x,y)=(x^2+y^2)^{\frac{1}{2}}$. Therefore, $(V,0)$ and $(\mathbb{R}^2,0)$ are bi-Lipschitz homeomorphic, as well.

Now, it is easy to see that there is no homeomorphism from V'_{Nash} to ${\mathbb{R}^2}'_{Nash}$ sending $\eta_V^{-1}(\mathrm{Sing}(V))$ over $\eta_{\mathbb{R}^2}^{-1}(\mathrm{Sing}(\mathbb{R}^2))$, since $\eta_V^{-1}(\mathrm{Sing}(V))\neq\emptyset$. □

In particular, the above proposition says that to give a positive answer to Question B_{Nash} in the case of complex analytic sets, we really need to use the complex structure of the sets.

Acknowledgements. The author would like to thank Javier Fernández de Bobadilla for telling how to get a negative answer to Question B_{Rel} and Alexandre Fernandes for the discussions about Question B_{Nash}. The author would also like to thank the anonymous referee for his useful comments.

References

[1] N. A'Campo, *Le nombre de Lefschetz d'une monodromie* (French), Nederl. Akad. Wetensch. Proc. Ser. A **76** = Indag. Math. **35** (1973), 113–118. MR0320364

[2] E. Artal Bartolo, J. Fernández de Bobadilla, I. Luengo, and A. Melle-Hernández, *Milnor number of weighted-Lê-Yomdin singularities*, Int. Math. Res. Not. IMRN **22** (2010), 4301–4318, DOI 10.1093/imrn/rnq041. MR2737772

[3] L. Birbrair, *Lipschitz geometry of curves and surfaces definable in O-minimal structures*, Illinois J. Math. **52** (2008), no. 4, 1325–1353. MR2595771

[4] L. Birbrair, A. Fernandes, and W. D. Neumann, *Separating sets, metric tangent cone and applications for complex algebraic germs*, Selecta Math. (N.S.) **16** (2010), no. 3, 377–391, DOI 10.1007/s00029-010-0024-0. MR2734336

[5] L. Birbrair, A. Fernandes, D. T. Lê, and J. E. Sampaio, *Lipschitz regular complex algebraic sets are smooth*, Proc. Amer. Math. Soc. **144** (2016), no. 3, 983–987, DOI 10.1090/proc/12783. MR3447652

[6] L. Birbrair and A. Grabrielov, *Ambient Lipschitz Equivalence of Real Surface Singularities*, International Mathematics Research Notices, **2019** (2019), no. 20, 6347–6361.

[7] E. V. Brieskorn, *Examples of singular normal complex spaces which are topological manifolds*, Proc. Nat. Acad. Sci. U.S.A. **55** (1966), 1395–1397, DOI 10.1073/pnas.55.6.1395. MR0198497

[8] A. Dimca, *Singularities and topology of hypersurfaces*, Universitext, Springer-Verlag, New York, 1992. MR1194180

[9] C. Eyral, *Zariski's multiplicity question—a survey*, New Zealand J. Math. **36** (2007), 253–276. MR2476643

[10] A. Fernandes and J. E. Sampaio, *Multiplicity of analytic hypersurface singularities under bi-Lipschitz homeomorphisms*, J. Topol. **9** (2016), no. 3, 927–933, DOI 10.1112/jtopol/jtw014. MR3551843

[11] A. Fernandes and J. Edson Sampaio, *On Lipschitz rigidity of complex analytic sets*. J. Geom. Anal., vol. online (2019), 1–13. `https://doi.org/10.1007/s12220-019-00162-x`.

[12] J. Fernández de Bobadilla, *Answers to some equisingularity questions*, Invent. Math. **161** (2005), no. 3, 657–675, DOI 10.1007/s00222-005-0441-4. MR2181471

[13] J. Fernández de Bobadilla, A. Fernandes, and J. Edson Sampaio, *Multiplicity and degree as bi-Lipschitz invariants for complex sets*. Jornal of Topology, vol. 11 (2018), 957–965.

[14] D. Kerner, H. M. Pedersen, and M. A. S. Ruas, *Lipschitz normal embeddings in the space of matrices*, Math. Z. **290** (2018), no. 1-2, 485–507, DOI 10.1007/s00209-017-2027-4. MR3848442

[15] H. C. King, *Topological type of isolated critical points*, Ann. Math. (2) **107** (1978), no. 2, 385–397, DOI 10.2307/1971121. MR0494153

[16] K. Kurdyka and G. Raby, *Densité des ensembles sous-analytiques* (French, with English summary), Ann. Inst. Fourier (Grenoble) **39** (1989), no. 3, 753–771. MR1030848

[17] Lê Dũng Tráng, *Calcul du nombre de cycles évanouissants d'une hypersurface complexe* (French, with English summary), Ann. Inst. Fourier (Grenoble) **23** (1973), no. 4, 261–270. MR0330501

[18] J. Milnor, *Singular points of complex hypersurfaces*, Annals of Mathematics Studies, No. 61, Princeton University Press, Princeton, N.J.; University of Tokyo Press, Tokyo, 1968. MR0239612

[19] J. Naumann and C. G. Simader, *Measure and Integration on Lipschitz-Manifolds*. Preprints aus dem Institut für Mathematik - 15, Mathematik-Preprints, 2011, 43 p. `https://doi.org/10.18452/2773`.

[20] D. Prill, *Cones in complex affine space are topologically singular*, Proc. Amer. Math. Soc. **18** (1967), 178–182, DOI 10.2307/2035251. MR206991

[21] O. Saeki, *Topological types of complex isolated hypersurface singularities*, Kodai Math. J. **12** (1989), no. 1, 23–29, DOI 10.2996/kmj/1138038986. MR987138

[22] J. Edson Sampaio, *Regularidade lipschitz, invariância da multiplicidade e a geometria dos cones tangentes de conjuntos analítico.* Ph.D. thesis, Universidade Federal Do Ceará (2015). http://www.repositorio.ufc.br/bitstream/riufc/12545/1/2015_tese_jesampaio.pdf

[23] J. E. Sampaio, *Bi-Lipschitz homeomorphic subanalytic sets have bi-Lipschitz homeomorphic tangent cones*, Selecta Math. (N.S.) **22** (2016), no. 2, 553–559, DOI 10.1007/s00029-015-0195-9. MR3477329

[24] J. Edson Sampaio, *Multiplicity, regularity and blow-spherical equivalence of complex analytic sets.* arXiv:1702.06213 [math.AG], preprint (2017).

[25] J. E. Sampaio, *On Zariski's multiplicity problem at infinity*, Proc. Amer. Math. Soc. **147** (2019), no. 4, 1367–1376, DOI 10.1090/proc/14351. MR3910404

[26] user103279 (https://mathoverflow.net/users/103279/user103279). *Topological invariance of the relative multiplicities*, URL (version: 2017-01-13): https://mathoverflow.net/q/259520.

[27] G. Valette, *Multiplicity mod 2 as a metric invariant*, Discrete Comput. Geom. **43** (2010), no. 3, 663–679, DOI 10.1007/s00454-009-9205-z. MR2587843

[28] O. Zariski, *On the Topology of Algebroid Singularities*, Amer. J. Math. **54** (1932), no. 3, 453–465, DOI 10.2307/2370887. MR1507926

[29] O. Zariski, *Some open questions in the theory of singularities*, Bull. Amer. Math. Soc. **77** (1971), 481–491, DOI 10.1090/S0002-9904-1971-12729-5. MR277533

(J. Edson Sampaio) (1) Departamento de Matemática, Universidade Federal do Ceará, Rua Campus do Pici, s/n, Bloco 914, Pici, 60440-900, Fortaleza-CE, Brazil. E-mail: edsonsampaio@mat.ufc.br
(2) BCAM - Basque Center for Applied Mathematics, Mazarredo, 14 E48009 Bilbao, Basque Country - Spain. E-mail: esampaio@bcamath.org

Contemporary Mathematics
Volume **742**, 2020
https://doi.org/10.1090/conm/742/14946

Higher dimensional Stokes structures are rare

Jean-Baptiste Teyssier

The purpose of this paper is to explore the geometry of the moduli of Stokes torsors. Stokes torsors are algebraic structures encoding the Stokes phenomenon for linear differential equations. By contrast to the Stokes matrices, Stokes torsors don't involve any choice and are available in any dimension. Moduli of *local* Stokes torsors were constructed in dimension 1 by Babbit and Varadarajan [**BV89**] following a method due to Deligne. In higher dimension, moduli of *local* Stokes torsors were constructed in [**Tey19**]. Although the works of Martinet and Ramis [**MR91**] and Loday-Richaud [**Lod94**] provide a concrete description of these moduli in dimension 1, very little is known in higher dimension. Our first result says that non trivial local Stokes torsors are rare in dimension > 1.

THEOREM 1. *Let $\mathcal{N}$ be a good split meromorphic flat bundle in a neighbourhood of the origin in $\mathbb{C}^n$ for some $n \geqslant 2$. Suppose that the pole locus D of $\mathcal{N}$ has at least two components. Suppose that the eigenvalues of the monodromy of $\mathcal{N}$ along the components of D are generic. Then, there are no non trivial torsors under the Stokes sheaf of $\mathcal{N}$.*

Good meromorphic flat bundles are ubiquitous in the theory of linear systems of differential equations. According to a fundamental result of Kedlaya [**Ked10**, **Ked11**] and Mochizuki [**Moc09**][**Moc11**], any meromorphic flat bundle becomes good after a pull-back by a suitable composition of blow-ups above the pole locus. Note that Theorem 1 has no counterpart in dimension 1. In dimension 1 indeed, the moduli of torsors under the Stokes sheaf of $\mathcal{N}$ does not depend on the monodromy of $\mathcal{N}$. From [**Tey18**, 2.2.1], we deduce the following rigidity theorem refining [**Tey19**, Th. 3]

THEOREM 2. *Let $\mathcal{N}$ be a good split meromorphic flat bundle in a neighbourhood of the origin 0 in $\mathbb{C}^n$ for some $n \geqslant 2$. Suppose that the pole locus D of $\mathcal{N}$ has at least two components. Suppose that the eigenvalues of the monodromy of $\mathcal{N}$ along the components of D are generic. Then, $\mathcal{N}$ itself is the only germ of meromorphic flat bundle at 0 formally isomorphic to $\mathcal{N}$ at 0.*

The local Stokes sheaf admits a global variant. In the global case, moduli of Stokes torsors were constructed in any dimension in [**Tey18**]. In this note, we show the following

THEOREM 3. *Let X be a smooth complex algebraic variety. Let D be a normal crossing divisor in X. Let $\mathcal{M}$ be a rank 2 good meromorphic connection on X with* *poles along D. Then, the moduli of torsors under the Stokes sheaf of $\mathcal{M}$ along D is an affine space.*

As an application of Theorem 3, we construct non trivial examples of meromorphic connections with poles along the fibre of some abelian scheme. This answers a question asked to the author by Y. André. In general, we give a conjectural description of the global moduli of Stokes torsors predicting (at least in the case where D is smooth) that they are affine spaces.

Acknowledgement. I thank Y. André, P. Boalch, F. Loray and C. Sabbah for interesting discussions. I thank the anonymous referee for numerous constructive remarks on a first draft of this manuscript. This work has been funded by the long term structural funding-Methusalem grant of the Flemish Government. We thank KU Leuven for providing outstanding working conditions. This paper benefited from a one month stay at the Hausdorff Research Institute for Mathematics, Bonn. We thank the Hausdorff Institute for providing outstanding working conditions.

1. Recollection on Stokes torsors

1.1. Torsors. Let M be a manifold. Let $\mathcal{G}$ be a sheaf of groups on M. We recall that a torsor under $\mathcal{G}$ is a sheaf $\mathcal{F}$ on M endowed with a left action of $\mathcal{G}$ such that there exists a cover $\mathcal{U}$ by open subsets of M such that for every $U \in \mathcal{U}$, there exists an isomorphism of sheaves $\mathcal{F}_{|U} \simeq \mathcal{G}_{|U}$ commuting with the action of $\mathcal{G}$, where $\mathcal{G}$ acts on itself by left multiplication. It is a standard fact that the isomorphism classes of $\mathcal{G}$-torsors are in bijection with $H^1(M, \mathcal{G})$, the set of non abelian cohomology classes of $\mathcal{G}$.

1.2. Geometric setup. Let X be a smooth complex algebraic variety of dimension n. Let D be a normal crossing divisor in X. For a quasi-coherent sheaf $\mathcal{F}$ on X, we denote by $\mathcal{F}_{|D}$ the sheaf of germs of sections of $\mathcal{F}$ along D. Let $D_1, \ldots, D_m$ be the irreducible components of D. For $I \subset [\![1, m]\!]$, set

$$D_I := \bigcap_{i \in I} D_i \text{ and } D_I^\circ := D_I \backslash \bigcup_{i \notin I} D_i$$

1.3. Functions with asymptotic expansion along D. For $i = 1, \ldots, m$, let $\widetilde{X}_i \longrightarrow X$ be the real blow-up of X along D_i. Let $p : \widetilde{X} \longrightarrow X$ be the fibre product of the $\widetilde{X}_i$, $i = 1, \ldots, m$ above X. For every subset $A \subset D$, put $\partial A := p^{-1}(A)$. Let $\iota_A : \partial A \longrightarrow \partial D$ be the canonical inclusion.

Let $\mathcal{A}$ be the sheaf of functions on ∂D admitting an asymptotic expansion along D **[Sab00]**. For a closed subset Z in D, let $\mathcal{A}_{\widehat{Z}}$ be the completion of $\mathcal{A}$ along the pull-back by p of the ideal sheaf of Z. Put $\mathcal{A}^{<Z} := \operatorname{Ker}(\mathcal{A} \longrightarrow \mathcal{A}_{\widehat{Z}})$. When $Z = D$, the sheaf $\mathcal{A}^{<D}$ can be concretely described locally as follows (see proposition 1.1.11 from **[Sab00]** for a proof). Let $(x_1, \ldots, x_n)$ be local coordinates centred at $0 \in D$ such that D is defined around 0 by $x_1 \cdots x_l = 0$ for some $l \in [\![1, m]\!]$. Then, the germ of $\mathcal{A}^{<D}$ at $\theta \in \partial 0$ is given by those holomorphic functions u defined over the trace on $X \backslash D$ of a neighbourhood Ω of θ in $\widetilde{X}$, and such that for every compact $K \subset \Omega$, for every $N := (N_1, \ldots, N_l) \in \mathbb{N}^l$, there exists a constant $C_{K,N} > 0$ satisfying

$$|u(x)| \leqslant C_{K,N} |x_1|^{N_1} \cdots |x_l|^{N_l} \text{ for every } x \in K \cap (X \backslash D)$$

1.4. Stokes torsors and the functor of relative Stokes torsors. Let $\mathcal{M}$ be a good meromorphic connection defined in a neighbourhood of D and with poles along D. We set

$$\partial\mathcal{M} = \mathcal{A} \otimes_{p_D^{-1}\mathcal{O}_{X|D}} p^{-1}\mathcal{M}$$

Let $\mathcal{D}_X$ be the sheaf of differential operators on X. The sheaf $\mathcal{A}$ is endowed with an action of $p^{-1}\mathcal{D}_{X|D}$. Hence, so does $\partial\mathcal{M}$. We can thus form the De Rham complex of $\mathcal{M}$ with coefficients in $\mathcal{A}$ as

$$\partial\mathcal{M} \longrightarrow \partial\mathcal{M} \otimes_{p^{-1}\mathcal{O}_{X|D}} p^{-1}\Omega^1_{X|D} \longrightarrow \cdots \longrightarrow \partial\mathcal{M} \otimes_{p^{-1}\mathcal{O}_{X|D}} p^{-1}\Omega^n_{X|D}$$

It is denoted by $\mathrm{DR}\,\partial\mathcal{M}$. Similarly, we denote by $\mathrm{DR}^{<D}\mathcal{M}$ the De Rham complex of $\mathcal{M}$ with coefficients in $\mathcal{A}^{<D}$.

Let Z be a closed subset of D. Let $\mathrm{St}^{<Z}_{\mathcal{M}}$ be the subsheaf of $\mathcal{H}^0\,\mathrm{DR}\,\partial\,\mathrm{End}\,\mathcal{M}$ of sections asymptotic to the Identity along Z, that is of the form $\mathrm{Id}+f$ where f has coefficients in $\mathcal{A}^{<Z}$. The sheaf $\mathrm{St}^{<Z}_{\mathcal{M}}$ is a sheaf of complex unipotent algebraic groups on ∂Z. This is *the Stokes sheaf of $\mathcal{M}$ along Z*. In particular, $\mathrm{St}^{<D}_{\mathcal{M}}$ is a sheaf on ∂D and for every point $P \in D$, the sheaf $\mathrm{St}^{<P}_{\mathcal{M}}$ is a sheaf on ∂P. The sheaf $\mathrm{St}^{<P}_{\mathcal{M}}$ will also be denoted by $\mathrm{St}_{\mathcal{M}}$ when there is no ambiguity on the point under consideration. This abuse of language is done for consistency with the notations from [**Tey19**].

Since $\mathrm{St}^{<Z}_{\mathcal{M}}$ is a sheaf of complex algebraic groups, for every $R \in \mathbb{C}$-alg, the sheaf of R-points of $\mathrm{St}^{<Z}_{\mathcal{M}}$ is a well-defined sheaf of groups on ∂Z. It is denoted by $\mathrm{St}^{<Z}_{\mathcal{M}}(R)$. This is the Stokes sheaf of $\mathcal{M}$ along Z relative to R. Torsors under $\mathrm{St}^{<Z}_{\mathcal{M}}(R)$ are the *Stokes torsors along Z relative to R*. For every subset $A \subset Z$, let $H^1(\partial A, \mathrm{St}^{<Z}_{\mathcal{M}})$ be the functor

$$\begin{array}{ccc} \mathbb{C}\text{-alg} & \longrightarrow & \mathrm{Set} \\ R & \longrightarrow & H^1(\partial A, \mathrm{St}^{<Z}_{\mathcal{M}}(R)) \end{array}$$

From [**Tey19**, Th. 1], the functor $H^1(\partial P, \mathrm{St}^{<P}_{\mathcal{M}})$ is an affine scheme of finite type over $\mathbb{C}$ for every $P \in D$. From [**Tey18**, Th. 6], the functor $H^1(\partial D, \mathrm{St}^{<D}_{\mathcal{M}})$ is an affine scheme of finite type over $\mathbb{C}$.

From [**Tey18**, 1.6.2], the presheaf of functors $R^1p_*\,\mathrm{St}^{<D}_{\mathcal{M}}$ defined as

$$\begin{array}{ccc} \mathrm{Open}(D) & \longrightarrow & \mathrm{Set} \\ U & \longrightarrow & H^1(\partial U, \mathrm{St}^{<D}_{\mathcal{M}}) \end{array}$$

is a sheaf of affine schemes of finite type over $\mathbb{C}$. Its stalk at $P \in D$ is $H^1(\partial P, \mathrm{St}^{<D}_{\mathcal{M}})$. As proved in [**Tey18**, 1.7.1], the sheaf $R^1p_*\,\mathrm{St}^{<D}_{\mathcal{M}}$ is constructible. More precisely, for every $I \subset [\![1, m]\!]$, the sheaf $R^1p_*\,\mathrm{St}^{<D}_{\mathcal{M}}$ is locally constant on D_I°.

2. The level filtration and applications

2.1. Geometric setup. In this section, we put $X = \mathbb{C}^n$ and D is given by $x_1 \cdots x_m = 0$. In particular, we have

$$\widetilde{X} \simeq ([0, +\infty[\times S^1)^m \times \mathbb{C}^{n-m}$$

and the map $p : \widetilde{X} \longrightarrow X$ reads

$$((r_k, z_k)_k, y) \longrightarrow ((r_k z_k)_k, y)$$

In particular, $\mathbb{T} := \partial 0$ is a torus. Let $i : \mathbb{T} \longrightarrow \widetilde{X}$ and $j : X\backslash D \longrightarrow \widetilde{X}$ be the canonical inclusions. Let $\pi : \mathbb{R}^m \longrightarrow \mathbb{T}$ be the canonical projection.

2.2. Irregular values and truncation. In this paragraph, we follow part I, Chapter 2 from [**Moc11**]. We endow $\mathbb{Z}^m$ with the order given by $\mathbf{m} \leqslant \mathbf{m}'$ if and only if $m_i \leqslant m_i'$ for every $i = 1, \dots m$. For $a \in \mathcal{O}_{\mathbb{C}^n}(*D)/\mathcal{O}_{\mathbb{C}^n}$, we write $a = \sum_{\mathbf{m}\in\mathbb{Z}^m} a_{\mathbf{m}} z^{\mathbf{m}}$ and denote by $\operatorname{ord} a$ the minimum of

$$\{\mathbf{m} \in \mathbb{Z}^m_{\leqslant 0} \text{ such that } a_{\mathbf{m}} \neq 0\}$$

when it exists. Let $\mathcal{I}$ be a good set of irregular values with poles contained in D. By definition, $\mathcal{I}$ is a finite subset of $\mathcal{O}_{\mathbb{C}^n,0}(*D)/\mathcal{O}_{\mathbb{C}^n,0}$ such that

- For every non zero $a \in \mathcal{I}$, $\operatorname{ord} a$ exists and $a_{\operatorname{ord} a}$ is invertible in a neighbourhood of 0.
- For every distinct $a, b \in \mathcal{I}$, $\operatorname{ord} a - b$ exists and $(a-b)_{\operatorname{ord} a-b}$ is invertible in a neighbourhood of 0.
- The set $\Phi(\mathcal{I}) := \{\operatorname{ord} a - b, a, b \in \mathcal{I} \text{ distinct}\}$ is totally ordered.

The elements of $\Phi(\mathcal{I})$ are the *levels of* $\mathcal{I}$. In particular, the set $\{\operatorname{ord} a, a \in \mathcal{I}\}$ is totally ordered. Let $\mathbf{m}(0) \in \mathbb{Z}^m_{\leqslant 0}$ be its minimum. Let $(\mathbf{m}(0), \dots, \mathbf{m}(L), \mathbf{m}(L+1))$ be an auxiliary sequence for $\mathcal{I}$. This means that $\mathbf{m}(i+1) = \mathbf{m}(i) + (0, \dots, 1, \dots, 0)$ with 1 located in position $h_i \leqslant m$, that $\Phi(\mathcal{I}) \subset \{\mathbf{m}(0), \dots, \mathbf{m}(L+1)\}$ and that $\mathbf{m}(L+1) = 0$ by convention. We set for every $a \in \mathcal{I}$ and every $i = 0, \dots, L+1$,

$$\xi_{\mathbf{m}(i)}(a) := \sum_{\mathbf{n} \ngeqslant \mathbf{m}(i)} a_{\mathbf{n}} z^{\mathbf{n}}$$

and $a_{\geqslant \mathbf{m}(i)} := a - \xi_{\mathbf{m}(i)}(a)$.

2.3. Good unramified split bundle. For every $a \in \mathcal{I}$, set

$$\mathcal{E}^a = (\mathcal{O}_{\mathbb{C}^n,0}(*D), d - da)$$

We fix once for all a germ of split unramified good meromorphic flat bundle of rank r with poles along D

$$\mathcal{N} := \bigoplus_{a\in\mathcal{I}} \mathcal{E}^a \otimes \mathcal{R}_a$$

where the $\mathcal{R}_a$ are regular. For $i = 0, \dots, L+1$, we set $\mathcal{I}(i) := \xi_{\mathbf{m}(i)}(\mathcal{I})$ and

$$\mathcal{N}(i) := \bigoplus_{a\in\mathcal{I}} \mathcal{E}^{\xi_{\mathbf{m}(i)}(a)} \otimes \mathcal{R}_a$$

The levels of $\mathcal{N}(i)$ belong to $\{\mathbf{m}(0), \dots, \mathbf{m}(i-1)\}$. For $\alpha \in \mathcal{I}(i)$, we set

$$\mathcal{N}_\alpha := \bigoplus_{a\in\mathcal{I}, \xi_{\mathbf{m}(i)}(a)=\alpha} \mathcal{E}^a \otimes \mathcal{R}_a$$

The levels of $\mathcal{N}_\alpha$ belong to $\{\mathbf{m}(i), \dots, \mathbf{m}(L+1)\}$.

For $a \in \mathcal{I}$, let $\mathcal{L}_a(\mathcal{N})$ be the local system of flat sections for $\mathcal{R}_a$ on $\mathbb{C}^n\backslash D$. We put $\widetilde{\mathcal{L}}_a(\mathcal{N}) = i^* j_* \mathcal{L}_a(\mathcal{N})$.

2.4. The level filtration. In this subsection, we abuse notations by noting $\mathrm{St}_{\mathcal{N}}$ for $\mathrm{St}_{\mathcal{N}}^{<0}$. We recall the definition of the level filtration on $\mathrm{St}_{\mathcal{N}}$. It is a straightforward generalization of [**BV89**, II 3.2.1]. We include it for the reader's convenience due to a lack of reference in the higher dimensional case. For $i = 0, \dots, L+1$, let us set

$$\mathrm{St}_{\mathcal{N}}^{i} := \{g \in \mathrm{St}_{\mathcal{N}} \,|e^{a}(g - \mathrm{id}) \text{ has rapid decay for every } a \text{ with } \operatorname{ord} a > \mathbf{m}(i-1)\}$$

The sheaf $\mathrm{St}_{\mathcal{N}}^{i}$ is a sheaf of normal algebraic subgroups of $\mathrm{St}_{\mathcal{N}}$. Let us define three diagonal matrices $M := \mathrm{Diag}(e^{a}, a \in \mathcal{I})$, $M_{\leqslant} := \mathrm{Diag}(e^{\xi_{\mathbf{m}(i)}(a)}, a \in \mathcal{I})$ and $M_{\geqslant} := \mathrm{Diag}(e^{a_{\geqslant \mathbf{m}(i)}}, a \in \mathcal{I})$. The sheaf $\mathrm{St}_{\mathcal{N}}^{i}$ admits the following Stokes theoretic description:

LEMMA 2.4.1. *The map*

$$\begin{array}{rcl} \varphi : \mathrm{St}_{\mathcal{N}(i)} & \longrightarrow & \mathrm{St}_{\mathcal{N}} \\ s & \longrightarrow & e^{M_{\geqslant}} s e^{-M_{\geqslant}} \end{array}$$

induces an isomorphism between $\mathrm{St}_{\mathcal{N}(i)}$ *and* $\mathrm{St}_{\mathcal{N}}^{i}$.

PROOF. The statement is local. Hence, it is enough to work on an open set $\mathcal{S}$ contained in a product of strict open intervals. For such an open set, a choice of fundamental matrix F of flat sections for $\bigoplus_{a\in\mathcal{I}} \mathcal{R}_a$ yields a commutative diagram with injective arrows

$$\begin{array}{ccc} \Gamma(\mathcal{S}, \mathrm{St}_{\mathcal{N}(i)}) & \xrightarrow{\ \varphi\ } & \Gamma(\mathcal{S}, \mathrm{St}_{\mathcal{N}}) \\ & \searrow^{\widetilde{\iota}} & \downarrow{\scriptstyle \iota} \\ & & \mathrm{GL}_r \end{array} \tag{2.4.2}$$

where ι is given by $s \longrightarrow e^{-M}F^{-1}sFe^{M}$ and where $\widetilde{\iota}$ is given by $s \longrightarrow e^{-M_{\leqslant}}F^{-1}sFe^{M_{\leqslant}}$. By definition, $\iota(\Gamma(\mathcal{S}, \mathrm{St}_{\mathcal{N}}))$ is the subgroup of elements $g \in \mathrm{GL}_r$ such that for every $a, b \in \mathcal{I}$,

$$\begin{cases} g_{aa} = \mathrm{id} & \\ g_{ab} = 0 & \text{if } a \neq b \text{ and } a \not<_{\mathcal{S}} b \end{cases}$$

Hence, $\widetilde{\iota}(\Gamma(\mathcal{S}, \mathrm{St}_{\mathcal{N}(i)}))$ is the subgroup of elements of $g \in GL_r$ such that for every $a, b \in \mathcal{I}$,

$$\begin{cases} g_{aa} = \mathrm{id} & \\ g_{ab} = 0 & \text{if } \xi_{\mathbf{m}(i)}(a) \neq \xi_{\mathbf{m}(i)}(b) \text{ and } \xi_{\mathbf{m}(i)}(a) \not<_{\mathcal{S}} \xi_{\mathbf{m}(i)}(b) \\ g_{ab} = 0 & \text{if } a \neq b \text{ and } \xi_{\mathbf{m}(i)}(a) = \xi_{\mathbf{m}(i)}(b) \end{cases}$$

Note that if $a, b \in \mathcal{I}$ with $\xi_{\mathbf{m}(i)}(a) \neq \xi_{\mathbf{m}(i)}(b)$, then

$$a \not<_{\mathcal{S}} b \text{ if and only if } \xi_{\mathbf{m}(i)}(a) \not<_{\mathcal{S}} \xi_{\mathbf{m}(i)}(b)$$

Thus, $\widetilde{\iota}(\Gamma(\mathcal{S}, \mathrm{St}_{\mathcal{N}(i)}))$ is the subgroup of elements $g \in \iota(\Gamma(\mathcal{S}, \mathrm{St}_{\mathcal{N}}))$ such that for every $a, b \in \mathcal{I}$,

$$g_{ab} = 0 \text{ if } \xi_{\mathbf{m}(i)}(a) = \xi_{\mathbf{m}(i)}(b) \text{ and } a \neq b$$

Let $s \in \Gamma(\mathcal{S}, \mathrm{St}_{\mathcal{N}}^{i})$, and let $a, b \in \mathcal{I}$ with $a \neq b$. If $\xi_{\mathbf{m}(i)}(a) = \xi_{\mathbf{m}(i)}(b)$, then

$$\iota(s)_{ab} = e^{b-a}F_a^{-1}s_{ab}F_b = F_a^{-1}(e^{b_{\geqslant \mathbf{m}(i)} - a_{\geqslant \mathbf{m}(i)}}s_{ab})F_b$$

By definition, $e^{b_{\geqslant \mathbf{m}(i)} - a_{\geqslant \mathbf{m}(i)}}s_{ab}$ has rapid decay. Since F_a and F_b have moderate growth at 0, we deduce that the constant matrix $\iota(s)_{ab}$ has rapid decay. Hence,

$\iota(s)_{ab} = 0$. Thus $\mathrm{St}^i_{\mathcal{N}} \subset \varphi(\mathrm{St}_{\mathcal{N}(i)})$. On the other hand, let $s \in \varphi(\mathrm{St}_{\mathcal{N}(i)})$ and let c with $\operatorname{ord} c > \mathbf{m}(i-1)$. We have to show that for every $a, b \in \mathcal{I}$ with $a \neq b$,

$$e^c s_{ab} = e^{c+a-b} F_a \iota(s)_{ab} F_b^{-1}$$

has rapid decay. We can suppose $\xi_{\mathbf{m}(i)}(a) <_{\mathcal{S}} \xi_{\mathbf{m}(i)}(b)$. In particular $a <_{\mathcal{S}} b$. Since the leading term of $c + a - b$ is the leading term of $a - b$, the exponential e^{c+a-b} has rapid decay on $\mathcal{S}$. Thus, so does $e^c s_{ab}$. Hence, $s \in \mathrm{St}^i_{\mathcal{N}}$ and we deduce $\mathrm{St}^i_{\mathcal{N}} = \varphi(\mathrm{St}_{\mathcal{N}(i)})$. □

2.5. Quotients of the level filtration.

LEMMA 2.5.1. *There is a split exact sequence of sheaves of algebraic groups*

$$1 \longrightarrow \mathrm{St}_{\mathcal{N}(i)} \xrightarrow{\varphi} \mathrm{St}_{\mathcal{N}} \xrightarrow{\psi} \prod\nolimits_{\alpha \in \mathcal{I}(i)} \mathrm{St}_{\mathcal{N}_\alpha} \longrightarrow 1$$

In particular, $\mathrm{Gr}^i \mathrm{St}_{\mathcal{N}} := \mathrm{St}^{i+1}_{\mathcal{N}} / \mathrm{St}^i_{\mathcal{N}} \simeq \prod\nolimits_{\alpha \in \mathcal{I}(i)} \mathrm{St}_{\mathcal{N}(i+1)_\alpha}$

PROOF. Let us define

$$\begin{aligned} \psi : \mathrm{St}_{\mathcal{N}} &\longrightarrow \prod\nolimits_{\alpha \in \mathcal{I}(i)} \mathrm{St}_{\mathcal{N}_\alpha} \\ s &\longrightarrow \times (s_{ab})_{\substack{\xi_{\mathbf{m}(i)}(a)=\alpha \\ \xi_{\mathbf{m}(i)}(b)=\alpha}} \end{aligned}$$

From the local description of $\mathrm{St}_{\mathcal{N}(i)}$ given in the proof of 2.4.1, we see that the only a priori non obvious thing to prove is the fact that ψ is a group homomorphism. Let $\mathcal{S}$ be an open set of $\mathbb{T}$, let $s, t \in \Gamma(\mathcal{S}, \mathrm{St}_{\mathcal{N}})$, let $\alpha \in \mathcal{I}(i)$ and let $a, b \in \mathcal{I}$ such that $\xi_{\mathbf{m}(i)}(a) = \xi_{\mathbf{m}(i)}(b) = \alpha$. Let us denote by ψ_α the component of ψ associated to α. Then

$$(\psi_\alpha(st))_{ab} = \sum_{c \in \mathcal{I}} s_{ac} t_{cb} = \sum_{\substack{c \in \mathcal{I} \\ a \leqslant_{\mathcal{S}} c \leqslant_{\mathcal{S}} b}} s_{ac} t_{cb}$$

If $\xi_{\mathbf{m}(i)}(c) \neq \alpha$, the leading coefficient of $c - a$ is that of $\xi_{\mathbf{m}(i)}(c) - \xi_{\mathbf{m}(i)}(a) = \xi_{\mathbf{m}(i)}(c) - \alpha$. Hence, $a \leqslant_{\mathcal{S}} c$ if and only if $\alpha <_{\mathcal{S}} \xi_{\mathbf{m}(i)}(c)$. Similarly, $c \leqslant_{\mathcal{S}} b$ if and only if $\xi_{\mathbf{m}(i)}(c) <_{\mathcal{S}} \alpha$. Hence, for $\xi_{\mathbf{m}(i)}(c) \neq \alpha$, the condition $a \leqslant_{\mathcal{S}} c \leqslant_{\mathcal{S}} b$ is empty. Thus

$$(\psi_\sigma(st))_{ab} = \sum_{\substack{c \in \mathcal{I} \\ \xi_{\mathbf{m}(i)}(c)=\alpha}} s_{ac} t_{cb} = (\psi_\alpha(s)\psi_\alpha(t))_{ab}$$

□

2.6. Action of the fundamental group in the local one level case. We consider in this paragraph the case where $\mathcal{N}$ has a unique level $\mathbf{m}$ and we suppose that $\mathcal{N}$ is not regular, that is $\mathbf{m} \in \mathbb{Z}^m_{\leqslant 0} \backslash \{0\}$. Let $\iota : C \longrightarrow \mathbb{C}^n$ be a smooth curve passing through 0 and not contained in D. Let $\widetilde{C}$ be the real-blow up of C at 0. Let S^1_C be the boundary of $\widetilde{C}$. Following [**Sab12**, 8.b p120], the map ι lifts as a map $\widetilde{\iota} : \widetilde{C} \longrightarrow \widetilde{X}$. Let us suppose that the restricted map $S^1_C \longrightarrow \mathbb{T}$ is injective. Let $H_{\mathbf{m}}$ be the hyperplane of $\mathbb{R}^m$ defined by $\sum_{i=1}^m m_i x_i = 0$. For an interval I of S^1_C, set $\mathbb{T}(H, I) := \pi(H) + I$ and

$$\mathbb{T}(\mathbf{m}, I) := \mathbb{T}(H_{\mathbf{m}}, I)$$

For $x \in \mathbb{T}$, the translation t_x by x provides an isomorphism

$$\pi_1(\mathbb{T}(\mathbf{m}, \pi(0)), \pi(0)) \longrightarrow \pi_1(\mathbb{T}(\mathbf{m}, x), x)$$

Hence, $\pi_1(\mathbb{T}(\mathbf{m}, \pi(0)), \pi(0))$ acts on $\mathrm{St}_{\mathcal{N}_C} \simeq (\mathrm{St}_{\mathcal{N}})_{|S^1_C}$ via the parallel transport. We deduce that $\pi_1(\mathbb{T}(\mathbf{m}, \pi(0)), \pi(0))$ acts on $H^1(S^1_C, \mathrm{St}_{\mathcal{N}_C})$. To simplify notations, we denote by $H^1(S^1_C, \mathrm{St}_{\mathcal{N}_C})^{\pi_1}$ the invariants for this action.

LEMMA 2.6.1. *For every cover $\underline{I}$ of S^1_C adapted to $\mathcal{N}_C$, the morphisms in the commutative triangle*

$$(2.6.2) \qquad \begin{array}{ccc} H^1(\mathbb{T}, \mathrm{St}_{\mathcal{N}}) & \xrightarrow{\quad \mathbf{res}_C \quad} & H^1(S^1_C, \mathrm{St}_{\mathcal{N}_C})^{\pi_1} \\ & {\scriptstyle f}\nwarrow \quad \nearrow & \\ & Z^1(\mathbb{T}(\mathbf{m}, \underline{I}), \mathrm{St}_{\mathcal{N}}) & \end{array}$$

are isomorphisms. In particular, $H^1(\mathbb{T}, \mathrm{St}_{\mathcal{N}})$ is an affine space.

For a single level connection in dimension 1, an *adapted cover* is a finite cover $\underline{I}$ by open intervals such that for $I_1, I_2, I_3 \in \underline{I}$ distinct, we have $I_1 \cap I_2 \cap I_3 = \varnothing$, and such that for every $I \in \underline{I}$ and every pair of distinct irregular values $\{a, b\}$, the interval I contains exactly one Stokes direction associated to $a - b$.

PROOF. We first show that f is surjective. Let $\mathcal{T} \in H^1(\mathbb{T}, \mathrm{St}_{\mathcal{N}})$. Let $x = (x_1, \ldots, x_m) \in \pi(H_{\mathbf{m}})$. If C is the curve $t \longrightarrow (c_1(t), \ldots, c_n(t))$, let C_x be the curve given by $t \longrightarrow (x_1 c_1(t), \ldots, x_m c_m(t), \ldots, c_n(t))$. By construction $S^1_{C_x} = x + S^1_C$. Since $\mathcal{N}$ has only one level, the same holds for $\mathcal{N}_{C_x}$. Since $\underline{I}$ is an adapted cover for $\mathcal{N}_C$, the translated cover $x + \underline{I}$ is an adapted cover for $\mathcal{N}_C$. From **[BV89]**, the restriction $\mathcal{T}_{x+S^1_C} \in H^1(x + S^1_C, \mathrm{St}_{\mathcal{N}_{C_x}})$ of $\mathcal{T}$ to $x + S^1_C$ admits a unique trivialisation $t_I(x)$ on each $x + I \in \underline{I}$, and

$$(2.6.3) \qquad Z^1(\underline{I}, \mathrm{St}_{\mathcal{N}_C}) \xrightarrow{\ \sim\ } H^1(S^1_C, \mathrm{St}_{\mathcal{N}_C})$$

Since $\mathrm{St}_{\mathcal{N}}$ has no non trivial section on $x + I$ for every $x \in \pi(H_{\mathbf{m}})$ and every $I \in \underline{I}$, the sections $(t_I(x))_{x \in \pi(H_{\mathbf{m}})}$ with I fixed glue into a section of $\mathcal{T}$ on $\mathbb{T}(\mathbf{m}, I)$. Hence, the cocycle corresponding to $\mathcal{T}_{S^1_C}$ via (2.6.3) extends uniquely into a cocycle for $\mathcal{T}$ relative to the cover $\mathbb{T}(\mathbf{m}, \underline{I})$. In particular, f is surjective.

Let us consider the commutative diagram

$$(2.6.4) \qquad \begin{array}{ccc} Z^1(\mathbb{T}(\mathbf{m}, \underline{I}), \mathrm{St}_{\mathcal{N}}) & \xrightarrow{\ f\ } & H^1(\mathbb{T}, \mathrm{St}_{\mathcal{N}}) \\ \downarrow & & \downarrow \\ Z^1(\underline{I}, \mathrm{St}_{\mathcal{N}_C}) & \xrightarrow{\ \sim\ } & H^1(S^1_C, \mathrm{St}_{\mathcal{N}_C}) \end{array}$$

Every section of $\mathrm{St}_{\mathcal{N}}$ on a connected open set is determined by its germ at a point. Hence, the left vertical arrow of (2.6.4) is injective. We deduce that f is bijective. Tautologically, the image of the left vertical arrow is formed by those $g \in Z^1(\underline{I}, \mathrm{St}_{\mathcal{N}_C})$ extending to $\mathbb{T}(\mathbf{m}, \underline{I})$. These are exactly the invariants under the action of $\pi_1(\mathbb{T}(\mathbf{m}, \pi(0)), \pi(0))$. Taking the invariants under $\pi_1(\mathbb{T}(\mathbf{m}, \pi(0)), \pi(0))$

thus yields a commutative diagram

$$(2.6.5)\qquad \begin{array}{ccc} Z^1(\mathbb{T}(\mathbf{m},\underline{I}),\mathrm{St}_{\mathcal{N}}) & \xrightarrow{f} & H^1(\mathbb{T},\mathrm{St}_{\mathcal{N}}) \\ \wr\downarrow & & \downarrow \mathbf{res}_C \\ Z^1(\underline{I},\mathrm{St}_{\mathcal{N}_C})^{\pi_1} & \xrightarrow{\sim} & H^1(S^1_C,\mathrm{St}_{\mathcal{N}_C})^{\pi_1} \end{array}$$

We deduce that $\mathbf{res}_C$ is bijective. To conclude, we observe that $\mathrm{St}_{\mathcal{N}}$ being a sheaf of unipotent algebraic groups, the scheme $\Gamma(U,\mathrm{St}_{\mathcal{N}})$ is an affine space for every open subset $U\subset\mathbb{T}$. Since $Z^1(\mathbb{T}(\mathbf{m},\underline{I}),\mathrm{St}_{\mathcal{N}})$ is a product of such schemes, it is also an affine space. □

2.7. Proof of Theorem 1. Let $\mathcal{N}$ be a good split meromorphic flat bundle in a neighbourhood of the origin in $\mathbb{C}^n$ for some $n\geqslant 2$ as in paragraph 2.3. Suppose that the pole locus D of $\mathcal{N}$ has at least two components. Suppose that the eigenvalues of the monodromy of $\mathcal{N}$ along the components of D are generic. We want to show that $H^1(\mathbb{T},\mathrm{St}_{\mathcal{N}})$ is reduced to the trivial torsor. We argue recursively on the number of levels of $\mathcal{N}$. The notations from section 2 will be in use. Suppose that $\mathcal{N}$ has only one level $\mathbf{m}$. Let Δ be the diagonal curve of $\mathbb{C}^n$. Let $\underline{I}$ be a cover of S^1_Δ adapted to $\mathcal{N}_\Delta$. Fix $I_1,I_2\in\underline{I}$, put $J=I_1\cap I_2$ and suppose without loss of generality that $\pi(0)\in J$. Note that $\mathbb{T}(\mathbf{m},J)$ is homotopic to the product of $m-1$ circles. By assumption, $m>1$. Hence, $\pi_1(\mathbb{T}(\mathbf{m},J),\pi(0))$ is non trivial. Pick $\gamma\in\pi_1(\mathbb{T}(\mathbf{m},J),\pi(0))$ non trivial and let $\mathbf{n}=(n_1,\dots,n_m)\in\mathbb{Z}^m$ be the coordinates of γ in the canonical basis of $\pi_1(\mathbb{T},\pi(0))$. For $a\in\mathcal{I}$ and $i=1,\dots,m$, let $\Sigma_a(i)$ be the set of eigenvalues for the monodromy of $\mathcal{L}_a(\mathcal{N})$ around D_i. Then, the eigenvalues for the monodromy of $\widetilde{\mathcal{L}}_b(\mathcal{N})^*\otimes\widetilde{\mathcal{L}}_a(\mathcal{N})$ along γ are the

$$\{\prod_{i=1}^m\left(\frac{\lambda_i}{\mu_i}\right)^{n_i},\lambda_i\in\Sigma_a(i),\mu_i\in\Sigma_b(i),i=1,\dots,m\}$$

Since $\mathbf{n}\neq 0$, the zero locus of the polynomial

$$\prod_{i=1}^m x_i^{n_i}-\prod_{i=1}^m y_i^{n_i}$$

is a strict closed subset in $\mathbb{C}^{2m}$. Hence, for a generic choice of the $\Sigma_a(i)$, the restriction of $\widetilde{\mathcal{L}}_b(\mathcal{N})^*\otimes\widetilde{\mathcal{L}}_a(\mathcal{N})=\widetilde{\mathcal{L}}_{a-b}(\operatorname{End}\mathcal{N})$, $a\neq b$ to $\mathbb{T}(\mathbf{m},J)$ cannot have non zero global sections. Hence, the Identity is the only section of $\mathrm{St}_{\mathcal{N}}$ on $\mathbb{T}(\mathbf{m},J)$. Thus, Theorem 1 in the one level case is a consequence of lemma 2.6.1.

Suppose that $\mathcal{N}$ has at least two levels. Let $(\mathbf{m}(0),\dots,\mathbf{m}(L),\mathbf{m}(L+1))$ be an auxiliary sequence for $\mathcal{I}$. Then, there is an index i such that $\mathcal{N}(i)$ has only one level and such that the number of levels of $\mathrm{St}_{\mathcal{N}_\alpha}$ is strictly less than number of levels of $\mathcal{N}$ for every $\alpha\in\mathcal{I}(i)$. Since the $\mathcal{N}_\alpha$ are direct summands of $\mathcal{N}$, they have generic monodromy eigenvalues along the irreducible components of D. By recursion hypothesis applied to the $\mathcal{N}_\alpha$, we obtain that the right term in the exact sequence of pointed sets

$$H^1(\mathbb{T},\mathrm{St}_{\mathcal{N}(i)})\longrightarrow H^1(\mathbb{T},\mathrm{St}_{\mathcal{N}})\longrightarrow\prod\nolimits_{\alpha\in\mathcal{I}(i)}H^1(\mathbb{T},\mathcal{N}_\alpha)$$

deduced from lemma 2.5.1 is trivial. Hence $H^1(\mathbb{T},\mathrm{St}_{\mathcal{N}})\simeq H^1(\mathbb{T},\mathrm{St}_{\mathcal{N}(i)})$. Note that the eigenvalues of the monodromy of $\mathcal{N}(i)$ along the irreducible components of

D are generic. Hence, the one level case treated above implies that $H^1(\mathbb{T}, \mathrm{St}_{\mathcal{N}(i)})$ is a point. This finishes the proof of Theorem 1.

3. Moduli of Stokes torsors and the relative Stokes group

3.1. Roadmap. The goal of this section is to describe the global moduli of Stokes torsors constructed in [**Tey18**] when the irregular values have only one level. To do this, we compare a relative version of the Stokes group from [**MR91**][**Lod94**] with the relative non abelian cohomology of the Stokes sheaf. For the problem raised by this comparison in the multi-level case, we refer to the discussion in 3.6. Note that over a smooth base (corresponding in this paper to the case where D is smooth), relative Stokes groups appeared in the one level case in [**JMU81**] and in more generality in [**Boa02**]. In particular, over a smooth base, they were already considered in the multi-level case in [**Boa14**].

3.2. Loday-Richaud's theorem. In this subsection, the notations from section 1 will be in use and we will suppose that X is a neighbourhood of the origin 0 in the affine line. Let $\mathcal{M}$ be an unramified meromorphic connection on X with poles at 0. Let $\mathcal{I}$ be the set of irregular values of $\mathcal{M}$ at 0. Let $p : \widetilde{X} \longrightarrow X$ be the real blow-up of X at 0. For $a, b \in \mathcal{I}$, the function $G_{a,b} := (a-b)/|a-b|$ induces a C^∞-function $\partial G_{a,b}$ on $\partial 0$. The *anti-Stokes directions of* (a,b) are the directions of $z \in \partial 0$ such that $\partial G_{a,b}(z) \in \mathbb{R}^-$. Let $H^<$ be the set of all anti-Stokes directions, for $a, b \in \mathcal{I}$ distinct. For $z \in H^<$, we put

$$\mathbb{S}to_{\mathcal{M},z} = \{g \in \mathrm{St}_{\mathcal{M},z} \text{ such that for every } a, b \in \mathcal{I} \text{ distinct}, g_{ab} = 0 \text{ unless } z \in H^<_{ab}\}$$

and

$$\mathbb{S}to_{\mathcal{M}} := \prod_{z \in H^<} \mathbb{S}to_{\mathcal{M},z}$$

The group $\mathbb{S}to_{\mathcal{M}}$ is the *Stokes group of* $\mathcal{M}$. For a possibly ramified connection $\mathcal{M}$, we define the Stokes group of $\mathcal{M}$ via Galois descent from the unramified case. Note that an element of the Stokes group $g = (g_z)_{z \in H^<}$ induces a well-defined Stokes torsor. To see this, let $N \geqslant 1$ be an integer and let $(z_i)_{i \in \mathbb{Z}/N\mathbb{Z}}$ be the anti-Stokes directions of $\mathcal{M}$ ordered in an ascending order with respect to the clockwise orientation on $\partial 0$. Pick $i \in \mathbb{Z}/N\mathbb{Z}$. Since $\mathrm{St}_{\mathcal{M}}$ is constructible, g_{z_i} extends into a section $\widetilde{g}_{z_i}$ of $\mathrm{St}_{\mathcal{M}}$ on a small interval $]z_i - \epsilon, z_i + \epsilon[$ for $\epsilon > 0$. At the cost of shrinking ϵ, we can suppose that

(1) The intervals $I_i = (]z_i - \epsilon, z_{i+1} + \epsilon[)_{i \in \mathbb{Z}/N\mathbb{Z}}$ contain exactly two anti-Stokes directions, namely z_i and z_{i+1}.

(2) We have $I_i \cap I_j \cap I_k = \varnothing$ for every $i, j, k \in \mathbb{Z}/N\mathbb{Z}$ distinct.

In particular, $(\widetilde{g}_{z_i})_{i \in \mathbb{Z}/N\mathbb{Z}}$ defines a 1-cocycle for $\mathrm{St}_{\mathcal{M}}$ with respect to the cover $(I_i)_{i \in \mathbb{Z}/N\mathbb{Z}}$. Let $\mathrm{LR}_0(g)$ be the associated Stokes torsor on $\partial 0$. It is independent of ϵ for ϵ small enough. The following theorem is due to Loday-Richaud [**Lod94**, II.2.1]

THEOREM 4. *The map*

$$\mathrm{LR}_0 : \mathbb{S}to_{\mathcal{M}} \longrightarrow H^1(\partial 0, \mathrm{St}_{\mathcal{M}}) \tag{3.2.1}$$

is an isomorphism.

3.3. Relative Stokes groups. In this subsection, we introduce a relative version of the Stokes group. The notations from section 1 will be in use. Let X be a smooth complex algebraic variety and let D be a normal crossing divisor in X. Let $\mathcal{M}$ be a good meromorphic connection on X with poles along D. Let $\mathcal{I}$ be the sheaf of irregular values of $\mathcal{M}$. We first suppose that $\mathcal{M}$ is unramified. In particular, $\mathcal{I}$ is a subsheaf of $\mathcal{O}_X(*D)/\mathcal{O}_X$. Let $p : \widetilde{X} \longrightarrow X$ be the real blow-up of X along D. For $a, b \in \mathcal{I}$, the function $G_{a,b} := (a-b)/|a-b|$ induces a C^∞-function $\partial G_{a,b}$ on ∂D. The *anti-Stokes hyperplanes of* (a,b) are the irreducible components of

$$H^{<}_{ab} := \{z \in \partial D \text{ such that } \partial G_{a,b}(z) \in \mathbb{R}^-\}$$

The set $H^{<}_{ab}$ is a smooth C^∞-hypersurface in ∂D. Let $H^{<}$ be the union of all the $H^{<}_{ab}$, $a, b \in \mathcal{I}$ distinct. Let Z be a closed subset of D. Let $\mathbb{S}to^{<Z}_{\mathcal{M}}$ be the sheaf on $\partial Z \cap H^{<D}$ whose germ at z is

$$\mathbb{S}to^{<Z}_{\mathcal{M},z} = \{g \in \mathrm{St}^{<Z}_{\mathcal{M},z} \text{ such that for every } a, b \in \mathcal{I} \text{ distinct}, g_{ab} = 0 \text{ unless } z \in H^{<}_{ab}\}$$

We call $p_*\mathbb{S}to^{<Z}_{\mathcal{M}}$ the *relative Stokes group of* $\mathcal{M}$ *along* Z. Note that when Z is a point $P \in D$, the sheaf $\mathbb{S}to^{<P}_{\mathcal{M}}$ will also be denoted by $\mathbb{S}to_{\mathcal{M}}$ when there is no ambiguity on the point under consideration.

For a possibly ramified connection $\mathcal{M}$, we define the relative Stokes group of $\mathcal{M}$ via Galois descent from the unramified case.

LEMMA 3.3.1. *If D is smooth, the sheaf $p_*\mathbb{S}to^{<D}_{\mathcal{M}}$ is a local system on D.*

PROOF. For every $a, b \in \mathcal{I}$, the Stokes hyperplanes of (a,b) are parallel to the anti-Stokes hyperplanes of (a,b). Hence, $H^{<}_{ab}$ does not meet any Stokes hyperplane of (a,b). Thus, for any $z \in \partial D$ and any $g \in \mathbb{S}to^{<D}_{\mathcal{M},z}$, the germ g extends uniquely on a small product $\Delta \times I$ containing z, where Δ is a disc in D centred at $p(z)$ and where I is an interval of S^1. Both Δ and I depend only on z and not on g. This concludes the proof of lemma 3.3.1. □

3.4. Comparison over a point in the one level case.

PROPOSITION 3.4.1. *Let X be a smooth complex algebraic variety and let D be a normal crossing divisor in X. Let $P \in D$. Let $\mathcal{M}$ be a good meromorphic connection on X with poles along D. Suppose that $\mathcal{M}$ has a single level. Then, there is a canonical isomorphism*

$$\mathrm{LR}_P : \Gamma(\partial P, \mathbb{S}to_{\mathcal{M}}) \longrightarrow H^1(\partial P, \mathrm{St}_{\mathcal{M}}) \tag{3.4.2}$$

PROOF. By Galois descent, we can suppose that $\mathcal{M}$ is unramified. Let $\mathbf{m} \in \mathbb{Z}^m_{\leqslant 0}\backslash\{0\}$ be the level of $\mathcal{M}$. We are going to construct a cover of ∂P similar to that used by Loday-Richaud in Theorem 4. Let $(x_1, \dots, x_n)$ be local coordinates centred at P such that D is defined by $x_1 \cdots x_m = 0$ in a neighbourhood of P in X. For $a, b \in \mathcal{I}$ distinct, we have

$$a - b = f_{ab}x^{\mathbf{m}}$$

with $f_{ab}(0) \neq 0$. Put $-g_{ab}^{-1} = f_{ab}/|f_{ab}|$. Then, the map $\partial G_{a,b}$ reads

$$((r_k, z_k)_{1 \leqslant k \leqslant m}, (x_k)_{m+1 \leqslant k \leqslant n}) \longrightarrow -g_{ab}(r_k z_k, x_k)^{-1} z^{\mathbf{m}} \tag{3.4.3}$$

If $\phi_m : \partial P \longrightarrow S^1$ is the map $z \longrightarrow z^{\mathbf{m}}$, then

$$H^{<}_{ab} \cap \partial P = \{z \in \partial D \text{ with } \phi_m(z) = g_{ab}(0)\}$$

Let us re-index the g_{ab} with $\mathbb{Z}/N\mathbb{Z}$ in such a way that the $(g_i(0))_{i\in\mathbb{Z}/N\mathbb{Z}}$ are in an ascending order with respect to the clockwise orientation on S^1. Then, at the cost of shrinking ϵ, the connected components of the

$$\phi_m^{-1}(]g_i(0)-\epsilon, g_{i+1}(0)+\epsilon[), i\in\mathbb{Z}/N\mathbb{Z}$$

form a cover $\mathcal{S}=(S_i)_{i\in\mathbb{Z}/N'\mathbb{Z}}$ of ∂P by open multi-sectors such that

(1) The open set S_i contains exactly two anti-Stokes hyperplanes.

(2) The open set $S_i\cap S_{i+1}$ contains exactly one anti-Stokes hyperplane for every $i\in\mathbb{Z}/N'\mathbb{Z}$.

(3) We have $S_i\cap S_j\cap S_k=\varnothing$ for every $i,j,k\in\mathbb{Z}/N'\mathbb{Z}$ distinct.

At the cost of shrinking ϵ, a section $g\in\Gamma(\partial P,\mathbb{S}to_{\mathcal{M}})$ extends uniquely into a 1-cocycle for $\mathrm{St}_{\mathcal{M}}$ with respect to the cover $\mathcal{S}$. Taking the associated Stokes torsor defines the map LR_P from (3.4.2). We now have to show that this map is an isomorphism.

Let C be a smooth curve passing through P and not contained in D. Then, we have $(\mathrm{St}_{\mathcal{M}})_{|S^1_C}\simeq\mathrm{St}_{\mathcal{M}_C}$ and $\mathcal{M}_C$ admits only one level. Thus, restriction to S^1_C provides a commutative diagram

$$\begin{array}{ccc}\Gamma(\partial P,\mathbb{S}to_{\mathcal{M}}) & \longrightarrow & \Gamma(S^1_C,\mathbb{S}to_{\mathcal{M}_C})\\ {\scriptstyle \mathrm{LR}_P}\downarrow & & \downarrow\\ H^1(\partial P,\mathrm{St}_{\mathcal{M}}) & \longrightarrow & H^1(S^1_C,\mathrm{St}_{\mathcal{M}_C})\end{array}\tag{3.4.4}$$

From Loday-Richaud's theorem, the right vertical map of (3.4.4) is an isomorphism. Taking the invariants under the action of $\pi_1(\mathbb{T}(\mathbf{m},\pi(0)),\pi(0))$ on the right part of the diagram (3.4.4) gives a commutative diagram

$$\begin{array}{ccc}\Gamma(\partial P,\mathbb{S}to_{\mathcal{M}}) & \xrightarrow{\sim} & \Gamma(S^1_C,\mathbb{S}to_{\mathcal{M}_C})^{\pi_1}\\ {\scriptstyle \mathrm{LR}_P}\downarrow & & \downarrow\wr\\ H^1(\partial P,\mathrm{St}_{\mathcal{M}}) & \xrightarrow[\mathbf{res}_C]{} & H^1(S^1_C,\mathrm{St}_{\mathcal{M}_C})^{\pi_1}\end{array}\tag{3.4.5}$$

From lemma 2.6.1, the map $\mathbf{res}_C$ is an isomorphism. Hence, LR_P is an isomorphism. This concludes the proof of proposition 3.4.1. □

3.5. Proof of Theorem 3. We are going to prove something slightly stronger than Theorem 3, that is the following

THEOREM 5. *Let X be a smooth complex algebraic variety and let D be a normal crossing divisor in X. Let $\mathcal{M}$ be a good meromorphic connection on X with poles along D. Suppose that the Stokes hyperplanes of $\mathcal{M}$ don't meet and that locally along D, the connection $\mathcal{M}$ admits only one level. Then, $H^1(\partial D,\mathrm{St}^{<D}_{\mathcal{M}})$ is an affine space.*

PROOF. Let $\mathcal{U}$ be a cover of D by small enough open subsets of D on which the level of $\mathcal{M}$ is defined. Since $R^1p_*\mathrm{St}^{<D}_{\mathcal{M}}$ is a sheaf, we have

$$H^1(\partial D,\mathrm{St}^{<D}_{\mathcal{M}})\simeq\lim_{U\in\mathcal{U}}H^1(\partial U,\mathrm{St}^{<D}_{\mathcal{M}})\tag{3.5.1}$$

Let $U \in \mathcal{U}$. Since $\mathcal{M}$ has locally only one level, if two distinct irregular values $a_0, b_0 \in \Gamma(U, \mathcal{I})$ are such that $a_0 - b_0$ does not have poles along every irreducible component of U, then so does $a - b$ for every $a, b \in \Gamma(U, \mathcal{I})$ distinct. Thus, $\mathrm{St}_{\mathcal{M}}^{<D}$ is trivial on U. So $H^1(\partial U, \mathrm{St}_{\mathcal{M}}^{<D})$ is trivial. Hence the limit (3.5.1) can be taken over the open subsets $U \in \mathcal{U}$ for which for every $a, b \in \Gamma(U, \mathcal{I})$ distinct, the difference $a-b$ has poles along every irreducible component of U. In particular we can suppose that $\mathcal{M}$ is *very good*, that is for every open set $V \subset D$, for every $a, b \in \Gamma(V, \mathcal{I})$ distinct, the difference $a - b$ has poles along every irreducible component of V. In that case, for every $P \in D$, we have

$$(\mathrm{St}_{\mathcal{M}}^{<D})_{|\partial P} \simeq \mathrm{St}_{\mathcal{M}}^{<P} \text{ and } (\mathbb{S}to_{\mathcal{M}}^{<D})_{|\partial P} \simeq \mathbb{S}to_{\mathcal{M}}^{<P} \tag{3.5.2}$$

In that case, we show the following more precise result

PROPOSITION 3.5.3. *Let X be a smooth complex algebraic variety and let D be a normal crossing divisor in X. Let $\mathcal{M}$ be a very good meromorphic connection on X with poles along D. Suppose that the Stokes hyperplanes of $\mathcal{M}$ don't meet and that locally along D, the connection $\mathcal{M}$ admits only one level. Then, there is an isomorphism of schemes*

$$\Gamma(\partial D, \mathbb{S}to_{\mathcal{M}}^{<D}) \longrightarrow H^1(\partial D, \mathrm{St}_{\mathcal{M}}^{<D}) \tag{3.5.4}$$

In particular, $H^1(\partial D, \mathrm{St}_{\mathcal{M}}^{<D})$ is an affine scheme.

Note that $H^1(\partial D, \mathrm{St}_{\mathcal{M}}^{<D})$ is the global section space of the sheaf $R^1p_* \mathrm{St}_{\mathcal{M}}^{<D}$ and that $\Gamma(\partial D, \mathbb{S}to_{\mathcal{M}}^{<D})$ is the global section space of $p_*\mathbb{S}to_{\mathcal{M}}^{<D}$. Hence, it is enough to prove that the sheaves $R^1p_* \mathrm{St}_{\mathcal{M}}^{<D}$ and $p_*\mathbb{S}to_{\mathcal{M}}^{<D}$ are isomorphic. We are going to construct local isomorphisms between these sheaves and then observe that they glue into a global isomorphism.

As recalled in paragraph 1.4, the sheaf $R^1p_* \mathrm{St}_{\mathcal{M}}^{<D}$ is constructible. Hence, there exists a cover $\mathcal{U}$ of D by open subsets of D such that for every $U \in \mathcal{U}$, there exists a point $P \in U$ such that the restriction of torsors

$$H^1(\partial U, \mathrm{St}_{\mathcal{M}}^{<D}) \longrightarrow H^1(\partial P, \mathrm{St}_{\mathcal{M}}^{<D}) \tag{3.5.5}$$

is an isomorphism of schemes. Let $U \in \mathcal{U}$ and let $P \in U$ as above. Let $(x_1, \dots, x_n)$ be local coordinates centred at P such that D is defined by $x_1 \cdots x_m = 0$ in U. At the cost of shrinking $\mathcal{U}$, we can suppose that $\mathcal{M}$ has a unique level $\mathbf{m} \in \mathbb{Z}_{<0}^m$ on U. For $a, b \in \Gamma(U, \mathcal{I})$ distinct, we have

$$a - b = f_{ab}x^{\mathbf{m}}$$

with $f_{ab}(0) \neq 0$. Put $-g_{ab}^{-1} = f_{ab}/|f_{ab}|$. Then, the map $\partial G_{a,b}$ reads

$$((r_k, z_k)_{1\leqslant k\leqslant m}, (x_k)_{m+1\leqslant k\leqslant n}) \longrightarrow -g_{ab}(r_k z_k, x_k)^{-1} z^{\mathbf{m}} \tag{3.5.6}$$

If $\phi_{\mathbf{m}} : \partial U \longrightarrow S^1$ is the map $((r_k, z_k)_{1\leqslant k\leqslant m}, (x_k)_{m+1\leqslant k\leqslant n}) \longrightarrow z^{\mathbf{m}}$, then

$$H_{ab}^{<} \cap \partial U = \{z \in \partial U \text{ with } \phi_{\mathbf{m}}(z) = g_{ab}(z)\}$$

Let us re-index the g_{ab} with $\mathbb{Z}/N\mathbb{Z}$ in such a way that the $(g_i(0))_{i\in\mathbb{Z}/N\mathbb{Z}}$ are in an ascending order with respect to the clockwise orientation on S^1. Note that the g_{ab} are continuous. Hence, at the cost of shrinking $\mathcal{U}$, we can suppose that there exists $\epsilon > 0$ such that g_i maps ∂U into $]g_i(0) - \epsilon, g_i(0) + \epsilon[$ for every $i \in \mathbb{Z}/N\mathbb{Z}$. Since the

Stokes hyperplanes of $\mathcal{M}$ don't meet, the $(g_i(0))_{i\in\mathbb{Z}/N\mathbb{Z}}$ are distinct. Hence, at the cost of shrinking ϵ, the connected components of the

$$\phi_{\mathbf{m}}^{-1}(]g_i(0)-\epsilon, g_{i+1}(0)+\epsilon[), i\in\mathbb{Z}/N\mathbb{Z}$$

form a cover $\mathcal{S}(U)=(S(U)_i)_{i\in\mathbb{Z}/N'\mathbb{Z}}$ of ∂U such that

(1) The open set $S(U)_i$ contains exactly two anti-Stokes hyperplanes.

(2) The open set $S(U)_i\cap S(U)_{i+1}$ contains exactly one anti-Stokes hyperplane for every $i\in\mathbb{Z}/N'\mathbb{Z}$.

(3) We have $S(U)_i\cap S(U)_j\cap S(U)_k=\varnothing$ for every $i,j,k\in\mathbb{Z}/N'\mathbb{Z}$ distinct.

At the cost of shrinking ϵ, any section $g\in\Gamma(\partial U,\mathbb{S}to_{\mathcal{M}}^{<D})$ extends uniquely into a 1-cocycle for $\mathrm{St}_{\mathcal{M}}^{<D}$ with respect to the cover $\mathcal{S}(U)$. Taking the associated Stokes torsor defines a map

$$\mathrm{LR}_U:\Gamma(\partial U,\mathbb{S}to_{\mathcal{M}}^{<D})\longrightarrow H^1(\partial U,\mathrm{St}_{\mathcal{M}}^{<D})$$

The restriction to ∂P induces a commutative diagram

$$\begin{array}{ccc}\Gamma(\partial U,\mathbb{S}to_{\mathcal{M}}^{<D}) & \longrightarrow & \Gamma(\partial P,\mathbb{S}to_{\mathcal{M}}^{<D})\\ {\scriptstyle \mathrm{LR}_U}\downarrow & & \downarrow{\scriptstyle \mathrm{LR}_P}\\ H^1(\partial U,\mathrm{St}_{\mathcal{M}}^{<D}) & \longrightarrow & H^1(\partial P,\mathrm{St}_{\mathcal{M}}^{<D})\end{array} \tag{3.5.7}$$

From our choice of cover, the bottom arrow in (3.5.7) is an isomorphism. At the cost of refining the cover $\mathcal{U}$, we can suppose that the top arrow in (3.5.7) is an isomorphism. Since the relation (3.5.2) holds, proposition 3.4.1 implies that LR_U is an isomorphism.

If $U,U'\in\mathcal{U}$, observe that the trace of the covers $\mathcal{S}(U)$ and $\mathcal{S}(U')$ on $\partial(U\cap U')$ have a common refinement $\mathcal{V}$ such that the following diagram commutes

$$\begin{array}{ccccc} & & Z^1(\mathcal{S}(U)\cap\partial U',\mathrm{St}_{\mathcal{M}}^{<D}) & & \\ & \nearrow & & \searrow & \\ \Gamma(\partial(U\cap U'),\mathbb{S}to_{\mathcal{M}}^{<D}) & & & & Z^1(\mathcal{V},\mathrm{St}_{\mathcal{M}}^{<D})\\ & \searrow & & \nearrow & \\ & & Z^1(\mathcal{S}(U')\cap\partial U,\mathrm{St}_{\mathcal{M}}^{<D}) & & \end{array}$$

Hence, LR_U and $\mathrm{LR}_{U'}$ coincide on $U\cap U'$ for every $U,U'\in\mathcal{U}$. Thus, the $(\mathrm{LR}_U)_{U\in\mathcal{U}}$ glue into an isomorphism between $p_*\mathbb{S}to_{\mathcal{M}}$ and $R^1p_*\,\mathrm{St}_{\mathcal{M}}^{<D}$. This concludes the proof of proposition 3.5.3.

□

3.6. A conjecture in the multi-level case. In this subsection, we restrict to the case where D is smooth. The question whether $R^1p_*\,\mathrm{St}_{\mathcal{M}}^{<D}$ and $p_*\mathbb{S}to_{\mathcal{M}}^{<D}$ are isomorphic is fruitful since it would imply that *when D is smooth, the moduli of Stokes torsors are affine spaces.* We thus formulate the following

Conjecture. *Let X be a smooth complex algebraic variety. Let D be a smooth divisor in X. Let $\mathcal{M}$ be a good meromorphic connection on X with poles along D. Let $p : \widetilde{X} \longrightarrow X$ be the real blow-up of X along D. Then, the local systems $R^1 p_* \operatorname{St}_{\mathcal{M}}^{<D}$ and $p_* \mathbb{S}to_{\mathcal{M}}^{<D}$ are isomorphic.*

In the several level case, the difficulty comes from the fact that the parallel transports for $R^1 p_* \operatorname{St}_{\mathcal{M}}^{<D}$ and $p_* \mathbb{S}to_{\mathcal{M}}^{<D}$ produce different cocycles that are not equal on the nose, but might be cohomologous. The picture below illustrates this phenomenon. The picture on the left features part of our initial element of the

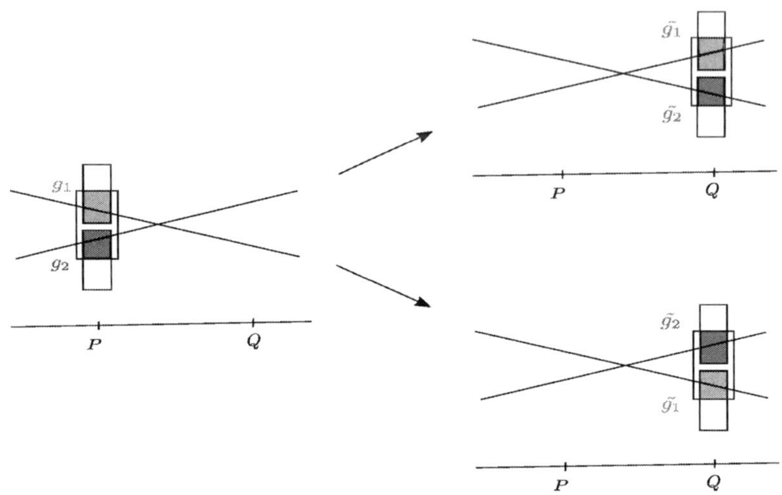

Stokes group above P. In this situation, two anti-Stokes hyperplanes L_1 and L_2 intersect once along the path joining P to Q. Let us call x the intersection point. Since the anti-Stokes hyperplanes are parallel to the Stokes hyperplanes, there is a neighbourhood Ω of x in ∂D not meeting any Stokes line coming from the differences of irregular values giving rise to L_1 and L_2. In particular, g_1 and g_2 extend uniquely into sections $\widetilde{g_1}$ and $\widetilde{g_2}$ of $\operatorname{St}_{\mathcal{M}}^{<D}$ over Ω. When applying the parallel transport for $R^1 p_* \operatorname{St}_{\mathcal{M}}^{<D}$, we end up with the cocycle in the upper right picture. The bottom right picture represents the effect of the parallel transport for $p_* \mathbb{S}to_{\mathcal{M}}^{<D}$. Finally, one passes from one cocycle to the other by permuting $\widetilde{g_1}$ and $\widetilde{g_2}$. Since the Stokes sheaf is not commutative, it is not a priori clear that these cocycles are cohomologous.

4. Application

The role of this paragraph is to provide non trivial examples of meromorphic connections with poles along the fibre of an abelian scheme over a curve. This answers a question asked to the author by Y. André. Trivial examples would be split connections, connections coming from the base curve and their tensor product with regular singular connexions on our abelian scheme. We also want to exclude Stokes structures coming from sub-connections of the previous form. We thus prove the following

Proposition 4.0.1. *Let $\mathcal{C}$ be a complex elliptic curve. Put $\mathcal{A} = \mathcal{C} \times \mathbb{A}^1_{\mathbb{C}}$ and let $p : \mathcal{A} \longrightarrow \mathbb{A}^1_{\mathbb{C}}$ be the second projection. Let $\mathcal{A}_0$ be the fibre of p above 0. Then,* *there exists a rank 4 meromorphic connection $\mathcal{M}$ defined in a neighbourhood of $\mathcal{A}_0$ and with poles along $\mathcal{A}_0$ such that*

(1) *The connection $\mathcal{M}$ is not of the form $\mathcal{M}_1 \otimes p^*\mathcal{M}_2$ where $\mathcal{M}_1$ is a regular connection on $\mathcal{A}$ with poles along $\mathcal{A}_0$ and where $\mathcal{M}_2$ is a meromorphic connection on S with poles at 0.*

(2) *The connection $\mathcal{M}$ underlies a non trivial Stokes structure.*

(3) *The Stokes structure of any strict sub-connection of $\mathcal{M}$ is trivial.*

PROOF. Let (γ_1, γ_2) be a basis for the fundamental group of $\mathcal{C}(\mathbb{C})$. Let γ_3 be a basis for the fundamental group of $\mathbb{C}^*$. Let $A_1, A_2 \in \mathrm{GL}_2(\mathbb{C})$ be matrices such that A_1 has a unique eigenvalue λ_1 and A_2 has a unique eigenvalue λ_2 with $\lambda_1 \neq \lambda_2$ and such that $\ker({}^tA_2 - A_1)$ is one dimensional. For example, take z_1, z_2, z_3, z_4 non zero complex numbers such that

$$A_1 = \begin{pmatrix} z_1 & z_3 \\ z_2 & z_4 \end{pmatrix}$$

has a unique non zero eigenvalue distinct from z_1, and put

$$A_2 = \begin{pmatrix} z_1 & z_2 \\ 0 & z_1 \end{pmatrix}$$

For $i = 1, 2$, let $\rho_i : \pi_1(\mathcal{A}\backslash\mathcal{A}_0(\mathbb{C})) \longrightarrow \mathrm{GL}_2(\mathbb{C})$ be the representation associating A_i to γ_1 and γ_2 and Id to γ_3. Let $\mathcal{R}_i$ be the regular meromorphic connection on $\mathcal{A}$ with poles along $\mathcal{A}_0$ induced by ρ_i. Let $a \in \mathcal{O}_{\mathbb{A}^1_{\mathbb{C}}}(*0)$ with a pole of order $d > 0$ at 0. Put

$$\mathcal{N} := p^*\mathcal{E}^a \otimes \mathcal{R}_1 \oplus \mathcal{R}_2 \tag{4.0.2}$$

Since $\mathcal{R}_1$ and $\mathcal{R}_2$ are non isomorphic, $\mathcal{N}$ satisfies the condition (1). From Theorem 5, the moduli space $H^1(\partial\mathcal{A}_0, \mathrm{St}_{\mathcal{N}}^{<\mathcal{A}_0})$ is a complex affine space. From [**Tey18**, 2.4.4], we have

$$\dim H^1(\partial\mathcal{A}_0, \mathrm{St}_{\mathcal{N}}^{<\mathcal{A}_0}) = \dim H^1(\mathcal{A}_0, \mathrm{Irr}^*_{\mathcal{A}_0} \mathrm{End}\,\mathcal{N})$$

From our choice for $\mathcal{R}_1$ and $\mathcal{R}_2$, note that $\mathrm{End}\,\mathcal{N}$ contains $p^*\mathcal{E}^a$ and $p^*\mathcal{E}^{-a}$ as direct summands. Hence,

$$\begin{aligned}\dim H^1(\partial\mathcal{A}_0, \mathrm{St}_{\mathcal{N}}^{<\mathcal{A}_0}) &\geqslant 2\dim H^1(\mathcal{A}_0, \mathrm{Irr}^*_{\mathcal{A}_0} p^*\mathcal{E}^a) \\ &\geqslant 2\dim H^1(\mathcal{A}_0, p^{-1}\mathrm{Irr}^*_{\mathcal{A}_0} \mathcal{E}^a) \\ &\geqslant 2\dim H^1(\mathcal{A}_0, \mathbb{C}^d[-1]) \\ &\geqslant 2d\end{aligned}$$

Hence, $H^1(\partial\mathcal{A}_0, \mathrm{St}_{\mathcal{N}}^{<\mathcal{A}_0})$ is a non trivial affine space. From [**Tey18**, 2.2.1], a point of $H^1(\partial\mathcal{A}_0, \mathrm{St}_{\mathcal{N}}^{<\mathcal{A}_0})$ distinct form the trivial torsor gives rise to a germ of meromorphic connection $\mathcal{M}$ with poles along $\mathcal{A}_0$ satisfying conditions (1) and (2). Indeed, the connection $\mathcal{M}$ satisfies (1) because $\mathcal{N}$ does, and by construction, $\mathcal{M}$ underlies a non trivial Stokes structure.

We are left to show that $\mathcal{M}$ satisfies condition (3). That is, for every strict sub-connection $\mathcal{M}'$ of $\mathcal{M}$, we have

$$H^1(\partial\mathcal{A}_0, \mathrm{St}_{\mathcal{M}'}^{<\mathcal{A}_0}) \simeq 0$$

Let $\mathcal{N}'$ be the split formal model of $\mathcal{M}'$ along $\mathcal{A}_0$. Then, $\mathcal{N}'$ identifies to a non trivial strict sub-connection of $\mathcal{N}$. For $i = 1, 2$, let χ_i be the rank one connection on $\mathcal{A}$ corresponding to the character of $\pi_1(\mathcal{A}\backslash\mathcal{A}_0(\mathbb{C})) \longrightarrow \mathbb{C}^*$ associating λ_i to γ_1 and γ_2, and 1 to γ_3. Then, the non trivial strict sub-connections of $\mathcal{N}$ are

$$p^*\mathcal{E}^a \otimes \chi_1, \chi_2, \mathcal{R}_2, p^*\mathcal{E}^a \otimes \mathcal{R}_1, p^*\mathcal{E}^a \otimes \chi_1 \oplus \chi_2, p^*\mathcal{E}^a \otimes \chi_1 \oplus \mathcal{R}_2, p^*\mathcal{E}^a \otimes \mathcal{R}_1 \oplus \chi_2$$

If $\mathcal{N}'$ is regular or is the twist of a regular connection by an exponential one, then $\mathcal{M}'$ does not underlie a non trivial Stokes structure. Hence, we can suppose that $\mathcal{N}'$ is one of the three last connections given in the above list. From Theorem 5, the moduli space $H^1(\partial\mathcal{A}_0, \mathrm{St}^{<\mathcal{A}_0}_{\mathcal{M}'})$ is a complex affine space. Let us show that it is trivial. From [**Tey18**, 2.4.4], we have

$$\dim H^1(\partial\mathcal{A}_0, \mathrm{St}^{<\mathcal{A}_0}_{\mathcal{M}'}) = \dim H^1(\mathcal{A}_0, \mathrm{Irr}^*_{\mathcal{A}_0} \mathrm{End}\, \mathcal{M}')$$

From [**Sab17**, 1.2], there is a non canonical isomorphism

$$\mathrm{Irr}^*_{\mathcal{A}_0} \mathrm{End}\, \mathcal{M}' \simeq \mathrm{Irr}^*_{\mathcal{A}_0} \mathrm{End}\, \mathcal{N}'$$

Hence, we are left to show that

$$H^1(\mathcal{A}_0, \mathrm{Irr}^*_{\mathcal{A}_0} \mathrm{End}\, \mathcal{N}') \simeq 0$$

Since χ_1 and χ_2 are distinct, it is enough to show that for any non trivial character $\chi : \pi_1(\mathcal{A}_0(\mathbb{C})) \longrightarrow \mathbb{C}^*$, we have

$$H^1(\mathcal{A}_0, \mathrm{Irr}^*_{\mathcal{A}_0}(p^*\mathcal{E}^a \otimes \chi)) \simeq 0$$

where we abuse notation by denoting by χ the pull-back to $\mathcal{A}$ of the rank one connection on $\mathcal{A}_0$ induced by χ. Let $p_{\mathcal{A}_0} : \widetilde{\mathcal{A}} \longrightarrow \mathcal{A}$ be the real blow-up of $\mathcal{A}$ along $\mathcal{A}_0$. From [**Sab17**, 3.2], we have

$$\mathrm{Irr}^*_{\mathcal{A}_0}(p^*\mathcal{E}^a \otimes \chi) \simeq Rp_{\mathcal{A}_0 *}\mathcal{H}^0\, \mathrm{DR}^{<\mathcal{A}_0}(p^*\mathcal{E}^{-a} \otimes \chi^{-1})$$

Thus, we have to show that for any non trivial character $\chi : \pi_1(\mathcal{A}_0(\mathbb{C})) \longrightarrow \mathbb{C}^*$, we have

$$H^1(\partial\mathcal{A}_0, \mathcal{H}^0\, \mathrm{DR}^{<\mathcal{A}_0}(p^*\mathcal{E}^a \otimes \chi)) \simeq 0$$

Let $p_0 : \widetilde{\mathbb{C}} \longrightarrow \mathbb{C}$ be the real blow-up of $\mathbb{C}$ at 0. Observe that $\partial\mathcal{A}_0 \simeq \mathcal{A}_0 \times \partial 0$. Let U be the open subset of $\partial 0$ consisting in the directions along which e^a has rapid decay at 0. Note that $F := \partial 0 \backslash U$ is a union of closed intervals. Let $j : \mathcal{A}_0 \times U \longrightarrow \partial\mathcal{A}_0$ and let $i : \mathcal{A}_0 \times F \longrightarrow \partial\mathcal{A}_0$ be the canonical inclusions. Let L_χ be the local system on $\mathcal{A}_0$ induced by χ. Let $\mathcal{L}_\chi$ be the pull-back of L_χ via $\partial\mathcal{A}_0 \longrightarrow \mathcal{A}_0$. Then, we have

$$\mathcal{H}^0\, \mathrm{DR}^{<\mathcal{A}_0}(p^*\mathcal{E}^a \otimes \chi) \simeq j_! j^{-1}\mathcal{L}_\chi$$

Since χ is not trivial, the sheaf $i^{-1}\mathcal{L}_\chi$ does not have non trivial global sections. Hence, the long exact sequence in cohomology induced by

$$0 \longrightarrow j_! j^{-1}\mathcal{L}_\chi \longrightarrow \mathcal{L}_\chi \longrightarrow i_* i^{-1}\mathcal{L}_\chi \longrightarrow 0$$

gives rise to an injective morphism $H^1(\partial\mathcal{A}_0, j_! j^{-1}\mathcal{L}_\chi) \longrightarrow H^1(\partial\mathcal{A}_0, \mathcal{L}_\chi)$. We are thus left to show that $\mathcal{L}_\chi$ is acyclic. Note that

$$R\Gamma(\partial\mathcal{A}_0, \mathcal{L}_\chi) \simeq R\Gamma(\partial 0, R\pi_*\mathcal{L}_\chi)$$

where $\pi : \partial\mathcal{A}_0 \longrightarrow \partial 0$ is the natural projection. Hence, it is enough to prove that $R\pi_*\mathcal{L}_\chi$ is zero. By proper base change theorem, the fibres of $R\pi_*\mathcal{L}_\chi$ identify with $R\Gamma(\mathcal{A}_0, L_\chi)$. We get the desired vanishing from the fact that non trivial rank 1 local systems on elliptic curves are acyclic.

□

References

[Boa02] P. P. Boalch, *G-bundles, isomonodromy, and quantum Weyl groups*, Int. Math. Res. Not. **22** (2002), 1129–1166, DOI 10.1155/S1073792802111081. MR1904670

[Boa14] Philip P. Boalch, *Geometry and braiding of Stokes data; Fission and wild character varieties*, Annals of Maths. **179** (2014), 301–365.

[BV89] D. G. Babbitt and V. S. Varadarajan, *Local moduli for meromorphic differential equations* (English, with French summary), Astérisque **169-170** (1989), 217. MR1014083

[JMU81] M. Jimbo, T. Miwa, and K. Ueno, *Monodromy preserving deformation of linear ordinary differential equations with rational coefficients. I. General theory and τ-function*, Phys. D **2** (1981), no. 2, 306–352, DOI 10.1016/0167-2789(81)90013-0. MR630674

[Ked10] K. S. Kedlaya, *Good formal structures for flat meromorphic connections, I: surfaces*, Duke Math. J. **154** (2010), no. 2, 343–418, DOI 10.1215/00127094-2010-041. MR2682186

[Ked11] K. S. Kedlaya, *Good formal structures for flat meromorphic connections, II: excellent schemes*, J. Amer. Math. Soc. **24** (2011), no. 1, 183–229, DOI 10.1090/S0894-0347-2010-00681-9. MR2726603

[Lod94] M. Loday-Richaud, *Stokes phenomenon, multisummability and differential Galois groups* (English, with English and French summaries), Ann. Inst. Fourier (Grenoble) **44** (1994), no. 3, 849–906. MR1303886

[Moc09] T. Mochizuki, *Good formal structure for meromorphic flat connections on smooth projective surfaces*, Algebraic analysis and around, Adv. Stud. Pure Math., vol. 54, Math. Soc. Japan, Tokyo, 2009, pp. 223–253, DOI 10.2969/aspm/05410223. MR2499558

[Moc11] T. Mochizuki, *Wild harmonic bundles and wild pure twistor D-modules* (English, with English and French summaries), Astérisque **340** (2011), x+607. MR2919903

[MR91] J. Martinet and J.-P. Ramis, *Elementary acceleration and multisummability. I*, Ann. Inst. H. Poincaré Phys. Théor. **54** (1991), no. 4, 331–401. MR1128863

[Sab00] C. Sabbah, *Équations différentielles à points singuliers irréguliers et phénomène de Stokes en dimension 2* (French, with English and French summaries), Astérisque **263** (2000), viii+190. MR1741802

[Sab12] C. Sabbah, *Introduction to Stokes structures*, Lecture Notes in Mathematics, vol. 2060, Springer, Heidelberg, 2013. MR2978128

[Sab17] C. Sabbah, *A remark on the irregularity complex*, J. Singul. **16** (2017), 101–114. MR3665434

[Tey18] J.-B. Teyssier, *Moduli of Stokes torsors and singularities of differential equations. Preprint*, January 2018.

[Tey19] J.-B. Teyssier, *Skeletons and moduli of Stokes torsors*, Annales Scientifiques de l'Ecole Normale Supérieure **52** (2019).

Current address: Institut de Mathématiques de Jussieu. 4 place Jussieu. Paris.
Email address: jean-baptiste.teyssier@imj-prg.fr

Selected Published Titles in This Series

742 **Francisco-Jesús Castro-Jiménez, David Bradley Massey, Bernard Teissier, and Meral Tosun, Editors,** A Panorama of Singularities, 2020

740 **Alina Bucur and David Zureick-Brown, Editors,** Analytic Methods in Arithmetic Geometry, 2019

739 **Yaiza Canzani, Linan Chen, and Dmitry Jakobson, Editors,** Probabilistic Methods in Geometry, Topology and Spectral Theory, 2019

738 **Shrikrishna G. Dani, Surender K. Jain, Jugal K. Verma, and Meenakshi P. Wasadikar, Editors,** Contributions in Algebra and Algebraic Geometry, 2019

737 **Fernanda Botelho, Editor,** Recent Trends in Operator Theory and Applications, 2019

736 **Jane Hawkins, Rachel L. Rossetti, and Jim Wiseman, Editors,** Dynamical Systems and Random Processes, 2019

735 **Yanir A. Rubinstein and Bernard Shiffman, Editors,** Advances in Complex Geometry, 2019

734 **Peter Kuchment and Evgeny Semenov, Editors,** Differential Equations, Mathematical Physics, and Applications, 2019

733 **Peter Kuchment and Evgeny Semenov, Editors,** Functional Analysis and Geometry, 2019

732 **Samuele Anni, Jay Jorgenson, Lejla Smajlović, and Lynne Walling, Editors,** Automorphic Forms and Related Topics, 2019

731 **Robert G. Niemeyer, Erin P. J. Pearse, John A. Rock, and Tony Samuel, Editors,** Horizons of Fractal Geometry and Complex Dimensions, 2019

730 **Alberto Facchini, Lorna Gregory, Sonia L'Innocente, and Marcus Tressl, Editors,** Model Theory of Modules, Algebras and Categories, 2019

729 **Daniel G. Davis, Hans-Werner Henn, J. F. Jardine, Mark W. Johnson, and Charles Rezk, Editors,** Homotopy Theory: Tools and Applications, 2019

728 **Nicolás Andruskiewitsch and Dmitri Nikshych, Editors,** Tensor Categories and Hopf Algebras, 2019

727 **André Leroy, Christian Lomp, Sergio López-Permouth, and Frédérique Oggier, Editors,** Rings, Modules and Codes, 2019

726 **Eugene Plotkin, Editor,** Groups, Algebras and Identities, 2019

725 **Shijun Zheng, Marius Beceanu, Jerry Bona, Geng Chen, Tuoc Van Phan, and Avy Soffer, Editors,** Nonlinear Dispersive Waves and Fluids, 2019

724 **Lubjana Beshaj and Tony Shaska, Editors,** Algebraic Curves and Their Applications, 2019

723 **Donatella Danielli, Arshak Petrosyan, and Camelia A. Pop, Editors,** New Developments in the Analysis of Nonlocal Operators, 2019

722 **Yves Aubry, Everett W. Howe, and Christophe Ritzenthaler, Editors,** Arithmetic Geometry: Computation and Applications, 2019

721 **Petr Vojtěchovský, Murray R. Bremner, J. Scott Carter, Anthony B. Evans, John Huerta, Michael K. Kinyon, G. Eric Moorhouse, and Jonathan D. H. Smith, Editors,** Nonassociative Mathematics and its Applications, 2019

720 **Alexandre Girouard, Editor,** Spectral Theory and Applications, 2018

719 **Florian Sobieczky, Editor,** Unimodularity in Randomly Generated Graphs, 2018

718 **David Ayala, Daniel S. Freed, and Ryan E. Grady, Editors,** Topology and Quantum Theory in Interaction, 2018

717 **Federico Bonetto, David Borthwick, Evans Harrell, and Michael Loss, Editors,** Mathematical Problems in Quantum Physics, 2018

For a complete list of titles in this series, visit the
AMS Bookstore at **www.ams.org/bookstore/conmseries/**.